# Data Cleaning and Exploration with Machine Learning

Get to grips with machine learning techniques to achieve sparkling-clean data quickly

**Michael Walker**

BIRMINGHAM—MUMBAI

# Data Cleaning and Exploration with Machine Learning

**Publishing Product Manager**: Ali Abidi

**Senior Editor**: David Sugarman

**Content Development Editor**: Manikandan Kurup

**Technical Editor**: Rahul Limbachiya

**Copy Editor**: Safis Editing

**Project Coordinator**: Farheen Fathima

**Proofreader**: Safis Editing

**Indexer**: Hemangini Bari

**Production Designer**: Alishon Mendonca

**Marketing Coordinators**: Shifa Ansari and Abeer Riyaz Dawe

First published: August 2022

Production reference: 1290722

Published by Packt Publishing Ltd.

Livery Place

35 Livery Street

Birmingham

B3 2PB, UK.

ISBN 978-1-80324-167-8

www.packt.com

# Contributors

## About the author

**Michael Walker** has worked as a data analyst for over 30 years at a variety of educational institutions. He has also taught data science, research methods, statistics, and computer programming to undergraduates since 2006. He is currently the Chief Information Officer at College Unbound in Providence, Rhode Island.

# About the reviewers

**Kalyana Bedhu** is an engineering leader for data science at Microsoft. Kalyana has over 20 years of industry experience in data analytics across various companies such as Ericsson, Sony, Bosch Fidelity, and Oracle, among others. Kalyana was an early practitioner of data science at Ericsson, setting up a data science lab and building up competence in solving some practical data science problems. He played a pivotal role in transforming a central IT organization that dealt with most of the enterprise business intelligence, data, and analytical systems, into an AI and data science engine. Kalyana is a recipient of patents, a speaker, and has authored award-winning papers and data science courses.

*Thanks to Packt and the author for the opportunity to review this book.*

**Divya Sardana** serves as the lead AI/ ML engineer at Nike. Previously, she was a senior data scientist at Teradata Corp. She holds a Ph.D. in computer science from the University of Cincinnati, OH. She has experience working on end-to-end machine learning and deep learning problems involving techniques such as regression and classification. She has further experience in moving developed models to production and ensuring scalability. Her interests include solving complex big data and machine learning/deep learning problems in real-world domains. She is actively involved in the peer review of journals and books in the area of machine learning. She has served as a session chair at machine learning conferences such as ICMLA 2021 and BDA 2021.

# Table of Contents

# 3
# Identifying and Fixing Missing Values

# Section 2 – Preprocessing, Feature Selection, and Sampling

# 4
# Encoding, Transforming, and Scaling Features

# 5
# Feature Selection

# 6
# Preparing for Model Evaluation

# Section 3 – Modeling Continuous Targets with Supervised Learning

## 7
## Linear Regression Models

## 8
## Support Vector Regression

## 9
## K-Nearest Neighbors, Decision Tree, Random Forest, and Gradient Boosted Regression

# Section 4 – Modeling Dichotomous and Multiclass Targets with Supervised Learning

## 10
## Logistic Regression

## 11
## Decision Trees and Random Forest Classification

## 12
## K-Nearest Neighbors for Classification

# 13

## Support Vector Machine Classification

# 14

## Naïve Bayes Classification

# Section 5 – Clustering and Dimensionality Reduction with Unsupervised Learning

# 15

## Principal Component Analysis

# 16
## K-Means and DBSCAN Clustering

## Index

## Other Books You May Enjoy

# Preface

The work that researchers do to prepare data for analysis – extraction, transformation, cleaning, and exploration – has not changed fundamentally with the increased popularity of machine learning tools. When we prepared data for multivariate analyses 30 years ago, we were every bit as concerned with missing values, outliers, the shape of the distribution of our variables, and how variables correlate, as we are when we use machine learning algorithms now. Although it is true that widespread use of the same libraries for machine learning (scikit-learn, TensorFlow, PyTorch, and others) does encourage greater uniformity in approach, good data cleaning and exploration practices are largely unchanged.

How we talk about machine learning is still very much algorithm-focused; just choose the right model and organization-changing insights will follow. But we have to make room for the same kind of learning from data that we have been engaged in over the last few decades, where the predictions we make from data, our modeling of relationships in the data, and our cleaning and exploration of that data are very much part of the conversation. Getting our models right has as much to do with gleaning as much information as we can from a histogram or a confusion matrix as from carefully tuning hyperparameters.

Similarly, the work that data analysts and scientists do does not progress neatly from cleaning, to exploration, to preprocessing, to modeling, to evaluation. We have potential models in mind at each step of the process, regularly updating our previous models. For example, we may initially think that we will be using logistic regression to model a particular binary target but then recognize when we see the distribution of features that we might need to at least try using random forest classification. We will discuss implications for modeling throughout this text, even when explaining relatively routine data cleaning tasks. We will also explore the use of machine learning tools early in the process to help us identify anomalies, impute values, and select features.

This points to another change in the workflow of data analysts and scientists over the last decade – less emphasis on *the one model* and greater acceptance of model building as an iterative process. A project might require multiple machine learning algorithms – for example, principal component analysis to reduce dimensions (the number of features) and then logistic regression for classification.

That being said, there is one key difference in our approach to data cleaning, exploration, and modeling as machine learning tools guide more of our work – an increased emphasis on prediction over an understanding of the underlying data. We are more concerned with how well our features (also known as independent variables, inputs, or predictors) predict our targets (dependent variables, outputs, responses) than with the relationships between features and the underlying structure of our data. I point out throughout the first two sections of this book how that alters our focus somewhat, even when we are cleaning and exploring our data.

# Who this book is for

I had multiple audiences in mind as I wrote this book, but I most consistently thought about a dear friend of mine who bought a Transact-SQL book 30 years ago and instantly developed great confidence in her database work, ultimately building a career around those skills. I would love it if someone just starting their career as a data scientist or analyst worked through this book and had a similar experience as my friend. More than anything else, I want you to feel good and excited about what you can do as a result of reading this book.

I also hope this book will be a useful reference for folks who have been doing this kind of work for a while. Here, I imagine someone opening the book and wondering to themselves, *what are good values to use in my grid search for my logistic regression model?*

In keeping with the hands-on nature of this text, every bit of output is reproducible with code in this book. I also stuck to a rule throughout, even when it was challenging. Every section, except for the conceptual sections, starts with raw data largely unchanged from the original downloaded file. You go from data file to model in each section. If you have forgotten how a particular object was created, all you will ever need to do is turn back a page or two to see.

Readers who have some knowledge of pandas and NumPy will have an easier time with some code blocks, as will folks with some knowledge of Python and scikit-learn. None of that is essential though. There are just some sections you might want to pause over longer. If you need additional instruction on doing data work with Python, my *Python Data Cleaning Cookbook* is a good companion book I think.

# What this book covers

*Chapter 1, Examining the Distribution of Features and Targets*, explores using common NumPy and pandas techniques to get a better sense of the attributes of our data. We will generate summary statistics, such as `mean`, `min`, and `max`, and standard deviation, and count the number of missings. We will also create visualizations of key features, including histograms and boxplots, to give us a better sense of the distribution of each feature than we can get by just looking at summary statistics. We will hint at the implications of feature distribution for data transformation, encoding and scaling, and the modeling that we will be doing in subsequent chapters with the same data.

*Chapter 2, Examining Bivariate and Multivariate Relationships between Features and Targets*, focuses on the correlation between possible features and target variables. We will use pandas methods for bivariate analysis, and Matplotlib for visualizations. We will discuss the implications of what we find for feature engineering and modeling. We also use multivariate techniques in this chapter to understand the relationship between features.

*Chapter 3, Identifying and Fixing Missing Values*, goes over techniques for identifying missing values for each feature or target, and for identifying observations where values for a large number of the features are absent. We will explore strategies for imputing values, such as setting values to the overall mean, to the mean for a given category, or forward filling. We will also examine multivariate techniques for imputing values for missings and discuss when they are appropriate.

*Chapter 4, Encoding, Transforming, and Scaling Features*, covers a range of feature engineering techniques. We will use tools to drop redundant or highly correlated features. We will explore the most common kinds of encoding – one-hot, ordinal, and hashing encoding. We will also use transformations to improve the distribution of our features. Finally, we will use common binning and scaling approaches to address skew, kurtosis, and outliers, and to adjust for features with widely different ranges.

*Chapter 5, Feature Selection* will go over a number of feature selection methods, from filter, to wrapper, to embedded methods. We will explore how they work with categorical and continuous targets. For wrapper and embedded methods, we consider how well they work with different algorithms.

*Chapter 6, Preparing for Model Evaluation*, will see us build our first full-fledged pipeline, separating our data into testing and training datasets, and learning how to do preprocessing without data leakage. We will implement cross-validation with k-fold and look more closely into assessing the performance of our models.

*Chapter 7, Linear Regression Models*, is the first of several chapters on building regression models with an old favorite of many data scientists, linear regression. We will run a classical linear model while also examining the qualities of a feature space that make it a good candidate for a linear model. We will explore how to improve linear models, when necessary, with regularization and transformations. We will look into stochastic gradient descent as an alternative to **ordinary least square (OLS)** optimization. We will also learn how to do hyperparameter tuning with grid searches.

*Chapter 8, Support Vector Regression*, discusses key support vector machine concepts and how they can be applied to regression problems. In particular, we will examine how concepts such as epsilon-insensitive tubes and soft margins can give us the flexibility to get the best fit possible, given our data and domain-related challenges. We will also explore, for the first time but definitely not the last, the very handy kernel trick, which allows us to model nonlinear relationships without transformations or increasing the number of features.

*Chapter 9, K-Nearest Neighbors, Decision Tree, Random Forest, and Gradient Boosted Regression*, explores some of the most popular non-parametric regression algorithms. We will discuss the advantages of each algorithm, when you might want to choose one over the other, and possible modeling challenges. These challenges include how to avoid underfitting and overfitting with careful adjusting of hyperparameters.

*Chapter 10, Logistic Regression*, is the first of several chapters on building classification models with logistic regression, an efficient algorithm with low bias. We will carefully examine the assumptions of logistic regression and discuss the attributes of a dataset and a modeling problem that make logistic regression a good choice. We will use regularization to address high variance or when we have a number of highly correlated predictors. We will extend the algorithm to multiclass problems with multinomial logistic regression. We will also discuss how to handle class imbalance for the first, but not the last, time.

*Chapter 11, Decision Trees and Random Forest Classification*, returns to the decision tree and random forest algorithms that were introduced in *Chapter 9, K-Nearest Neighbors, Decision Tree, Random Forest, and Gradient Boosted Regression*, this time dealing with classification problems. This gives us another opportunity to learn how to construct and interpret decision trees. We will adjust key hyperparameters, including the depth of trees, to avoid overfitting. We will then explore random forest and gradient boosted decision trees as good, lower variance alternatives to decision trees.

*Chapter 12, K-Nearest Neighbors for Classification*, returns to **k-nearest neighbors (KNNs)** to handle both binary and multiclass modeling problems. We will discuss and demonstrate the advantages of KNN – how easy it is to build a no-frills model and the limited number of hyperparameters to adjust. By the end of the chapter, we will know both – how to do KNN and when we should consider it for our modeling.

*Chapter 13, Support Vector Machine Classification*, explores different strategies for implementing **support vector classification** (**SVC**). We will use linear SVC, which can perform very well when our classes are linearly separable. We will then examine how to use the kernel trick to extend SVC to cases where the classes are not linearly separable. Finally, we will use one-versus-one and one-versus-rest classification to handle targets with more than two values.

*Chapter 14, Naïve Bayes Classification*, discusses the fundamental assumptions of naïve Bayes in this chapter and how the algorithm is used to tackle some of the modeling challenges we have already explored, as well as some new ones, such as text classification. We will consider when naïve Bayes is a good option and when it is not. We will also examine the interpretation of naïve Bayes models.

*Chapter 15, Principal Component Analysis*, examines **principal component analysis** (**PCA**), including how it works and when we might want to use it. We will learn how to interpret the components created from PCA, including how each feature contributes to each component and how much of the variance is explained. We will learn how to visualize components and how to use components in subsequent analyses. We will also examine how to use kernels for PCA and when that might give us better results.

*Chapter 16, K-Means and DBSCAN Clustering*, explores two popular clustering techniques, k-means and **Density-based spatial clustering of applications with noise** (**DBSCAN**). We will discuss the strengths of each approach and develop a sense of when to choose one clustering algorithm over the other. We will also learn how to evaluate our clusters and how to change hyperparameters to improve our model.

# To get the most out of this book

To run the code in this book, you will need to have installed a scientific distribution of Python, such as Anaconda. All code was tested with scikit-learn versions 0.24.2 and 1.0.2.

# Download the example code files

You can download the example code files for this book from GitHub at `https://github.com/PacktPublishing/Data-Cleaning-and-Exploration-with-Machine-Learning`. If there's an update to the code, it will be updated in the GitHub repository.

We also have other code bundles from our rich catalog of books and videos available at `https://github.com/PacktPublishing/`. Check them out!

# Download the color images

We also provide a PDF file that has color images of the screenshots and diagrams used in this book. You can download it here: `https://packt.link/aLE6J`.

# Conventions used

There are a number of text conventions used throughout this book.

`Code in text`: Indicates code words in text, database table names, folder names, filenames, file extensions, pathnames, dummy URLs, user input, and Twitter handles. Here is an example: "For learning purposes, we have provided two example `mlruns` artifacts and the `huggingface` cache folder in the GitHub repository under the `chapter08` folder."

A block of code is set as follows:

```
client = boto3.client('sagemaker-runtime')
response = client.invoke_endpoint(
        EndpointName=app_name,
        ContentType=content_type,
        Accept=accept,
        Body=payload
        )
```

When we wish to draw your attention to a particular part of a code block, the relevant lines or items are set in bold:

```
loaded_model = mlflow.pyfunc.spark_udf(
    spark,
    model_uri=logged_model,
    result_type=StringType())
```

Any command-line input or output is written as follows:

```
mlflow models serve -m models:/inference_pipeline_model/6
```

**Bold**: Indicates a new term, an important word, or words that you see onscreen. For instance, words in menus or dialog boxes appear in **bold**. Here is an example: "To execute the code in this cell, you can just click on **Run Cell** in the top-right drop-down menu."

> **Tips or Important Notes**
> Appear like this.

# Get in touch

Feedback from our readers is always welcome.

**General feedback**: If you have questions about any aspect of this book, email us at customercare@packtpub.com and mention the book title in the subject of your message.

**Errata**: Although we have taken every care to ensure the accuracy of our content, mistakes do happen. If you have found a mistake in this book, we would be grateful if you would report this to us. Please visit www.packtpub.com/support/errata and fill in the form.

**Piracy**: If you come across any illegal copies of our works in any form on the internet, we would be grateful if you would provide us with the location address or website name. Please contact us at copyright@packt.com with a link to the material.

**If you are interested in becoming an author**: If there is a topic that you have expertise in and you are interested in either writing or contributing to a book, please visit authors.packtpub.com.

# Share Your Thoughts

Once you've read *Data Cleaning and Exploration with Machine Learning*, we'd love to hear your thoughts! Scan the QR code below to go straight to the Amazon review page for this book and share your feedback.

https://packt.link/r/1-803-24167-5

Your review is important to us and the tech community and will help us make sure we're delivering excellent quality content.

# Section 1 –
# Data Cleaning and
# Machine Learning
# Algorithms

I try to avoid thinking about different parts of the model building process sequentially, to see myself as cleaning data, then preprocessing, and so on until I have done model validation. I do not want to think about that process as involving phases that ever end. We start with data cleaning in this section, but I hope the chapters in this section convey that we are always looking ahead, anticipating modeling challenges as we clean data; and that we also typically reflect back on the data cleaning we have done when we evaluate our models.

To some extent, the clean and dirty metaphor hides the nuance in preparing data for subsequent analysis. The real concern is how representative our instances and attributes (observations and variables) are of phenomena of interest. This can always be improved, and easily made worse without care. One thing is for certain though. There is nothing we can do in any other part of the model building process that will make right something important we have gotten wrong during data cleaning.

The first three chapters of this book are about getting our data as right as we can. To do that we have to have a good sense of how all variables, features and targets, are distributed. There are three questions we should ask ourselves before we do any formal analysis: 1) Are we confident that we know the full range of values, and the shape of the distribution, of every variable of interest? 2) Do we have a good idea of the bivariate relationship between variables, how each moves with others? 3) How successful are our attempts to fix potential problems, such as outliers and missing values? The chapters in this section provide the tools you need to answer these questions.

This section comprises the following chapters:

- *Chapter 1, Examining the Distribution of Features and Targets*
- *Chapter 2, Examining Bivariate and Multivariate Relationships between Features and Targets*
- *Chapter 3, Identifying and Fixing Missing Values*

# 1
# Examining the Distribution of Features and Targets

Machine learning writing and instruction are often algorithm-focused. Sometimes, this gives the impression that all we have to do is choose the right model and that organization-changing insights will follow. But the best place to begin a machine learning project is with an understanding of how the features and targets we will use are distributed.

It is important to make room for the same kind of learning from data that has been central to our work as analysts for decades – studying the distribution of variables, identifying anomalies, and examining bivariate relationships – even as we focus more and more on the accuracy of our predictions.

We will explore tools for doing so in the first three chapters of this book, while also considering implications for model building.

In this chapter, we will use common NumPy and pandas techniques to get a better sense of the attributes of our data. We want to know how key features are distributed before we do any predictive analyses. We also want to know the central tendency, shape, and spread of the distribution of each continuous feature and have a count for each value for categorical features. We will take advantage of very handy NumPy and pandas tools for generating summary statistics, such as the mean, min, and max, as well as standard deviation.

After that, we will create visualizations of key features, including histograms and boxplots, to give us a better sense of the distribution of each feature than we can get by just looking at summary statistics. We will hint at the implications of feature distribution for data transformation, encoding and scaling, and the modeling that we will be doing in subsequent chapters with the same data.

Specifically, in this chapter, we are going to cover the following topics:

- Subsetting data
- Generating frequencies for categorical features
- Generating summary statistics for continuous features
- Identifying extreme values and outliers in univariate analysis
- Using histograms, boxplots, and violin plots to examine the distribution of continuous features

# Technical requirements

This chapter will rely heavily on the pandas, NumPy, and Matplotlib libraries, but you don't require any prior knowledge of these. If you have installed Python from a scientific distribution, such as Anaconda or WinPython, then these libraries are probably already installed. If you need to install one of them to run the code in this chapter, you can run `pip install [package name]` from a terminal.

# Subsetting data

Almost every statistical modeling project I have worked on has required removing some data from the analysis. Often, this is because of missing values or outliers. Sometimes, there are theoretical reasons for limiting our analysis to a subset of the data. For example, we have weather data going back to 1600, but our analysis goals only involve changes in weather since 1900. Fortunately, the subsetting tools in pandas are quite powerful and flexible. We will work with data from the United States **National Longitudinal Survey (NLS)** of Youth in this section.

> **Note**
>
> The NLS of Youth is conducted by the United States Bureau of Labor Statistics. This survey started with a cohort of individuals in 1997 who were born between 1980 and 1985, with annual follow-ups each year through 2017. For this recipe, I pulled 89 variables on grades, employment, income, and attitudes toward government from the hundreds of data items on the survey. Separate files for SPSS, Stata, and SAS can be downloaded from the repository. The NLS data is available for public use at `https://www.nlsinfo.org/investigator/pages/search`.

Let's start subsetting the data using pandas:

1.  We will start by loading the NLS data. We also set an index:

    ```
    import pandas as pd
    import numpy as np
    nls97 = pd.read_csv("data/nls97.csv")
    nls97.set_index("personid", inplace=True)
    ```

2.  Let's select a few columns from the NLS data. The following code creates a new DataFrame that contains some demographic and employment data. A useful feature of pandas is that the new DataFrame retains the index of the old DataFrame, as shown here:

    ```
    democols = ['gender','birthyear','maritalstatus',
      'weeksworked16','wageincome','highestdegree']

    nls97demo = nls97[democols]
    nls97demo.index.name
    'personid'
    ```

3.  We can use slicing to select rows by position. `nls97demo[1000:1004]` selects every row, starting from the row indicated by the integer to the left of the colon (`1000`, in this case) up to, but not including, the row indicated by the integer to the right of the colon (`1004`). The row at `1000` is the 1,001st row because of zero-based indexing. Each row appears as a column in the output since we have transposed the resulting DataFrame:

    ```
    nls97demo[1000:1004].T
    personid        195884       195891       195970\
    gender          Male         Male         Female
    ```

```
birthyear        1981          1980            1982
maritalstatus NaN             Never-married Never-married
weeksworked16 NaN             53              53
wageincome      NaN           14,000          52,000
highestdegree 4.Bachelors   2.High School 4.Bachelors


personid         195996
gender           Female
birthyear        1980
maritalstatus    NaN
weeksworked16    NaN
wageincome       NaN
highestdegree    3.Associates
```

4.  We can also skip rows over the interval by setting a value for the step after the second colon. The default value for the step is 1. The value for the following step is 2, which means that every other row between 1000 and 1004 will be selected:

```
nls97demo[1000:1004:2].T
personid         195884        195970
gender           Male          Female
birthyear        1981          1982
maritalstatus    NaN           Never-married
weeksworked16    NaN           53
wageincome       NaN           52,000
highestdegree    4.Bachelors   4. Bachelors
```

5.  If we do not include a value to the left of the colon, row selection will start with the first row. Notice that this returns the same DataFrame as the head method does:

```
nls97demo[:3].T
personid         100061        100139         100284
gender           Female        Male           Male
birthyear        1980          1983           1984
maritalstatus    Married       Married        Never-
married
weeksworked16    48            53             47
wageincome       12,500        120,000        58,000
highestdegree    2.High School 2. High School 0.None
```

```
nls97demo.head(3).T
personid            100061          100139          100284
gender              Female          Male            Male
birthyear           1980            1983            1984
maritalstatus       Married         Married         Never-
married
weeksworked16       48              53              47
wageincome          12,500          120,000         58,000
highestdegree       2.High School   2.High School   0. None
```

6.  If we use a negative number, -n, to the left of the colon, the last n rows of the
    DataFrame will be returned. This returns the same DataFrame as the `tail`
    method does:

```
nls97demo[-3:].T
personid            999543          999698          999963
gender              Female          Female          Female
birthyear           1984            1983            1982
maritalstatus       Divorced        Never-married   Married
weeksworked16       0               0               53
wageincome          NaN             NaN             50,000
highestdegree       2.High School   2.High School   4. Bachelors


nls97demo.tail(3).T
personid            999543          999698          999963
gender              Female          Female          Female
birthyear           1984            1983            1982
maritalstatus       Divorced        Never-married   Married
weeksworked16       0               0               53
wageincome          NaN             NaN             50,000
highestdegree       2.High School   2.High School   4. Bachelors
```

7.   We can select rows by index value using the `loc` accessor. Recall that for the `nls97demo` DataFrame, the index is `personid`. We can pass a list of the index labels to the `loc` accessor, such as `loc[[195884,195891,195970]]`, to get the rows associated with those labels. We can also pass a lower and upper bound of index labels, such as `loc[195884:195970]`, to retrieve the indicated rows:

```
nls97demo.loc[[195884,195891,195970]].T
personid         195884        195891         195970
gender           Male          Male           Female
birthyear        1981          1980           1982
maritalstatus    NaN           Never-married  Never-married
weeksworked16    NaN           53             53
wageincome       NaN           14,000         52,000
highestdegree    4.Bachelors   2.High School  4.Bachelors


nls97demo.loc[195884:195970].T
personid         195884        195891         195970
gender           Male          Male           Female
birthyear        1981          1980           1982
maritalstatus    NaN           Never-married  Never-married
weeksworked16    NaN           53             53
wageincome       NaN           14,000         52,000
highestdegree    4.Bachelors   2.High School  4.Bachelors
```

8.   To select rows by position, rather than by index label, we can use the `iloc` accessor. We can pass a list of position numbers, such as `iloc[[0,1,2]]`, to the accessor to get the rows at those positions. We can pass a range, such as `iloc[0:3]`, to get rows between the lower and upper bound, not including the row at the upper bound. We can also use the `iloc` accessor to select the last n rows. `iloc[-3:]` selects the last three rows:

```
nls97demo.iloc[[0,1,2]].T
personid         100061       100139       100284
gender           Female       Male         Male
birthyear        1980         1983         1984
maritalstatus    Married      Married      Never-
married
weeksworked16    48           53           47
wageincome       12,500       120,000      58,000
```

| highestdegree | 2.High School | 2.High School | 0. None |

```
nls97demo.iloc[0:3].T
```

| personid | 100061 | 100139 | 100284 |
| gender | Female | Male | Male |
| birthyear | 1980 | 1983 | 1984 |
| maritalstatus<br>married | Married | Married | Never- |
| weeksworked16 | 48 | 53 | 47 |
| wageincome | 12,500 | 120,000 | 58,000 |
| highestdegree | 2.High School | 2.High School | 0. None |

```
nls97demo.iloc[-3:].T
```

| personid | 999543 | 999698 | 999963 |
| gender | Female | Female | Female |
| birthyear | 1984 | 1983 | 1982 |
| maritalstatus | Divorced | Never-married | Married |
| weeksworked16 | 0 | 0 | 53 |
| wageincome | NaN | NaN | 50,000 |
| highestdegree | 2.High School | 2.High School | 4. Bachelors |

Often, we need to select rows based on a column value or the values of several columns. We can do this in pandas by using Boolean indexing. Here, we pass a vector of Boolean values (which can be a Series) to the `loc` accessor or the bracket operator. The Boolean vector needs to have the same index as the DataFrame.

9.  Let's try this using the `nightlyhrssleep` column on the NLS DataFrame. We want a Boolean Series that is `True` for people who sleep 6 or fewer hours a night (the 33rd percentile) and `False` if `nightlyhrssleep` is greater than 6 or is missing. `sleepcheckbool = nls97.nightlyhrssleep<=lowsleepthreshold` creates the boolean Series. If we display the first few values of `sleepcheckbool`, we will see that we are getting the expected values. We can also confirm that the `sleepcheckbool` index is equal to the `nls97` index:

```
nls97.nightlyhrssleep.head()
```

| personid | |
| 100061 | 6 |
| 100139 | 8 |

```
100284       7
100292       nan
100583       6
Name: nightlyhrssleep, dtype: float64


lowsleepthreshold = nls97.nightlyhrssleep.quantile(0.33)
lowsleepthreshold
6.0


sleepcheckbool = nls97.nightlyhrssleep<=lowsleepthreshold
sleepcheckbool.head()
personid
100061       True
100139       False
100284       False
100292       False
100583       True
Name: nightlyhrssleep, dtype: bool


sleepcheckbool.index.equals(nls97.index)
True
```

Since the `sleepcheckbool` Series has the same index as `nls97`, we can just pass it to the `loc` accessor to create a DataFrame containing people who sleep 6 hours or less a night. This is a little pandas magic here. It handles the index alignment for us:

```
lowsleep = nls97.loc[sleepcheckbool]
lowsleep.shape
(3067, 88)
```

10. We could have created the `lowsleep` subset of our data in one step, which is what we would typically do unless we need the Boolean Series for some other purpose:

```
lowsleep = nls97.loc[nls97.
nightlyhrssleep<=lowsleepthreshold]
lowsleep.shape
(3067, 88)
```

11. We can pass more complex conditions to the `loc` accessor and evaluate the values of multiple columns. For example, we can select rows where `nightlyhrssleep` is less than or equal to the threshold and `childathome` (number of children living at home) is greater than or equal to 3:

```
lowsleep3pluschildren = \
  nls97.loc[(nls97.nightlyhrssleep<=lowsleepthreshold)
    & (nls97.childathome>=3)]


lowsleep3pluschildren.shape
(623, 88)
```

Each condition in `nls97.loc[(nls97.nightlyhrssleep<=lowsleepthreshold) & (nls97.childathome>3)]` is placed in parentheses. An error will be generated if the parentheses are excluded. The `&` operator is the equivalent of `and` in standard Python, meaning that *both* conditions have to be `True` for the row to be selected. We could have used `|` for `or` if we wanted to select the row if *either* condition was `True`.

12. Finally, we can select rows and columns at the same time. The expression to the left of the comma selects rows, while the list to the right of the comma selects columns:

```
lowsleep3pluschildren = \
  nls97.loc[(nls97.nightlyhrssleep<=lowsleepthreshold)
    & (nls97.childathome>=3),
    ['nightlyhrssleep','childathome']]


lowsleep3pluschildren.shape
(623, 2)
```

We used three different tools to select columns and rows from a pandas DataFrame in the last two sections: the `[]` bracket operator and two pandas-specific accessors, `loc` and `iloc`. This will be a little confusing if you are new to pandas, but it becomes clear which tool to use in which situation after just a few months. If you came to pandas with a fair bit of Python and NumPy experience, you will likely find the `[]` operator most familiar. However, the pandas documentation recommends against using the `[]` operator for production code. The `loc` accessor is used for selecting rows by Boolean indexing or by index label, while the `iloc` accessor is used for selecting rows by row number.

This section was a brief primer on selecting columns and rows with pandas. Although we did not go into too much detail on this, most of what you need to know to subset data was covered, as well as everything you need to know to understand the pandas-specific material in the rest of this book. We will start putting some of that to work in the next two sections by creating frequencies and summary statistics for our features.

# Generating frequencies for categorical features

Categorical features can be either nominal or ordinal. **Nominal** features, such as gender, species name, or country, have a limited number of possible values, and are either strings or are numerical without having any intrinsic numerical meaning. For example, if country is represented by 1 for Afghanistan, 2 for Albania, and so on, the data is numerical but it does not make sense to perform arithmetic operations on those values.

**Ordinal** features also have a limited number of possible values but are different from nominal features in that the order of the values matters. A **Likert scale** rating (ranging from 1 for very unlikely to 5 for very likely) is an example of an ordinal feature. Nonetheless, arithmetic operations would not typically make sense because there is no uniform and meaningful distance between values.

Before we begin modeling, we want to have counts of all the possible values for the categorical features we may use. This is typically referred to as a one-way frequency distribution. Fortunately, pandas makes this very easy to do. We can quickly select columns from a pandas DataFrame and use the `value_counts` method to generate counts for each categorical value:

1.  Let's load the NLS data, create a DataFrame that contains just the first 20 columns of the data, and look at the data types:

```
nls97 = pd.read_csv("data/nls97.csv")
nls97.set_index("personid", inplace=True)
nls97abb = nls97.iloc[:,:20]
nls97abb.dtypes
gender                  object
birthmonth              int64
birthyear               int64
highestgradecompleted   float64
maritalstatus           object
childathome             float64
```

```
childnotathome             float64
wageincome                 float64
weeklyhrscomputer          object
weeklyhrstv                object
nightlyhrssleep            float64
satverbal                  float64
satmath                    float64
gpaoverall                 float64
gpaenglish                 float64
gpamath                    float64
gpascience                 float64
highestdegree              object
govprovidejobs             object
govpricecontrols           object
dtype: object
```

> **Note**
>
> Recall from the previous section how column and row selection works with
> the loc and iloc accessors. The colon to the left of the comma indicates that
> we want all the rows, while :20 to the right of the comma gets us the first 20
> columns.

2. All of the object type columns in the preceding code are categorical. We can use
   value_counts to see the counts for each value for maritalstatus. We can
   also use dropna=False to get value_counts to show the missing values (NaN):

```
nls97abb.maritalstatus.value_counts(dropna=False)
Married           3066
Never-married     2766
NaN               2312
Divorced          663
Separated         154
Widowed           23
Name: maritalstatus, dtype: int64
```

3.  If we just want the number of missing values, we can chain the `isnull` and `sum` methods. `isnull` returns a Boolean Series containing `True` values when `maritalstatus` is missing and `False` otherwise. `sum` then counts the number of `True` values, since it will interpret `True` values as 1 and `False` values as 0:

    ```
    nls97abb.maritalstatus.isnull().sum()
    2312
    ```

4.  You have probably noticed that the `maritalstatus` values were sorted by frequency by default. You can sort them alphabetically by values by sorting the index. We can do this by taking advantage of the fact that `value_counts` returns a Series with the values as the index:

    ```
    marstatcnt = nls97abb.maritalstatus.value_
    counts(dropna=False)
    type(marstatcnt)
    <class 'pandas.core.series.Series'>
    marstatcnt.index
    Index(['Married', 'Never-married', nan, 'Divorced',
    'Separated', 'Widowed'], dtype='object')
    ```

5.  To sort the index, we just need to call `sort_index`:

    ```
    marstatcnt.sort_index()
    Divorced              663
    Married              3066
    Never-married        2766
    Separated             154
    Widowed                23
    NaN                  2312
    Name: maritalstatus, dtype: int64
    ```

6.  Of course, we could have gotten the same results in one step with `nls97. maritalstatus.value_counts(dropna=False).sort_index()`. We can also show ratios instead of counts by setting `normalize` to `True`. In the following code, we can see that 34% of the responses were `Married` (notice that we did not set `dropna` to `True`, so missing values have been excluded):

    ```
    nls97.maritalstatus.\
        value_counts(normalize=True, dropna=False).\
            sort_index()
    ```

```
Divorced              0.07
Married               0.34
Never-married         0.31
Separated             0.02
Widowed               0.00
NaN                   0.26
Name: maritalstatus, dtype: float64
```

7. pandas has a category data type that can store data much more efficiently than the object data type when a column has a limited number of values. Since we already know that all of our object columns contain categorical data, we should convert those columns into the category data type. In the following code, we're creating a list that contains the column names for the object columns, catcols. Then, we're looping through those columns and using astype to change the data type to category:

```
catcols = nls97abb.select_dtypes(include=["object"]).
columns

for col in nls97abb[catcols].columns:
...        nls97abb[col] = nls97abb[col].astype('category')
...

nls97abb[catcols].dtypes
gender                    category
maritalstatus             category
weeklyhrscomputer         category
weeklyhrstv               category
highestdegree             category
govprovidejobs            category
govpricecontrols          category
dtype: object
```

8.  Let's check our category features for missing values. There are no missing values for `gender` and very few for `highestdegree`. But the overwhelming majority of values for `govprovidejobs` (the government should provide jobs) and `govpricecontrols` (the government should control prices) are missing. This means that those features probably won't be useful for most modeling:

```
nls97abb[catcols].isnull().sum()
gender                  0
maritalstatus           2312
weeklyhrscomputer       2274
weeklyhrstv             2273
highestdegree           31
govprovidejobs          7151
govpricecontrols        7125
dtype: int64
```

9.  We can generate frequencies for multiple features at once by passing a `value_counts` call to `apply`. We can use `filter` to select the columns that we want – in this case, all the columns with *gov* in their name. Note that the missing values for each feature have been omitted since we did not set `dropna` to `False`:

```
nls97abb.filter(like="gov").apply(pd.value_counts,
normalize=True)
```

|                    | govprovidejobs | govpricecontrols |
|--------------------|----------------|------------------|
| 1. Definitely      | 0.25           | 0.54             |
| 2. Probably        | 0.34           | 0.33             |
| 3. Probably not    | 0.25           | 0.09             |
| 4. Definitely not  | 0.16           | 0.04             |

10. We can use the same frequencies on a subset of our data. If, for example, we want to see the responses of only married people to the government role questions, we can do that subsetting by placing `nls97abb[nls97abb.maritalstatus=="Married"]` before `filter`:

```
nls97abb.loc[nls97abb.maritalstatus=="Married"].\
filter(like="gov").\
    apply(pd.value_counts, normalize=True)
```

|                | govprovidejobs | govpricecontrols |
|----------------|----------------|------------------|
| 1. Definitely  | 0.17           | 0.46             |
| 2. Probably    | 0.33           | 0.38             |

```
  3. Probably not              0.31                    0.11
  4. Definitely not            0.18                    0.05
```

11. Since, in this case, there were only two *gov* columns, it may have been easier to do the following:

```
nls97abb.loc[nls97abb.maritalstatus=="Married",
   ['govprovidejobs','govpricecontrols']].\
   apply(pd.value_counts, normalize=True)
                  govprovidejobs      govpricecontrols
1. Definitely            0.17                    0.46
2. Probably              0.33                    0.38
3. Probably not          0.31                    0.11
4. Definitely not        0.18                    0.05
```

Nonetheless, it will often be easier to use `filter` since it is not unusual to have to do the same cleaning or exploration task on groups of features with similar names.

There are times when we may want to model a continuous or discrete feature as categorical. The NLS DataFrame contains `highestgradecompleted`. A year increase from 5 to 6 may not be as important as that from 11 to 12 in terms of its impact on a target. Let's create a dichotomous feature instead – that is, 1 when the person has completed 12 or more grades, 0 if they have completed less than that, and missing when `highestgradecompleted` is missing.

12. We need to do a little bit of cleaning up first, though. `highestgradecompleted` has two logical missing values – an actual NaN value that pandas recognizes as missing and a 95 value that the survey designers intend for us to also treat as missing for most use cases. Let's use `replace` to fix that before moving on:

```
nls97abb.highestgradecompleted.\
   replace(95, np.nan, inplace=True)
```

13. We can use NumPy's `where` function to assign values to `highschoolgrad` based on the values of `highestgradecompleted`. If `highestgradecompleted` is null (`NaN`), we assign `NaN` to our new column, `highschoolgrad`. If the value for `highestgradecompleted` is not null, the next clause tests for a value less than 12, setting `highschoolgrad` to 0 if that is true, and to 1 otherwise. We can confirm that the new column, `highschoolgrad`, contains the values we want by using `groupby` to get the min and max values of `highestgradecompleted` at each level of `highschoolgrad`:

```
nls97abb['highschoolgrad'] = \
   np.where(nls97abb.highestgradecompleted.isnull(),np.
nan, \
   np.where(nls97abb.highestgradecompleted<12,0,1))

nls97abb.groupby(['highschoolgrad'], dropna=False) \
   ['highestgradecompleted'].agg(['min','max','size'])
```

|                 | min | max | size |
|-----------------|-----|-----|------|
| highschoolgrad  |     |     |      |
| 0               | 5   | 11  | 1231 |
| 1               | 12  | 20  | 5421 |
| nan             | nan | nan | 2332 |

```
 nls97abb['highschoolgrad'] = \
...   nls97abb['highschoolgrad'].astype('category')
```

While 12 makes conceptual sense as the threshold for classifying our new feature, `highschoolgrad`, this would present some modeling challenges if we intended to use `highschoolgrad` as a target. There is a pretty substantial class imbalance, with `highschoolgrad` equal to 1 class being more than 4 times the size of the 0 group. We should explore using more groups to represent `highestgradecompleted`.

14. One way to do this with pandas is with the `qcut` function. We can set the `q` parameter of `qcut` to 6 to create six groups that are as evenly distributed as possible. These groups are now closer to being balanced:

```
nls97abb['highgradegroup'] = \
   pd.qcut(nls97abb['highestgradecompleted'],
     q=6, labels=[1,2,3,4,5,6])

nls97abb.groupby(['highgradegroup'])
```

```
['highestgradecompleted'].\
    agg(['min','max','size'])
                    min         max        size
highgradegroup
1                     5          11        1231
2                    12          12        1389
3                    13          14        1288
4                    15          16        1413
5                    17          17         388
6                    18          20         943
nls97abb['highgradegroup'] = \
    nls97abb['highgradegroup'].astype('category')
```

15. Finally, I typically find it helpful to generate frequencies for all the categorical features and save that output so that I can refer to it later. I rerun that code whenever I make some change to the data that may change these frequencies. The following code iterates over all the columns that are of the category data type and runs value_counts:

```
freqout = open('views/frequencies.txt', 'w')
for col in nls97abb.select_dtypes(include=["category"]):
    print(col, "----------------------",
        "frequencies",
        nls97abb[col].value_counts(dropna=False).sort_
index(),
        "percentages",
        nls97abb[col].value_counts(normalize=True).\
        sort_index(),
        sep="\n\n", end="\n\n\n", file=freqout)

freqout.close()
```

These are the key techniques for generating one-way frequencies for the categorical features in your data. The real star of the show has been the value_counts method. We can use value_counts to create frequencies a Series at a time, use it with apply for multiple columns, or iterate over several columns and call value_counts each time. We have looked at examples of each in this section. Next, let's explore some techniques for examining the distribution of continuous features.

# Generating summary statistics for continuous and discrete features

Getting a feel for the distribution of continuous or discrete features is a little more complicated than it is for categorical features. A continuous feature can take an infinite number of values. An example of a continuous feature is weight, as someone can weigh 70 kilograms, or 70.1, or 70.01. Discrete features have a finite number of values, such as the number of birds sighted, or the number of apples purchased. One way of thinking about the difference is that a discrete feature is typically something that has been counted, while a continuous feature is usually captured by measurement, weighing, or timekeeping.

Continuous features will generally be stored as floating-point numbers unless they have been constrained to be whole numbers. In that case, they may be stored as integers. Age for individual humans, for example, is continuous but is usually truncated to an integer.

For most modeling purposes, continuous and discrete features are treated similarly. We would not model age as a categorical feature. We assume that the interval between ages has largely the same meaning between 25 and 26 as it has between 35 and 36, though this breaks down at the extremes. The interval between 1 and 2 years of age for humans is not at all like that between 71 and 72. Data analysts and scientists are usually skeptical of assumed linear relationships between continuous features and targets, though modeling is much easier when that is true.

To understand how a continuous feature (or discrete feature) is distributed, we must examine its central tendency, shape, and spread. Key summary statistics are mean and median for central tendency, skewness and kurtosis for shape, and range, interquartile range, variance, and standard deviation for spread. In this section, we will learn how to use pandas, supplemented by the **SciPy** library, to get these statistics. We will also discuss important implications for modeling.

We will work with COVID-19 data in this section. The dataset contains one row per country, with total cases and deaths through June 2021, as well as demographic data for each country.

> **Note**
>
> *Our World in Data* provides COVID-19 public use data at `https://ourworldindata.org/coronavirus-source-data`. The data that will be used in this section was downloaded on July 9, 2021. There are more columns in the data than I have included. I created the region column based on country.

Follow these steps to generate the summary statistics:

1.  Let's load the COVID .csv file into pandas, set the index, and look at the data.
    There are 221 rows and 16 columns. The index we set, iso_code, contains a
    unique value for each row. We use sample to view two countries randomly, rather
    than the first two (we set a value for random_state to get the same results each
    time we run the code):

```
import pandas as pd
import numpy as np
import scipy.stats as scistat
covidtotals = pd.read_csv("data/covidtotals.csv",
    parse_dates=['lastdate'])
covidtotals.set_index("iso_code", inplace=True)
covidtotals.shape
(221, 16)
covidtotals.index.nunique()
221
covidtotals.sample(2, random_state=6).T
```

| iso_code | ISL | CZE |
|---|---|---|
| lastdate | 2021-07-07 | 2021-07-07 |
| location | Iceland | Czechia |
| total_cases | 6,555 | 1,668,277 |
| total_deaths | 29 | 30,311 |
| total_cases_mill | 19,209 | 155,783 |
| total_deaths_mill | 85 | 2,830 |
| population | 341,250 | 10,708,982 |
| population_density | 3 | 137 |
| median_age | 37 | 43 |
| gdp_per_capita | 46,483 | 32,606 |
| aged_65_older | 14 | 19 |
| total_tests_thous | NaN | NaN |
| life_expectancy | 83 | 79 |
| hospital_beds_thous | 3 | 7 |
| diabetes_prevalence | 5 | 7 |
| region | Western Europe | Western Europe |

Just by looking at these two rows, we can see significant differences in cases and deaths between Iceland and Czechia, even in terms of population size. (`total_cases_mill` and `total_deaths_mill` divide cases and deaths per million of population, respectively.) Data analysts are very used to wondering if there is anything else in the data that may explain substantially higher cases and deaths in Czechia than in Iceland. In a sense, we are always engaging in feature selection.

2.  Let's take a look at the data types and number of non-null values for each column. Almost all of the columns are continuous or discrete. We have data on cases and deaths, as well as likely targets, for 192 and 185 rows, respectively. An important data cleaning task we'll have to do will be figuring out what, if anything, we can do about countries that have missing values for our targets. We'll discuss how to handle missing values later:

```
covidtotals.info()
<class 'pandas.core.frame.DataFrame'>
Index: 221 entries, AFG to ZWE
Data columns (total 16 columns):
 #    Column               Non-Null Count    Dtype
---   -------              ---------------   ---------------
 0    lastdate             221 non-null      datetime64[ns]
 1    location             221 non-null              object
 2    total_cases          192 non-null             float64
 3    total_deaths         185 non-null             float64
 4    total_cases_mill     192 non-null             float64
 5    total_deaths_mill    185 non-null             float64
 6    population           221 non-null             float64
 7    population_density   206 non-null             float64
 8    median_age           190 non-null             float64
 9    gdp_per_capita       193 non-null             float64
 10   aged_65_older        188 non-null             float64
 11   total_tests_thous     13 non-null             float64
 12   life_expectancy      217 non-null             float64
 13   hospital_beds_thous  170 non-null             float64
 14   diabetes_prevalence  200 non-null             float64
 15   region               221 non-null              object
dtypes: datetime64[ns](1), float64(13), object(2)
memory usage: 29.4+ KB
```

3. Now, we are ready to examine the distribution of some of the features. We can get most of the summary statistics we want by using the describe method. The mean and median (50%) are good indicators of the center of the distribution, each with its strengths. It is also good to notice substantial differences between the mean and median, as an indication of skewness. For example, we can see that the mean cases per million is almost twice the median, with 36.7 thousand compared to 19.5 thousand. This is a clear indicator of positive skew. This is also true for deaths per million.

The interquartile range is also quite large for cases and deaths, with the 75th percentile value being about 25 times larger than the 25th percentile value in both cases. We can compare that with the percentage of the population aged 65 and older and diabetes prevalence, where the 75th percentile is just four times or two times that of the 25th percentile, respectively. We can tell right away that those two possible features (aged_65_older and diabetes_prevalence) would have to do a lot of work to explain the huge variance in our targets:

```
keyvars = ['location','total_cases_mill','total_deaths_
mill',
...    'aged_65_older','diabetes_prevalence']
covidkeys = covidtotals[keyvars]
covidkeys.describe()
```

| | total_cases_mill | total_deaths_mill | aged_65_older | diabetes_prevalence |
|---|---|---|---|---|
| count | 192.00 | 185.00 | 188.00 | 200.00 |
| mean | 36,649.37 | 683.14 | 8.61 | 8.44 |
| std | 41,403.98 | 861.73 | 6.12 | 4.89 |
| min | 8.52 | 0.35 | 1.14 | 0.99 |
| 25% | 2,499.75 | 43.99 | 3.50 | 5.34 |
| 50% | 19,525.73 | 293.50 | 6.22 | 7.20 |
| 75% | 64,834.62 | 1,087.89 | 13.92 | 10.61 |
| max | 181,466.38 | 5,876.01 | 27.05 | 30.53 |

4.  I sometimes find it helpful to look at the decile values to get a better sense of the distribution. The `quantile` method can take a single value for quantile, such as `quantile(0.25)` for the 25th percentile, or a list or tuple, such as `quantile((0.25,0.5))` for the 25th and 50th percentiles. In the following code, we're using `arange` from NumPy (`np.arange(0.0, 1.1, 0.1)`) to generate an array that goes from 0.0 to 1.0 with a 0.1 increment. We would get the same result if we were to use `covidkeys.quantile([0.0,0.1,0.2,0.3,0.4,0.5,0.6,0.7,0.8,0.9,1.0])`:

```
covidkeys.quantile(np.arange(0.0, 1.1, 0.1))
            total_cases_mill   total_deaths_mill   aged_65_older
diabetes_prevalence
0.00            8.52                0.35               1.14          0.99
0.10          682.13               10.68               2.80          3.30
0.20        1,717.39               30.22               3.16          4.79
0.30        3,241.84               66.27               3.86          5.74
0.40        9,403.58              145.06               4.69          6.70
0.50       19,525.73              293.50               6.22          7.20
0.60       33,636.47              556.43               7.93          8.32
0.70       55,801.33              949.71              11.19         10.08
0.80       74,017.81            1,333.79              14.92         11.62
0.90       94,072.18            1,868.89              18.85         13.75
1.00      181,466.38            5,876.01              27.05         30.53
```

For cases, deaths, and diabetes prevalence, much of the range (the distance between the min and max values) is in the last 10% of the distribution. This is particularly true for deaths. This hints at possible modeling problems and invites us to take a close look at outliers, something we will do in the next section.

5.  Some machine learning algorithms assume that our features have normal (also referred to as Gaussian) distributions, that they are distributed symmetrically (have low skew), and that they have relatively normal tails (neither excessively high nor excessively low kurtosis). The statistics we have seen so far already suggest a high positive skew for our two likely targets – that is, total cases and deaths per million people in the population. Let's put a finer point on this by calculating both skew and kurtosis for some of the features. For a Gaussian distribution, we expect a value near 0 for skew and 3 for kurtosis. `total_deaths_mill` has values for skew and kurtosis that are worth noting, and the `total_cases_mill` and `aged_65_older` features have excessively low kurtosis (skinny tails):

```
covidkeys.skew()
total_cases_mill              1.21
```

```
total_deaths_mill        2.00
aged_65_older            0.84
diabetes_prevalence      1.52
dtype: float64
  covidkeys.kurtosis()


total_cases_mill         0.91
total_deaths_mill        6.58
aged_65_older           -0.56
diabetes_prevalence      3.31
dtype: float64
```

6.  We can also explicitly test each distribution's normality by looping over the
    features in the keyvars list and running a **Shapiro-Wilk** test on the distribution
    (scistat.shapiro(covidkeys[var].dropna())). Notice that we need
    to drop missing values with dropna for the test to run. p-values less than 0.05
    indicate that we can reject the null hypothesis of normal, which is the case for each
    of the four features:

```
for var in keyvars[1:]:
        stat, p = scistat.shapiro(covidkeys[var].dropna())
        print("feature=", var, "      p-value=", '{:.6f}'.
format(p))


feature= total_cases_mill        p-value= 0.000000
feature= total_deaths_mill       p-value= 0.000000
feature= aged_65_older           p-value= 0.000000
feature= diabetes_prevalence     p-value= 0.000000
```

These results should make us pause if we are considering parametric models such as linear
regression. None of the distributions approximates a normal distribution. However, this
is not determinative. It is not as simple as deciding that we should use certain models
when we have normally distributed features and non-parametric models (say, k-nearest
neighbors) when we do not.

We want to do additional data cleaning before we make any modeling decisions.
For example, we may decide to remove outliers or determine that it is appropriate
to transform the data. We will explore transformations, such as log and polynomial
transformations, in several chapters in this book.

This section showed you how to use pandas and SciPy to understand how continuous and discrete features are distributed, including their central tendency, shape, and spread. It makes sense to generate these statistics for any feature or target that might be included in our modeling. This also points us in the direction of more work we need to do to prepare our data for analysis. We need to identify missing values and outliers and figure out how we will handle them. We should also visualize the distribution of our continuous features. This rarely fails to yield additional insights. We will learn how to identify outliers in the next section and create visualizations in the following section.

# Identifying extreme values and outliers in univariate analysis

An outlier can be thought of as an observation with feature values, or relationships between feature values, that are so unusual that they cannot help explain relationships in the rest of the data. This matters for modeling because we cannot assume that the outliers will have a neutral impact on our parameter estimates. Sometimes, our models work so hard to construct parameter estimates that can account for patterns in outlier observations that we compromise the model's explanatory or predictive power for all other observations. Raise your hand if you have ever spent days trying to interpret a model only to discover that your coefficients and predictions completely changed once you removed a few outliers.

I should quickly add that there is no agreed-upon definition of an outlier. I offer the preceding definition for use in this book because it helps us distinguish between outliers, as I have described them, and extreme values. There is a fair bit of overlap between the two, but many extreme values are not outliers. This is because such values reflect a natural and explainable trend in a feature, or because they reflect the same relationship between features as is observed throughout the data. The reverse is also true. Some outliers are not extreme values. For example, a target value might be right in the middle of the distribution but have quite unexpected predictor values.

For our modeling, then, it is hard to say that a particular feature or target value is an outlier without referencing multivariate relationships. But it should at least raise a red flag when, in our univariate analysis, we see values well to the left or right of the center. This should prompt us to investigate the observation at that value further, including examining the values of other features. We will look at multivariate relationships in more detail in the next two chapters. Here, and in the next section on visualizations, we will focus on identifying extreme values and outliers when looking at one variable.

A good starting point for identifying an extreme value is to look at its distance from the middle of the distribution. One common method for doing that is to calculate each value's distance from the **interquartile range (IQR)**, which is the distance between the first quartile value and the third quartile value. We often flag any value that is more than 1.5 times the interquartile range above the third quartile or below the first quartile. We can use this method to identify outliers in the COVID-19 data.

Let's get started:

1.  Let's start by importing the libraries we will need. In addition to pandas and NumPy, we will use Matplotlib and statsmodels for the plots we will create. We will also load the COVID data and select the variables we need:

```
import pandas as pd
import numpy as np
import matplotlib.pyplot as plt
import statsmodels.api as sm

covidtotals = pd.read_csv("data/covidtotals.csv")
covidtotals.set_index("iso_code", inplace=True)
keyvars = ['location','total_cases_mill','total_deaths_
mill',
    'aged_65_older','diabetes_prevalence','gdp_per_capita']
covidkeys = covidtotals[keyvars]
```

2.  Let's take a look at `total_cases_mill`. We get the first and third quartile values and calculate the interquartile range, `1.5*(thirdq-firstq)`. Then, we calculate their thresholds to determine the high and low extreme values, which are `interquartilerange+thirdq` and `firstq-interquartilerange`, respectively (if you are familiar with boxplots, you will notice that this is the same calculation that's used for the whiskers of a boxplot; we will cover boxplots in the next section):

```
thirdq, firstq = covidkeys.total_cases_mill.
quantile(0.75), covidkeys.total_cases_mill.quantile(0.25)
interquartilerange = 1.5*(thirdq-firstq)
extvalhigh, extvallow = interquartilerange+thirdq,
firstq-interquartilerange
print(extvallow, extvalhigh, sep=" <--> ")
-91002.564625 <--> 158336.930375
```

3.  This calculation indicates that any value for `total_cases_mill` that's above 158,337 can be considered extreme. We can ignore extreme values on the low end because they would be negative:

```
covidtotals.loc[covidtotals.total_cases_
mill>extvalhigh].T
```

| iso_code | AND | MNE | SYC |
|---|---|---|---|
| lastdate | 2021-07-07 | 2021-07-07 | 2021-07-07 |
| location | Andorra | Montenegro | Seychelles |
| total_cases | 14,021 | 100,392 | 16,304 |
| total_deaths | 127 | 1,619 | 71 |
| total_cases_mill | 181,466 | 159,844 | 165,792 |
| total_deaths_mill | 1,644 | 2,578 | 722 |
| population | 77,265 | 628,062 | 98,340 |
| population_density | 164 | 46 | 208 |
| median_age | NaN | 39 | 36 |
| gdp_per_capita | NaN | 16,409 | 26,382 |
| aged_65_older | NaN | 15 | 9 |
| total_tests_thous | NaN | NaN | NaN |
| life_expectancy | 84 | 77 | 73 |
| hospital_beds_thous | NaN | 4 | 4 |
| diabetes_prevalence | 8 | 10 | 11 |
| region | Western Europe | Eastern Europe | East Africa |

4.  Andorra, Montenegro, and Seychelles all have `total_cases_mill` above the threshold amount. This invites us to explore other ways these countries might be exceptional, and whether our features can capture that. We will not dive deeply into multivariate analysis here since we will do that in the next chapter, but it is a good idea to start wrapping our brains around why these extreme values may or may not make sense. Having some means across the full dataset may help us here:

```
covidtotals.mean()
total_cases             963,933
total_deaths             21,631
total_cases_mill         36,649
total_deaths_mill           683
population           35,134,957
population_density          453
```

| | |
|---|---|
| median_age | 30 |
| gdp_per_capita | 19,141 |
| aged_65_older | 9 |
| total_tests_thous | 535 |
| life_expectancy | 73 |
| hospital_beds_thous | 3 |
| diabetes_prevalence | 8 |

The main difference between these three countries and others is that they have very low populations. Surprisingly, each has a much lower population density than average. That is the opposite of what you would expect and merits further consideration in the analysis we will do throughout this book.

---

**An Alternative to the IQR Calculation**

An alternative to using the interquartile range to identify an extreme value would be to use several standard deviations away from the mean, say 3. One drawback of this method is that it is a little more susceptible to extreme values than using the interquartile range.

---

I find it helpful to produce this kind of analysis for all the key targets and features in my data, so let's automate this method of identifying extreme values. We should also output the results to a file so that we can use them when we need them:

5.  Let's define a function, getextremevalues, that iterates over all of the columns of our DataFrame (except for the first one, which contains the location column), calculates the interquartile range for that column, selects all the observations with values above the high threshold or below the low threshold for that column, and then appends the results to a new DataFrame (dfout):

```
def getextremevalues(dfin):
    dfout = pd.DataFrame(columns=dfin.columns,
                         data=None)
    for col in dfin.columns[1:]:
      thirdq, firstq = dfin[col].quantile(0.75), \
        dfin[col].quantile(0.25)
      interquartilerange = 1.5*(thirdq-firstq)
      extvalhigh, extvallow = \
        interquartilerange+thirdq, \
        firstq-interquartilerange
```

```
          df = dfin.loc[(dfin[col]>extvalhigh) |
    (dfin[col]<extvallow)]
          df = df.assign(varname = col,
            threshlow = extvallow,
            threshhigh = extvalhigh)
          dfout = pd.concat([dfout, df])
      return dfout
```

6.  Now, we can pass our covidkeys DataFrame to the getextremevalues
    function to get a DataFrame that contains the extreme values for each column.
    Then, we can display the number of extreme values for each column, which tells us
    that there were four extreme values for the total deaths per million people in the
    population (total_deaths_mill). Now, we can output the data to an Excel file:

```
extremevalues = getextremevalues(covidkeys)
extremevalues.varname.value_counts()
gdp_per_capita            9
diabetes_prevalence       8
total_deaths_mill         4
total_cases_mill          3
Name: varname, dtype: int64
extremevalues.to_excel("views/extremevaluescases.xlsx")
```

7.  Let's take a closer look at the extreme deaths per million values. We can query the
    DataFrame we just created to get the threshhigh value for total_deaths_
    mill, which is 2654. We can also get other key features for those countries with
    the extreme values since we have included that data in the new DataFrame:

```
extremevalues.loc[extremevalues.varname=="total_deaths_
mill",
     'threshhigh'][0]
2653.752
extremevalues.loc[extremevalues.varname=="total_deaths_
mill",
      keyvars].sort_values(['total_deaths_mill'],
ascending=False)
```

```
          location                    total_cases_mill    total_
deaths_mill
```

| | | | |
|---|---|---|---|
| PER | Peru | 62,830 | 5,876 |
| HUN | Hungary | 83,676 | 3,105 |
| BIH | Bosnia and Herzegovina | 62,499 | 2,947 |
| CZE | Czechia | 155,783 | 2,830 |

| | _65_older | diabetes_prevalence | gdp_per_capita |
|---|---|---|---|
| PER | 7 | 6 | 12,237 |
| HUN | 19 | 8 | 26,778 |
| BIH | 17 | 10 | 11,714 |
| CZE | 19 | 7 | 32,606 |

Peru, Hungary, Bosnia and Herzegovina, and Czechia have total_deaths_mill above the extreme value threshold. One thing that stands out for three of these countries is how much above the average for percent of population 65 or older they are as well (the average for that feature, as we displayed in a preceding table, is 9). Although these are extreme values for deaths, the relationship between the elderly percentage of the population and COVID deaths may account for much of this and can do so without overfitting the model to these extreme cases. We will go through some strategies for teasing that out in the next chapter.

So far, we have discussed outliers and extreme values without referencing distribution shape. What we have implied so far is that an extreme value is a rare value – significantly rarer than the values near the center of the distribution. But this makes the most sense when the feature's distribution approaches normal. If, on the other hand, a feature had a uniform distribution, a very high value would be no rarer than any other value.

In practice, then, we think about extreme values or outliers relative to the distribution of the feature. **Quantile-quantile (Q-Q)** plots can improve our sense of that distribution by allowing us to view it graphically relative to a theoretical distribution: normal, uniform, log, or others. Let's take a look:

1.  Let's create a Q-Q plot of total cases per million that's relative to the normal distribution. We can use the statsmodels library for this:

    ```
    sm.qqplot(covidtotals[['total_cases_mill']]. \
        sort_values(['total_cases_mill']).dropna(),line='s')
    )
    plt.show()
    ```

This produces the following plot:

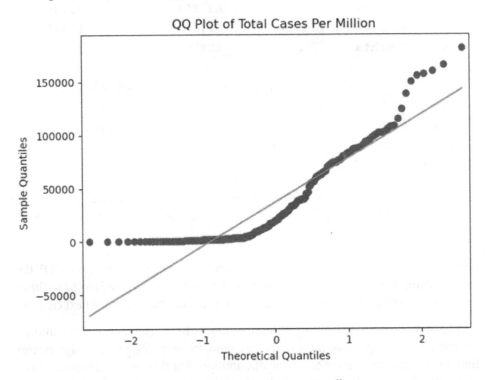

Figure 1.1 – Q-Q plot of total cases per million

This Q-Q plot makes it clear that the distribution of total cases across countries is not normal. We can see this by how much the data points deviate from the red line. It is a Q-Q plot that we would expect from a distribution with some positive skew. This is consistent with the summary statistics we have already calculated for the total cases feature. It further reinforces our developing sense, in that we will need to be cautious about parametric models and that we will probably have to account for more than just one or two outlier observations.

Let's look at a Q-Q plot for a feature with a distribution that is a little closer to normal. There isn't a great candidate in the COVID data, so we will work with data from the United States National Oceanic and Atmospheric Administration on land temperatures in 2019.

> **Data Note**
>
> The land temperature dataset contains the average temperature readings (in Celsius) in 2019 from over 12,000 stations across the world, though the majority of the stations are in the United States. The raw data was retrieved from the Global Historical Climatology Network integrated database. It has been made available for public use by the United States National Oceanic and Atmospheric Administration at `https://www.ncdc.noaa.gov/` `data-access/land-based-station-data/land-based-` `datasets/global-historical-climatology-network-` `monthly-version-2`.

2.  First, let's load the data into a pandas DataFrame and run some descriptive statistics on the temperature feature, `avgtemp`. We must add a few percentile statistics to the normal `describe` output to get a better sense of the range of values. `avgtemp` is the average temperature for the year at each of the 12,095 weather stations. The average temperature across all stations was 11.2 degrees Celsius. The median was 10.4. However, there were some very negative values, including 14 weather stations with an average temperature of less than -25. This contributes to a moderately negative skew, though both the skew and kurtosis are closer to what would be expected from a normal distribution:

```
landtemps = pd.read_csv("data/landtemps2019avgs.csv")
landtemps.avgtemp.describe(percentiles=[0.05, 0.1, 0.25,
0.5, 0.75, 0.9, 0.95])
count           12,095.0
mean                11.2
std                  8.6
min                -60.8
5%                  -0.7
10%                  1.7
25%                  5.4
50%                 10.4
75%                 16.9
90%                 23.1
95%                 27.0
max                 33.9
Name: avgtemp, dtype: float64
  landtemps.loc[landtemps.avgtemp<-25,'avgtemp'].count()
14
```

```
landtemps.avgtemp.skew()
-0.2678382583481768
landtemps.avgtemp.kurtosis()
2.1698313707061073
```

3.    Now, let's take a look at a Q-Q plot of the average temperature:

```
sm.qqplot(landtemps.avgtemp.sort_values().dropna(),
line='s')
plt.tight_layout()
plt.show()
```

This produces the following plot:

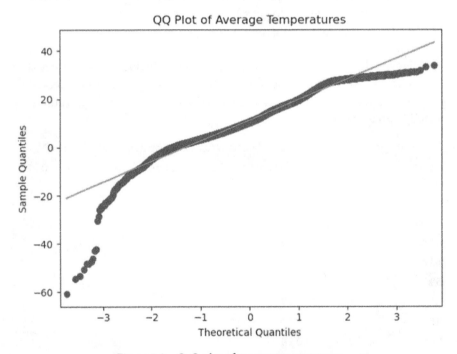

Figure 1.2 – Q-Q plot of average temperatures

Along most of the range, the distribution of average temperatures looks pretty close
to normal. The exceptions are the extremely low temperatures, contributing to a small
amount of negative skew. There is also some deviation from normal at the high end,
though this is much less of an issue (you may have noticed that Q-Q plots for features
with negative skew have an umbrella-like shape, while those with positive skews, such as
total cases, have more of a bowl-like shape).

We are off to a good start in our efforts to understand the distribution of possible features and targets and, to a related effort, to identify extreme values and outliers. This is important information to have at our fingertips when we construct, refine, and interpret our models. But there is more that we can do to improve our intuition about the data. A good next step is to construct visualizations of key features.

# Using histograms, boxplots, and violin plots to examine the distribution of features

We have already generated many of the numbers that would make up the data points of a histogram or boxplot. But we often improve our understanding of the data when we see it represented graphically. We see observations bunched around the mean, we notice the size of the tails, and we see what seem to be extreme values.

## Using histograms

Follow these steps to create a histogram:

1.  We will work with both the COVID data and the temperatures data in this section. In addition to the libraries we have worked with so far, we must import Seaborn to create some plots more easily than we could in Matplotlib:

```
import pandas as pd
import numpy as np
import matplotlib.pyplot as plt
import seaborn as sns

landtemps = pd.read_csv("data/landtemps2019avgs.csv")
covidtotals = pd.read_csv("data/covidtotals.csv", parse_
dates=["lastdate"])
covidtotals.set_index("iso_code", inplace=True)
```

2.  Now, let's create a simple histogram. We can use Matplotlib's `hist` method to create a histogram of total cases per million. We will also draw lines for the mean and median:

```
plt.hist(covidtotals['total_cases_mill'], bins=7)
plt.axvline(covidtotals.total_cases_mill.mean(),
color='red',
    linestyle='dashed', linewidth=1, label='mean')
```

```
plt.axvline(covidtotals.total_cases_mill.median(),
color='black',
    linestyle='dashed', linewidth=1, label='median')
plt.title("Total COVID Cases")
plt.xlabel('Cases per Million')
plt.ylabel("Number of Countries")
plt.legend()
plt.show()
```

This produces the following plot:

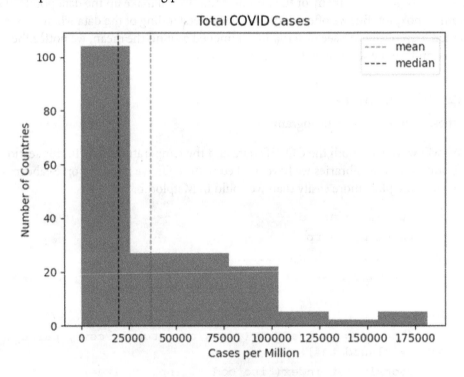

Figure 1.3 – Total COVID cases

One aspect of the total distribution that this histogram highlights is that most countries (more than 100 of the 192) are in the very first bin, between 0 cases per million and 25,000 cases per million. Here, we can see the positive skew, with the mean pulled to the right by extreme high values. This is consistent with what we discovered when we used Q-Q plots in the previous section.

3.  Let's create a histogram of average temperatures from the land temperatures dataset:

```
plt.hist(landtemps['avgtemp'])
plt.axvline(landtemps.avgtemp.mean(), color='red',
linestyle='dashed', linewidth=1, label='mean')
plt.axvline(landtemps.avgtemp.median(), color='black',
linestyle='dashed', linewidth=1, label='median')
plt.title("Average Land Temperatures")
plt.xlabel('Average Temperature')
plt.ylabel("Number of Weather Stations")
plt.legend()
plt.show()
```

This produces the following plot:

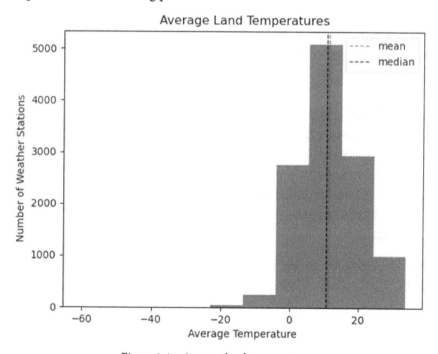

Figure 1.4 – Average land temperatures

The histogram for the average land temperatures from the land temperatures dataset looks quite different. Except for a few highly negative values, this distribution looks closer to normal. Here, we can see that the mean and the median are quite close and that the distribution looks fairly symmetrical.

4.   We should take a look at the observations at the extreme left of the distribution. They are all in Antarctica or the extreme north of Canada. Here, we have to wonder if it makes sense to include observations with such extreme values in the models we construct. However, it would be premature to make that determination based on these results alone. We will come back to this in the next chapter when we examine multivariate techniques for identifying outliers:

```
landtemps.loc[landtemps.avgtemp<-25,['station','country'
,'avgtemp']].\
...    sort_values(['avgtemp'], ascending=True)
```

|      | station | country | avgtemp |
|------|---------|---------|---------|
| 827  | DOME_PLATEAU_DOME_A | Antarctica | -60.8 |
| 830  | VOSTOK | Antarctica | -54.5 |
| 837  | DOME_FUJI | Antarctica | -53.4 |
| 844  | DOME_C_II | Antarctica | -50.5 |
| 853  | AMUNDSEN_SCOTT | Antarctica | -48.4 |
| 842  | NICO | Antarctica | -48.4 |
| 804  | HENRY | Antarctica | -47.3 |
| 838  | RELAY_STAT | Antarctica | -46.1 |
| 828  | DOME_PLATEAU_EAGLE | Antarctica | -43.0 |
| 819  | KOHNENEP9 | Antarctica | -42.4 |
| 1299 | FORT_ROSS | Canada | -30.3 |
| 1300 | GATESHEAD_ISLAND | Canada | -28.7 |
| 811  | BYRD_STATION | Antarctica | -25.8 |
| 816  | GILL | Antarctica | -25.5 |

An excellent way to visualize central tendency, spread, and outliers at the same time is with a boxplot.

# Using boxplots

**Boxplots** show us the interquartile range, with whiskers representing 1.5 times the interquartile range, and data points beyond that range that can be considered extreme values. If this calculation seems familiar, it's because it's the same one we used earlier in this chapter to identify extreme values! Let's get started:

1. We can use the Matplotlib `boxplot` method to create a boxplot of total cases per million people in the population. We can draw arrows to show the interquartile range (the first quartile, median, and third quartile) and the extreme value threshold. The three circles above the threshold can be considered extreme values. The line from the interquartile range to the extreme value threshold is typically referred to as the whisker. There are usually whiskers above and below the interquartile range, but the threshold value below the first quartile value would be negative in this case:

```
plt.boxplot(covidtotals.total_cases_mill.dropna(),
labels=['Total Cases per Million'])
plt.annotate('Extreme Value Threshold',
xy=(1.05,157000), xytext=(1.15,157000), size=7,
arrowprops=dict(facecolor='black', headwidth=2,
width=0.5, shrink=0.02))
plt.annotate('3rd quartile', xy=(1.08,64800),
xytext=(1.15,64800), size=7,
arrowprops=dict(facecolor='black', headwidth=2,
width=0.5, shrink=0.02))
plt.annotate('Median', xy=(1.08,19500),
xytext=(1.15,19500), size=7,
arrowprops=dict(facecolor='black', headwidth=2,
width=0.5, shrink=0.02))
plt.annotate('1st quartile',
xy=(1.08,2500), xytext=(1.15,2500), size=7,
arrowprops=dict(facecolor='black', headwidth=2,
width=0.5, shrink=0.02))
plt.title("Boxplot of Total Cases")
plt.show()
```

This produces the following plot:

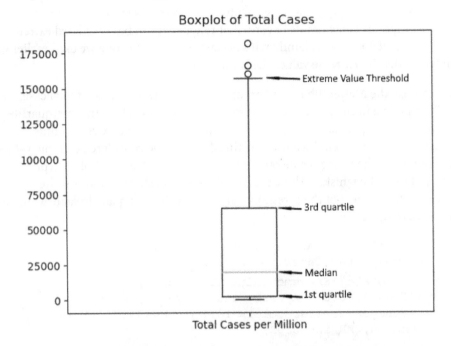

Figure 1.5 – Boxplot of total cases

It is helpful to take a closer look at the interquartile range, specifically where the median falls within the range. For this boxplot, the median is at the lower end of the range. This is what we see in distributions with positive skews.

2.  Now, let's create a boxplot for the average temperature. All of the extreme values are now at the low end of the distribution. Unsurprisingly, given what we have already seen with the average temperature feature, the median line is closer to the center of the interquartile range than with our previous boxplot (we will not annotate the plot this time – we only did this last time for explanatory purposes):

```
plt.boxplot(landtemps.avgtemp.dropna(), labels=['Boxplot
of Average Temperature'])
plt.title("Average Temperature")
plt.show()
```

This produces the following plot:

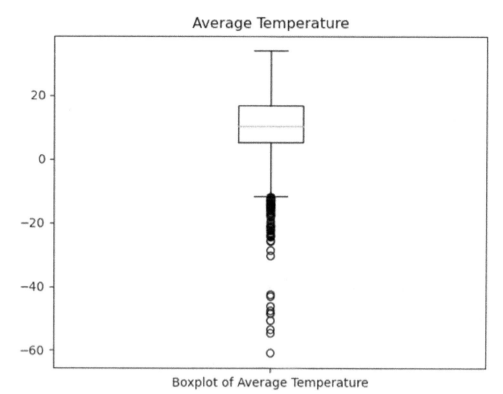

Figure 1.6 – Boxplot of average temperature

Histograms help us see the spread of a distribution, while boxplots make it easy to identify outliers. We can get a good sense of both the spread of the distribution and the outliers in one graphic with a violin plot.

## Using violin plots

Violin plots combine histograms and boxplots into one plot. They show the IQR, median, and whiskers, as well as the frequency of the observations at all the value ranges.

Let's get started:

1. We can use Seaborn to create violin plots of both the COVID cases per million and the average temperature features. I am using Seaborn here, rather than Matplotlib, because I prefer its default options for violin plots:

```
import seaborn as sns
fig = plt.figure()
fig.suptitle("Violin Plots of COVID Cases and Land
Temperatures")
```

```
ax1 = plt.subplot(2,1,1)
ax1.set_xlabel("Cases per Million")
sns.violinplot(data=covidtotals.total_cases_mill,
color="lightblue", orient="h")
ax1.set_yticklabels([])
ax2 = plt.subplot(2,1,2)
ax2.set_xlabel("Average Temperature")
sns.violinplot(data=landtemps.avgtemp, color="wheat",
orient="h")
ax2.set_yticklabels([])
plt.tight_layout()
plt.show()
```

This produces the following plot:

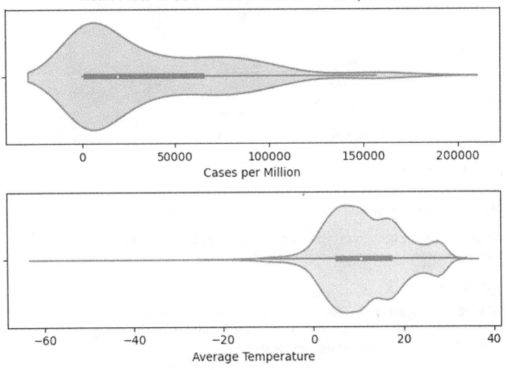

Figure 1.7 – Violin plots of COVID cases and land temperatures

The black bar with the white dot in the middle is the interquartile range, while the white dot represents the median. The height at each point (when the violin plot is horizontal) gives us the relative frequency. The thin black lines to the right of the interquartile range for cases per million, and to the right and left for the average temperature are the whiskers. The extreme values are shown in the part of the distribution beyond the whiskers.

If I am going to create just one plot for a numeric feature, I will create a violin plot. Violin plots allow me to see central tendency, shape, and spread all in one graphic. Space does not permit it here, but I usually like to create violin plots of all of my continuous features and save those to a PDF file for later reference.

## Summary

In this chapter, we looked at some common techniques for exploring data. We learned how to retrieve subsets of data when that is required for our analysis. We also used pandas methods to generate key statistics on features such as mean, interquartile range, and skew. This gave us a better sense of the central tendency, spread, and shape of the distribution of each feature. It also put us in a better position to identify outliers. Finally, we used the Matplotlib and Seaborn libraries to create histograms, boxplots, and violin plots. This yielded additional insights about the distribution of features, such as the length of the tail and divergence from the normal distribution.

Visualizations are a great supplement to the tools for univariate analysis that we have discussed in this chapter. Histograms, boxplots, and violin plots display the shape and spread of each feature's distribution. Graphically, they show what we may miss by examining a few summary statistics, such as where there is a bulge (or bulges) in the distribution and where the extreme values are. These visualizations will be every bit as helpful when we explore bivariate and multivariate relationships, which we will do in *Chapter 2, Examining Bivariate and Multivariate Relationships between Features and Targets*.

# 2

# Examining Bivariate and Multivariate Relationships between Features and Targets

In this chapter, we'll look at the correlation between possible features and target variables. Bivariate exploratory analysis, using crosstabs (two-way frequencies), correlations, scatter plots, and grouped boxplots can uncover key issues for modeling. Common issues include high correlation between features and non-linear relationships between features and the target variable. We will use pandas methods for bivariate analysis and Matplotlib for visualizations in this chapter. We will also discuss the implications of what we find in terms of feature engineering and modeling.

We will also use multivariate techniques to understand the relationship between features. This includes leaning on some machine learning algorithms to identify possibly problematic observations. After, we will provide tentative recommendations for eliminating certain observations from our modeling, as well as for transforming key features.

In this chapter, we will cover the following topics:

- Identifying outliers and extreme values in bivariate relationships
- Using scatter plots to view bivariate relationships between continuous features
- Using grouped boxplots to view bivariate relationships between continuous and categorical features
- Using linear regression to identify data points with significant influence
- Using K-nearest neighbors to find outliers
- Using Isolation Forest to find outliers

# Technical requirements

This chapter will rely heavily on the pandas and Matplotlib libraries, but you don't require any prior knowledge of these. If you have installed Python from a scientific distribution, such as Anaconda or WinPython, then these libraries have probably already been installed. We will also be using Seaborn for some of our graphics and the statsmodels library for some summary statistics. If you need to install any of the packages, you can do so by running `pip install [package name]` from a terminal window or Windows PowerShell. The code for this chapter can be found in this book's GitHub repository at `https://github.com/PacktPublishing/Data-Cleaning-and-Exploration-with-Machine-Learning`.

# Identifying outliers and extreme values in bivariate relationships

It is hard to develop a reliable model without having a good sense of the bivariate relationships in our data. We not only care about the relationship between particular features and target variables but also about how features move together. If features are highly correlated, then modeling their independent effect becomes tricky or unnecessary. This may be a challenge, even if the features are highly correlated over just a range of values.

Having a good understanding of bivariate relationships is also important for identifying outliers. A value might be unexpected, even if it is not an extreme value. This is because some values for a feature are unusual when a second feature has certain values. This is easy to illustrate when one feature is categorical and the other is continuous.

The following diagram illustrates the number of bird sightings per day over several years but shows different distributions for the two sites. One site has a (mean) sightings per day of 33, while the other has 52. (This is a fictional example that's been pulled from my *Python Data Cleaning Cookbook*.) The overall mean (not shown) is 42. What should we make of a value of 58 for daily sightings? Is it an outlier? This depends on which of the two sites was being observed. If there were 58 sightings in a day at site A, 58 would be an unusually high number. However, this wouldn't be true for site B, where 58 sightings would not be very different from the mean for that site:

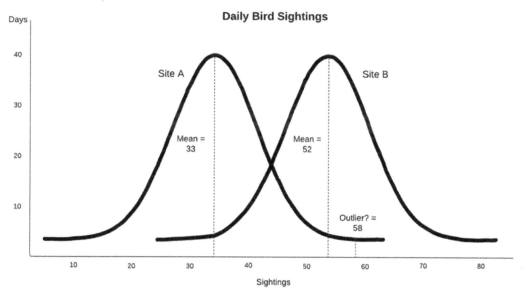

Figure 2.1 – Daily Bird Sightings

This hints at a useful rule of thumb: whenever a feature of interest is correlated with another feature, we should take that relationship into account when we're trying to identify outliers (or any modeling with that feature, actually). It is helpful to state this a little more precisely and extend it to cases where both features are continuous. If we assume a linear relationship between feature $x$ and feature $y$, we can describe that relationship with the familiar $y = mx + b$ equation, where $m$ is the slope and $b$ is the $y$-intercept. Then, we can expect the value of $y$ to be somewhere close to $x$ times the estimated slope, plus the $y$-intercept. Unexpected values are those that deviate substantially from this relationship, where the value of $y$ is much higher or lower than what would be predicted, given the value of $x$. This can be extended to multiple $x$, or predictor, variables.

In this section, we will learn how to identify outliers and unexpected values by examining the relationship a feature has with another feature. In subsequent sections of this chapter, we will use multivariate techniques to make additional improvements to our outlier detection.

We will work with data based on COVID-19 cases by country in this section. The dataset contains cases and deaths per million people in the population. We will treat both columns as possible targets. It also contains demographic data for each country, such as GDP per capita, median age, and diabetes prevalence. Let's get started:

> **Note**
>
> Our World in Data provides COVID-19 public-use data at `https://ourworldindata.org/coronavirus-source-data`. The dataset that's being used in this section was downloaded on July 9, 2021. There are more columns in the data than I have included. I created the `region` column based on country.

1. Let's start by loading the COVID-19 dataset and looking at how it is structured. We will also import the Matplotlib and Seaborn libraries since we will do a couple of visualizations:

```
import pandas as pd
import matplotlib.pyplot as plt
import seaborn as sns
covidtotals = pd.read_csv("data/covidtotals.csv")
covidtotals.set_index("iso_code", inplace=True)
covidtotals.info()
<class 'pandas.core.frame.DataFrame'>
Index: 221 entries, AFG to ZWE
Data columns (total 16 columns):
 #   Column             Non-Null Count    Dtype
---  ------             --------------    -------
 0   lastdate           221  non-null     object
 1   location           221  non-null     object
 2   total_cases        192  non-null     float64
 3   total_deaths       185  non-null     float64
 4   total_cases_mill   192  non-null     float64
 5   total_deaths_mill  185  non-null     float64
 6   population         221  non-null     float64
```

```
 7   population_density    206   non-null   float64
 8   median_age            190   non-null   float64
 9   gdp_per_capita        193   non-null   float64
10   aged_65_older         188   non-null   float64
11   total_tests_thous      13   non-null   float64
12   life_expectancy       217   non-null   float64
13   hospital_beds_thous   170   non-null   float64
14   diabetes_prevalence   200   non-null   float64
15   region                221   non-null   object
dtypes: float64(13), object(3)
memory usage: 29.4+ KB
```

2.  A great place to start with our examination of bivariate relationships is with correlations. First, let's create a DataFrame that contains a few key features:

```
totvars = ['location','total_cases_mill',
   'total_deaths_mill']
demovars = ['population_density','aged_65_older',
   'gdp_per_capita','life_expectancy',
   'diabetes_prevalence']
covidkeys = covidtotals.loc[:, totvars + demovars]
```

3.  Now, we can get the Pearson correlation matrix for these features. There is a strong positive correlation of 0.71 between cases and deaths per million. The percentage of the population that's aged 65 or older is positively correlated with cases and deaths, at 0.53 for both. Life expectancy is also highly correlated with cases per million. There seems to be at least some correlation of **gross domestic product (GDP)** per person with cases:

```
corrmatrix = covidkeys.corr(method="pearson")
corrmatrix
```

```
                      total_cases_mill   total_deaths_mill\
total_cases_mill              1.00                0.71
total_deaths_mill            0.71                1.00
population_density           0.04               -0.03
aged_65_older                0.53                0.53
gdp_per_capita               0.46                0.22
life_expectancy              0.57                0.46
diabetes_prevalence          0.02               -0.01
```

```
                    population_density  aged_65_older  gdp_
per_capita\
total_cases_mill            0.04             0.53        0.46
total_deaths_mill          -0.03             0.53        0.22
population_density          1.00             0.06        0.41
aged_65_older               0.06             1.00        0.49
gdp_per_capita              0.41             0.49        1.00
life_expectancy             0.23             0.73        0.68
diabetes_prevalence         0.01            -0.06        0.12
                    life_expectancy  diabetes_prevalence
total_cases_mill            0.57             0.02
total_deaths_mill           0.46            -0.01
population_density          0.23             0.01
aged_65_older               0.73            -0.06
gdp_per_capita              0.68             0.12
life_expectancy             1.00             0.19
diabetes_prevalence         0.19             1.00
```

It is worth noting the correlation between possible features, such as between life expectancy and GDP per capita (0.68) and life expectancy and those aged 65 or older (0.73).

4. It can be helpful to see the correlation matrix as a heat map. This can be done by passing the correlation matrix to the Seaborn `heatmap` method:

```
sns.heatmap(corrmatrix, xticklabels =
    corrmatrix.columns, yticklabels=corrmatrix.columns,
    cmap="coolwarm")
plt.title('Heat Map of Correlation Matrix')
plt.tight_layout()
plt.show()
```

This creates the following plot:

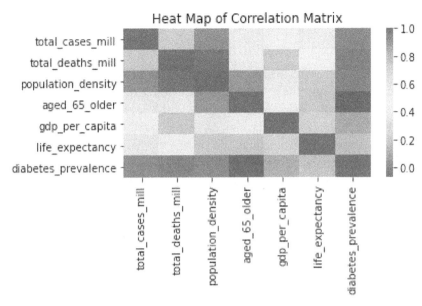

Figure 2.2 – Heat map of COVID data, with the strongest correlations in red and peach

We want to pay attention to the cells shown with warmer colors – in this case, mainly peach. I find that using a heat map helps me keep correlations in mind when modeling.

> **Note**
>
> All the color images contained in this book can be downloaded. Check the *Preface* of this book for the respective link.

5.  Let's take a closer look at the relationship between total cases per million and deaths per million. One way to get a better sense of this than with just a correlation coefficient is by comparing the high and low values for each and seeing how they move together. In the following code, we're using the `qcut` method to create a categorical feature with five values distributed relatively evenly, from very low to very high, for cases. We have done the same for deaths:

```
covidkeys['total_cases_q'] = \
  pd.qcut(covidkeys['total_cases_mill'],
  labels=['very low','low','medium','high',
  'very high'], q=5, precision=0)

covidkeys['total_deaths_q'] = \
  pd.qcut(covidkeys['total_deaths_mill'],
```

```
labels=['very low','low','medium','high',
'very high'], q=5, precision=0)
```

6.  We can use the `crosstab` function to view the number of countries for each quintile of cases and quintile of deaths. As we would expect, most of the countries are along the diagonal. There are 27 countries with very low cases and very low deaths, and 25 countries with very high cases and very high deaths. The interesting counts are those not on the diagonal, such as the four countries with very high cases but only medium deaths, nor the one with medium cases and very high deaths. Let's also look at the means of our features so that we can reference them later:

```
pd.crosstab(covidkeys.total_cases_q,
    covidkeys.total_deaths_q)
```

| total_deaths_q | very low | low | medium | high | very high |
|---|---|---|---|---|---|
| total_cases_q | | | | | |
| very low | 27 | 7 | 0 | 0 | 0 |
| low | 9 | 24 | 4 | 0 | 0 |
| medium | 1 | 6 | 23 | 6 | 1 |
| high | 0 | 0 | 6 | 21 | 11 |
| very high | 0 | 0 | 4 | 10 | 25 |

```
covidkeys.mean()
total_cases_mill        36,649
total_deaths_mill          683
population_density         453
aged_65_older                9
gdp_per_capita          19,141
life_expectancy             73
diabetes_prevalence          8
```

7.  Let's take a closer look at the countries away from the diagonal. Four countries – Cyprus, Kuwait, Maldives, and Qatar – have fewer deaths per million than average but well above average cases per million. Interestingly, all four countries are very small in terms of population; three of the four have population densities far below the average of 453; again, three of the four have people aged 65 or older percentages that are much lower than average:

```
covidtotals.loc[(covidkeys.total_cases_q=="very high")
    & (covidkeys.total_deaths_q=="medium")].T
```

| iso_code | CYP | KWT | MDV | QAT |
|---|---|---|---|---|
| lastdate | 2021-07-07 | 2021-07-07 | 2021-07-07 | 2021-07-07 |
| location | Cyprus | Kuwait | Maldives | Qatar |
| total_cases | 80,588 | 369,227 | 74,724 | 222,918 |
| total_deaths | 380 | 2,059 | 213 | 596 |
| total_cases_mill | 90,752 | 86,459 | 138,239 | 77,374 |
| total_deaths_mill | 428 | 482 | 394 | 207 |
| population | 888,005 | 4,270,563 | 540,542 | 2,881,060 |
| population_density | 128 | 232 | 1,454 | 227 |
| median_age | 37 | 34 | 31 | 32 |
| gdp_per_capita | 32,415 | 65,531 | 15,184 | 116,936 |
| aged_65_older | 13 | 2 | 4 | 1 |
| total_tests_thous | NaN | NaN | NaN | NaN |
| life_expectancy | 81 | 75 | 79 | 80 |
| hospital_beds_thous | 3 | 2 | NaN | 1 |
| diabetes_prevalence | 9 | 16 | 9 | 17 |
| region | Eastern Europe | West Asia | South Asia | West Asia |

8.  Let's take a closer look at the country with more deaths than we would have expected based on cases. For Mexico, the number of cases per million are well below average, while the number of deaths per million are quite a bit above average:

```
covidtotals.loc[(covidkeys. total_cases_q=="medium")
  & (covidkeys.total_deaths_q=="very high")].T
iso_code                           MEX
lastdate                    2021-07-07
location                        Mexico
total_cases                  2,558,369
total_deaths                   234,192
total_cases_mill                19,843
total_deaths_mill                1,816
population                  128,932,753
population_density                  66
median_age                          29
gdp_per_capita                  17,336
aged_65_older                        7
total_tests_thous                  NaN
life_expectancy                     75
hospital_beds_thous                  1
diabetes_prevalence                 13
region                   North America
```

Correlation coefficients and heat maps are a good place to start when we want to get a sense of the bivariate relationships in our dataset. However, it can be hard to visualize the relationship between continuous variables with just a correlation coefficient. This is particularly true when the relationship is not linear – that is, when it varies based on the ranges of a feature. We can often improve our understanding of the relationship between two features with a scatter plot. We will do that in the next section.

# Using scatter plots to view bivariate relationships between continuous features

In this section, we'll learn how to get a scatter plot of our data.

We can use scatter plots to get a more complete picture of the relationship between two features than what can be detected by a correlation coefficient alone. This is particularly useful when that relationship changes across certain ranges of the data. In this section, we will create scatter plots of some of the same features we examined in the previous section. Let's get started:

1.  It is helpful to plot a regression line through the data points. We can do this with Seaborn's `regplot` method. Let's load the COVID-19 data again, along with the Matplotlib and Seaborn libraries, and generate a scatter plot of `total_cases_mill` by `total_deaths_mill`:

```
import pandas as pd
import numpy as np
import matplotlib.pyplot as plt
import seaborn as sns
covidtotals = pd.read_csv("data/covidtotals.csv")
covidtotals.set_index("iso_code", inplace=True)
ax = sns.regplot(x="total_cases_mill",
  y="total_deaths_mill", data=covidtotals)
ax.set(xlabel="Cases Per Million", ylabel="Deaths Per
  Million", title="Total COVID Cases and Deaths by
  Country")
plt.show()
```

This produces the following plot:

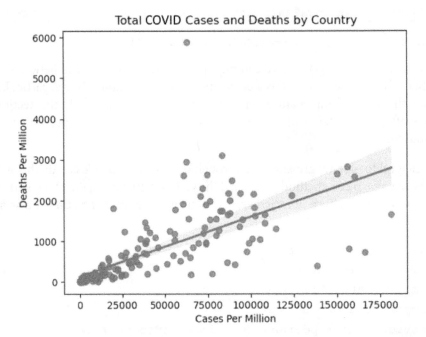

Figure 2.3 – Total COVID Cases and Deaths by Country

The regression line is an estimate of the relationship between cases per million and deaths per million. The slope of the line indicates how much we can expect deaths per million to increase with a 1-unit increase in cases per million. Those points on the scatter plot that are significantly above the regression line should be examined more closely.

2.  The country with deaths per million near 6,000 and cases per million below 75,000 is clearly an outlier. Let's take a closer look:

```
covidtotals.loc[(covidtotals.total_cases_mill<75000) \
  & (covidtotals.total_deaths_mill>5500)].T
iso_code                        PER
lastdate                 2021-07-07
location                       Peru
total_cases               2,071,637
total_deaths                193,743
total_cases_mill             62,830
total_deaths_mill             5,876
population               32,971,846
```

```
population_density                25
median_age                        29
gdp_per_capita                12,237
aged_65_older                      7
total_tests_thous                NaN
life_expectancy                   77
hospital_beds_thous                2
diabetes_prevalence                6
region                   South America
```

Here, we can see that the outlier country is Peru. Peru does have above-average cases per million, but its number of deaths per million is still much greater than would be expected given the number of cases. If we draw a line that's perpendicular to the *x* axis at 62,830, we can see that it crosses the regression line at about 1,000 deaths per million, which is far fewer than the 5,876 for Peru. The only other values in the data for Peru that also stand out as very different from the dataset averages are population density and GDP per person, both of which are substantially lower than average. Here, none of our features may help us explain the high number of deaths in Peru.

> **Note**
>
> When creating a scatter plot, it is common to put a feature or predictor variable on the *x* axis and a target variable on the *y* axis. If a regression line is drawn, then that represents the increase in the target that's been predicted by a 1-unit increase in the predictor. But scatter plots can also be used to examine the relationship between two predictors or two possible targets.

Looking back at how we defined an outlier in *Chapter 1*, *Examining the Distribution of Features and Targets*, an argument can be made that Peru is an outlier. But we still have more work to do before we can come to that conclusion. Peru is not the only country with points on the scatter plot far above or below the regression line. It is generally a good idea to investigate many of these points. Let's take a look:

1.  Creating scatter plots that contain most of the key continuous features can help us identify other possible outliers and better visualize the correlations we observed in the first section of this chapter. Let's create scatter plots of people who are aged 65 and older and GDP per capita with total cases per million:

    ```
    fig, axes = plt.subplots(1,2, sharey=True)
    sns.regplot(x=covidtotals.aged_65_older,
      y=covidtotals.total_cases_mill, ax=axes[0])
    ```

```
sns.regplot(x=covidtotals.gdp_per_capita,
  y=covidtotals.total_cases_mill, ax=axes[1])
axes[0].set_xlabel("Aged 65 or Older")
axes[0].set_ylabel("Cases Per Million")
axes[1].set_xlabel("GDP Per Capita")
axes[1].set_ylabel("")
plt.suptitle("Age 65 Plus and GDP with Cases Per
  Million")
plt.tight_layout()
fig.subplots_adjust(top=0.92)
plt.show()
```

This produces the following plot:

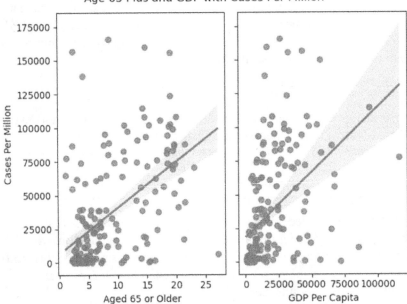

Figure 2.4 – Age 65 Plus and GDP with Cases Per Million

These scatter plots show that some countries that had very high cases per million had values close to what we would expect, given the age of the population or the GDP. These are extreme values, but not necessarily outliers as we have defined them.

It is possible to use scatter plots to illustrate the relationships between two features and a target, all in one graphic. Let's return to the land temperatures data that we worked with in the previous chapter to explore this.

> **Data Note**
>
> The land temperature dataset contains the average temperature readings (in Celsius) in 2019 from over 12,000 stations across the world, though the majority of the stations are in the United States. The dataset was retrieved from the Global Historical Climatology Network integrated database. It has been made available for public use by the United States National Oceanic and Atmospheric Administration at `https://www.ncdc.noaa.gov/data-access/land-based-station-data/land-based-datasets/global-historical-climatology-network-monthly-version-4`.

2.  We expect the average temperature at a weather station to be impacted by both latitude and elevation. Let's say that our previous analysis showed that elevation does not start having much of an impact on temperature until approximately the 1,000-meter mark. We can split the `landtemps` DataFrame into low- and high-elevation stations, with 1,000 meters as the threshold. In the following code, we can see that this gives us 9,538 low-elevation stations with an average temperature of 12.16 degrees Celsius, and 2,557 high-elevation stations with an average temperature of 7.58:

```
landtemps = pd.read_csv("data/landtemps2019avgs.csv")
low, high = landtemps.loc[landtemps.elevation<=1000],
  landtemps.loc[landtemps.elevation>1000]
low.shape[0], low.avgtemp.mean()
(9538, 12.161417937651676)
high.shape[0], high.avgtemp.mean()
(2557, 7.58321486951755)
```

3.  Now, we can visualize the relationship between elevation and latitude and temperature in one scatter plot:

```
plt.scatter(x="latabs", y="avgtemp", c="blue",
  data=low)
plt.scatter(x="latabs", y="avgtemp", c="red",
  data=high)
plt.legend(('low elevation', 'high elevation'))
plt.xlabel("Latitude (N or S)")
plt.ylabel("Average Temperature (Celsius)")
plt.title("Latitude and Average Temperature in 2019")
plt.show()
```

This produces the following scatter plot:

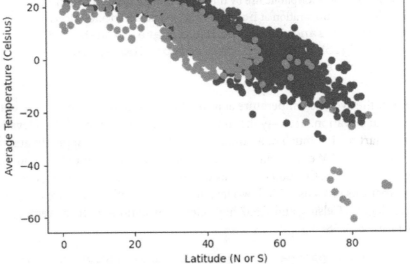

Figure 2.5 – Latitude and Average Temperature in 2019

Here, we can see that the temperatures gradually decrease as the distance from the equator (measured in latitude) increases. We can also see that high-elevation weather stations (those with red dots) are generally below low-elevation stations – that is, they have lower temperatures at similar latitudes.

4.  There also seems to be at least some difference in slope between high- and low-elevation stations. Temperatures appear to decline more quickly as latitude increases with high-elevation stations. We can draw two regression lines through the scatter plot – one for high and one for low-elevation stations – to get a clearer picture of this. To simplify the code a bit, let's create a categorical feature, elevation_group, for low- and high-elevation stations:

```
landtemps['elevation_group'] =
  np.where(landtemps.elevation<=1000,'low','high')
sns.lmplot(x="latabs", y="avgtemp",
  hue="elevation_group", palette=dict(low="blue",
  high="red"), legend_out=False, data=landtemps)
plt.xlabel("Latitude (N or S)")
plt.ylabel("Average Temperature")
```

```
plt.legend(('low elevation', 'high elevation'),
    loc='lower left')
plt.yticks(np.arange(-60, 40, step=20))
plt.title("Latitude and Average Temperature in 2019")
plt.tight_layout()
plt.show()
```

This produces the following plot:

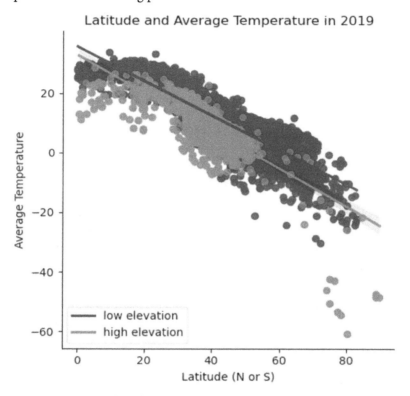

Figure 2.6 – Latitude and Average Temperature in 2019 with regression lines

Here, we can see the steeper negative slope for high-elevation stations.

5.  If we want to see a scatter plot with two continuous features and a continuous target, rather than forcing one of the features to be dichotomous, as we did in the previous example, we can take advantage of Matplotlib's 3D functionality:

```
fig = plt.figure()
plt.suptitle("Latitude, Temperature, and Elevation in
    2019")
ax = plt.axes(projection='3d')
```

```
ax.set_xlabel("Elevation")
ax.set_ylabel("Latitude")
ax.set_zlabel("Avg Temp")
ax.scatter3D(landtemps.elevation, landtemps.latabs,
    landtemps.avgtemp)
plt.show()
```

This produces the following three-dimensional scatter plot:

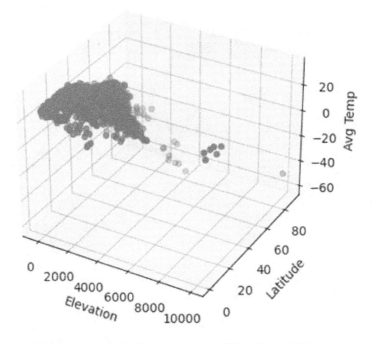

Figure 2.7 – Latitude, Temperature, and Elevation in 2019

Scatter plots are a go-to visualization for teasing out relationships between continuous features. We get a better sense of those relationships than correlation coefficients alone can reveal. However, we need a very different visualization if we are examining the relationship between a continuous feature and a categorical one. Grouped boxplots are useful in those cases. We will learn how to create grouped boxplots with Matplotlib in the next section.

# Using grouped boxplots to view bivariate relationships between continuous and categorical features

Grouped boxplots are an underappreciated visualization. They are helpful when we're examining the relationship between continuous and categorical features since they show how the distribution of a continuous feature can vary by the values of the categorical feature.

We can explore this by returning to the **National Longitudinal Survey (NLS)** data we worked with in the previous chapter. The NLS has one observation per survey respondent but collects annual data on education and employment (data for each year is captured in different columns).

> **Data Note**
>
> As stated in *Chapter 1*, *Examining the Distribution of Features and Targets*, the NLS of Youth is conducted by the United States Bureau of Labor Statistics. Separate files for SPSS, Stata, and SAS can be downloaded from the respective repository. The NLS data can be downloaded from `https://www.nlsinfo.org/investigator/pages/search`.

Follow these steps to create grouped bloxplots:

1. Among the many columns in the NLS DataFrame, there's `highestdegree` and `weeksworked17`, which represent the highest degree the respondent earned and the number of weeks the person worked in 2017, respectively. Let's look at the distribution of weeks worked for each value of the degree that was earned. First, we must define a function, `gettots`, to get the descriptive statistics we want. Then, we must pass a groupby series object, `groupby(['highestdegree'])` `['weeksworked17']`, to that function using `apply`:

> **Note**
>
> We will not go over how to use `groupby` or `apply` in this book. I have covered many examples of their use in my book, *Python Data Cleaning Cookbook*.

```
nls97 = pd.read_csv("data/nls97.csv")
nls97.set_index("personid", inplace=True)
def gettots(x):
  out = {}
```

```
out['min'] = x.min()
out['qr1'] = x.quantile(0.25)
out['med'] = x.median()
out['qr3'] = x.quantile(0.75)
out['max'] = x.max()
out['count'] = x.count()
return pd.Series(out)
```

```
nls97.groupby(['highestdegree'])['weeksworked17'].\
    apply(gettots).unstack()
```

| highestdegree | min | qr1 | med | qr3 | max | count |
|---|---|---|---|---|---|---|
| 0. None | 0 | 0 | 40 | 52 | 52 | 510 |
| 1. GED | 0 | 8 | 47 | 52 | 52 | 848 |
| 2. High School | 0 | 31 | 49 | 52 | 52 | 2,665 |
| 3. Associates | 0 | 42 | 49 | 52 | 52 | 593 |
| 4. Bachelors | 0 | 45 | 50 | 52 | 52 | 1,342 |
| 5. Masters | 0 | 46 | 50 | 52 | 52 | 538 |
| 6. PhD | 0 | 46 | 50 | 52 | 52 | 51 |
| 7. Professional | 0 | 47 | 50 | 52 | 52 | 97 |

Here, we can see how different the distribution of weeks worked for people with less than a high school degree is from that distribution for people with a bachelor's degree or more. For those with no degree, more than 25% had 0 weeks worked. For those with a bachelor's degree, even those at the 25th percentile worked 45 weeks during the year. The interquartile range covers the whole distribution for individuals with no degree (0 to 52), but only a small part of the range for individuals with bachelor's degrees (45 to 52).

We should also make note of the class imbalance for highestdegree. The counts get quite small after master's degrees and the counts for high school degrees are nearly twice that of the next largest group. We will likely need to collapse some categories before we do any modeling with this data.

2. Grouped boxplots make the differences in distributions even clearer. Let's create some with the same data. We will use Seaborn for this plot:

```
import seaborn as sns
myplt = sns.boxplot(x='highestdegree',
```

```
    y= 'weeksworked17' , data=nls97,
    order=sorted(nls97.highestdegree.dropna().unique()))
myplt.set_title("Boxplots of Weeks Worked by Highest
    Degree")
myplt.set_xlabel('Highest Degree Attained')
myplt.set_ylabel('Weeks Worked 2017')
myplt.set_xticklabels(myplt.get_xticklabels(),
    rotation=60, horizontalalignment='right')
plt.tight_layout()
plt.show()
```

This produces the following plot:

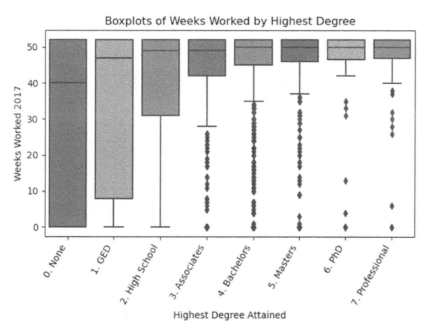

Figure 2.8 – Boxplots of Weeks Worked by Highest Degree

The grouped boxplots illustrate the dramatic difference in interquartile range for weeks worked by the degree earned. At the associate's degree level (a 2-year college degree in the United States) or above, there are values below the whiskers, represented by dots. Below the associate's degree level, the boxplots do not identify any outliers or extreme values. For example, a 0 weeks worked value is not an extreme value for someone with no degree, but it is for someone with an associate's degree or more.

3.  We can also use grouped boxplots to illustrate how the distribution of COVID-19 cases varies by region. Let's also add a swarmplot to view the data points since there aren't too many of them:

```
sns.boxplot(x='total_cases_mill', y='region',
    data=covidtotals)
sns.swarmplot(y="region", x="total_cases_mill",
    data=covidtotals, size=1.5, color=".3", linewidth=0)
plt.title("Boxplots of Total Cases Per Million by
    Region")
plt.xlabel("Cases Per Million")
plt.ylabel("Region")
plt.tight_layout()
plt.show()
```

This produces the following plot:

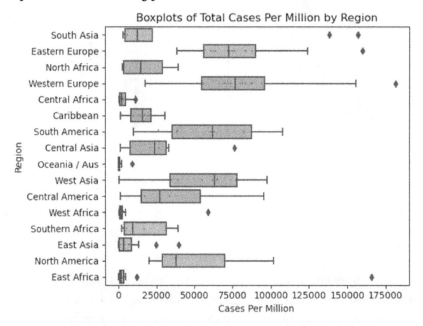

Figure 2.9 – Boxplots of Total Cases Per Million by Region

These grouped boxplots show just how much the median cases per million varies by region, from East Africa and East Asia on the low end to Eastern Europe and Western Europe on the high end. Extremely high values for East Asia are *below* the first quartile for Western Europe. We should probably avoid drawing too many conclusions beyond that since the counts for most regions (the number of countries) are fairly small.

So far in this chapter, we have focused mainly on bivariate relationships between features, as well as those between a feature and a target. The statistics and visualizations we have generated will inform the modeling we will do. We are already getting a sense of likely features, their influence on targets, and how the distributions of some features change with the values of another feature.

We will explore multivariate relationships in the remaining sections of this chapter. We want to have some sense of how multiple features move together before we begin our modeling. Do some features no longer matter once other features are included? Which observations pull on our parameter estimates more than others, and what are the implications for model fitting? Similarly, which observations are not like the others, because they either have invalid values or because they seem to be capturing a completely different phenomenon than the other observations? We will begin to answer those questions in the next three sections. Although we will not get any definitive answers until we construct our models, we can start making difficult modeling decisions by anticipating them.

# Using linear regression to identify data points with significant influence

It is not unusual to find that a few observations have a surprisingly high degree of influence on our model, our parameter estimates, and our predictions. This may or may not be desirable. Observations with significant influence may be unhelpful if they reflect a different social or natural process than the rest of the data does. For example, let's say we have a dataset of flying animals that migrate a great distance, and this is almost exclusively bird species, except for data on monarch butterflies. If we are using the wing architecture as a predictor of migration distance, the monarch butterfly data should probably be removed.

We should return to the distinction we made in the first section between an extreme value and an outlier. We mentioned that an outlier can be thought of as an observation with feature values, or relationships between feature values, that are so unusual that they cannot help explain relationships in the rest of the data. An extreme value, on the other hand, may reflect a natural and explainable trend in a feature, or the same relationship between features that has been observed throughout the data.

Distinguishing between an outlier and an extreme value matters most with observations that have a high influence on our model. A standard measure of influence in regression analysis is **Cook's Distance (Cook's D)**. This gives us a measure of how much our predictions would change if an observation were to be removed from the data.

Let's construct a relatively straightforward multivariate regression model in this section with the COVID-19 data we have been using, and then generate a Cook's D value for each observation:

1.  Let's load the COVID-19 data and the Matplotlib and statsmodels libraries:

    ```
    import pandas as pd
    import matplotlib.pyplot as plt
    import statsmodels.api as sm
    covidtotals = pd.read_csv("data/covidtotals.csv")
    covidtotals.set_index("iso_code", inplace=True)
    ```

2.  Now, let's look at the distribution of total cases per million in population and some possible predictors:

    ```
    xvars = ['population_density','aged_65_older',
      'gdp_per_capita','diabetes_prevalence']

    covidtotals[['total_cases_mill'] + xvars].\
      quantile(np.arange(0.0,1.05,0.25))
    ```
    | | total_cases_mill | population_density | aged_65_older\ |
    |---|---|---|---|
    | 0.00 | 8.52 | 0.14 | 1.14 |
    | 0.25 | 2,499.75 | 36.52 | 3.50 |
    | 0.50 | 19,525.73 | 87.25 | 6.22 |
    | 0.75 | 64,834.62 | 213.54 | 13.92 |
    | 1.00 | 181,466.38 | 20,546.77 | 27.05 |

    | | gdp_per_capita | diabetes_prevalence |
    |---|---|---|
    | 0.00 | 661.24 | 0.99 |
    | 0.25 | 3,823.19 | 5.34 |
    | 0.50 | 12,236.71 | 7.20 |
    | 0.75 | 27,216.44 | 10.61 |
    | 1.00 | 116,935.60 | 30.53 |

3. Next, let's define a function, getlm, that uses statsmodels to run a linear regression model and generate influence statistics, including Cook's D. This function takes a DataFrame, the name of the target column, and the column names for the features (it is customary to refer to a target as *y* and features as *X*).

We will use dropna to drop any observations where one of the features has a missing value. The function returns the estimated coefficients (along with pvalues), the influence measures for each observation, and the full regression results (lm):

```
def getlm(df, ycolname, xcolnames):
    df = df[[ycolname] + xcolnames].dropna()
    y = df[ycolname]
    X = df[xcolnames]
    X = sm.add_constant(X)
    lm = sm.OLS(y, X).fit()
    influence = lm.get_influence().summary_frame()
    coefficients = pd.DataFrame(zip(['constant'] +
        xcolnames, lm.params, lm.pvalues),
        columns=['features','params','pvalues'])
    return coefficients, influence, lm
```

4. Now, we can call the getlm function while specifying the total cases per million as the target and population density (people per square mile), age 65 plus the percentage, GDP per capita, and diabetes prevalence as predictors. Then, we can print the parameter estimates. Ordinarily, we would want to look at a full summary of the model, which can be generated with lm.summary(). We'll skip that here for ease of understanding:

```
coefficients, influence, lm = getlm(covidtotals,
    'total_cases_mill', xvars)
coefficients
```

| features | params | pvalues |
|---|---|---|
| 0  constant | -1,076.471 | 0.870 |
| 1  population_density | -6.906 | 0.030 |
| 2  aged_65_older | 2,713.918 | 0.000 |
| 3  gdp_per_capita | 0.532 | 0.001 |
| 4  diabetes_prevalence | 736.809 | 0.241 |

The coefficients for population density, age 65 plus, and GDP are all significant at the 95% level (have p-values less than 0.05). The result for population density is interesting since our bivariate analysis did not reveal a relationship between population density and cases per million. The coefficient indicates a 6.9-point reduction in cases per million, with a 1-point increase in people per square mile. Put more broadly, more crowded countries have fewer cases per million people once we control for the percentage of people that are 65 or older and their GDP per capita. This could be spurious, or it could be a relationship that can only be detected with multivariate analysis. (It could also be that population density is highly correlated with a feature that has a greater effect on cases per million, but that feature has been left out of the model. This would give us a biased coefficient estimate for population density.)

5.   We can use the influence DataFrame that we created in our call to `getlm` to take a closer look at those observations with a high Cook's D. One way of defining a high Cook's D is by using three times the mean value for Cook's D for all observations. Let's create a `covidtotalsoutliers` DataFrame with all the values above that threshold.

There were 13 countries with Cook's D values above the threshold. Let's print out the first five in descending order of the Cook's D value. Bahrain and Maldives are in the top quarter of the distribution for cases (see the descriptives we printed earlier in this section). They also have high population densities and low percentages of age 65 or older. All else being equal, we would expect lower cases per million for those two countries, given what our model says about the relationship between population density and age to cases. Bahrain does have a very high GDP per capita, however, which our model tells us is associated with high case numbers.

Singapore and Hong Kong have extremely high population densities and below-average cases per million, particularly Hong Kong. These two locations, alone, may account for the direction of the population density coefficient. They both also have very high GDP per capita values, which might be a drag on that coefficient. It may just be that our model should not include locations that are city-states:

```
influencethreshold = 3*influence.cooks_d.mean()
covidtotals = covidtotals.join(influence[['cooks_d']])
covidtotalsoutliers = \
  covidtotals.loc[covidtotals.cooks_d >
  influencethreshold]
covidtotalsoutliers.shape
(13, 17)
```

```
covidtotalsoutliers[['location','total_cases_mill',
   'cooks_d'] + xvars].sort_values(['cooks_d'],
   ascending=False).head()
```

|          | location | total_cases_mill | cooks_d | population_density\ |
|----------|----------|------------------|---------|---------------------|
| iso_code |          |                  |         |                     |
| BHR | Bahrain | 156,793.409 | 0.230 | 1,935.907 |
| SGP | Singapore | 10,709.116 | 0.200 | 7,915.731 |
| HKG | Hong Kong | 1,593.307 | 0.181 | 7,039.714 |
| JPN | Japan | 6,420.871 | 0.095 | 347.778 |
| MDV | Maldives | 138,239.027 | 0.069 | 1,454.433 |

|          | aged_65_older | gdp_per_capita | diabetes_prevalence |
|----------|---------------|----------------|---------------------|
| iso_code |               |                |                     |
| BHR | 2.372 | 43,290.705 | 16.520 |
| SGP | 12.922 | 85,535.383 | 10.990 |
| HKG | 16.303 | 56,054.920 | 8.330 |
| JPN | 27.049 | 39,002.223 | 5.720 |
| MDV | 4.120 | 15,183.616 | 9.190 |

6.  So, let's take a look at our regression model estimates if we remove Hong Kong and Singapore:

> **Note**
>
> I am not necessarily recommending this as an approach. We should look at each observation more carefully than we have done so far to determine whether it makes sense to exclude it from our analysis. We are removing the observations here just to demonstrate their effect on the model.

```
coefficients, influence, lm2 = \
   getlm(covidtotals.drop(['HKG','SGP']),
   'total_cases_mill', xvars)
coefficients
```

|   | features | params | pvalues |
|---|----------|--------|---------|
| 0 | constant | -2,864.219 | 0.653 |
| 1 | population_density | 26.989 | 0.005 |

| 2 | aged_65_older | 2,669.281 | 0.000 |
| 3 | gdp_per_capita | 0.553 | 0.000 |
| 4 | diabetes_prevalence | 319.262 | 0.605 |

The big change in the model is that the population density coefficient has now changed direction. This demonstrates how sensitive the population density estimate is to outlier observations whose feature and target values may not be generalizable to the rest of the data. In this case, that might be true for city-states such as Hong Kong and Singapore.

Generating influence measures with linear regression is a very useful technique, and it has the advantage that it is fairly easy to interpret, as we have seen. However, it does have one important disadvantage: it assumes a linear relationship between features, and that features are normally distributed. This is often not the case. We also needed to understand the relationships in the data enough to create *labels*, to identify total cases per million as the target. This is not always possible either. In the next two sections, we'll look at machine learning algorithms for outlier detection that do not make these assumptions.

# Using K-nearest neighbors to find outliers

Machine learning tools can help us identify observations that are unlike others when we have unlabeled data – that is, when there is no target or dependent variable. Even when selecting targets and features is relatively straightforward, it might be helpful to identify outliers without making any assumptions about relationships between features, or the distribution of features.

Although we typically use **K-nearest neighbors** (**KNN**) with labeled data, for classification or regression problems, we can use it to identify anomalous observations. These are observations where there is the greatest difference between their values and their nearest neighbors' values. KNN is a very popular algorithm because it is intuitive, makes few assumptions about the structure of the data, and is quite flexible. The main disadvantage of KNN is that it is not as efficient as many other approaches, particularly parametric techniques such as linear regression. We will discuss these advantages in much greater detail in *Chapter 9, K-Nearest Neighbors, Decision Tree, Random Forest, and Gradient Boosted Regression*, and *Chapter 12, K-Nearest Neighbors for Classification*.

We will use **PyOD**, short for **Python outlier detection**, to identify countries in the COVID-19 data that are significantly different from others. PyOD can use several algorithms to identify outliers, including KNN. Let's get started:

1.  First, we need to import the KNN module from PyOD and `StandardScaler` from the `sklearn` preprocessing utility functions. We also load the COVID-19 data:

    ```
    import pandas as pd
    from pyod.models.knn import KNN
    from sklearn.preprocessing import StandardScaler
    covidtotals = pd.read_csv("data/covidtotals.csv")
    covidtotals.set_index("iso_code", inplace=True)
    ```

    Next, we standardize the data, which is important when we have features with very different ranges, from over 100,000 for total cases per million and GDP per capita to less than 20 for diabetes prevalence and age 65 and older. We can use scikit-learn's standard scaler, which converts each feature value into a z-score, as follows:

    $$z_{ij} = (x_{ij} - u_j)/s_j$$

    Here, $x_{ij}$ is the value for the $i$th observation of the $j$th feature, $u_j$ is the mean for feature $j$, and $s_j$ is the standard deviation for that feature.

2.  We can use the scaler for just the features we will be including in our model, and then drop all observations that are missing values for one or more features:

    ```
    standardizer = StandardScaler()
    analysisvars =['location', 'total_cases_mill',
       'total_deaths_mill','population_density',
       'diabetes_prevalence', 'aged_65_older',
       'gdp_per_capita']
    covidanalysis =
       covidtotals.loc[:,analysisvars].dropna()
    covidanalysisstand =
       standardizer.fit_transform(covidanalysis.iloc[:,1:])
    ```

3.  Now, we can run the model and generate predictions and anomaly scores. First, we must set `contamination` to `0.1` to indicate that we want 10% of observations to be identified as outliers. This is pretty arbitrary but not a bad starting point. After using the `fit` method to run the KNN algorithm, we get predictions (1 if an outlier, 0 if an inlier) and an anomaly score, which is the basis of the prediction (in this case, the top 10% of anomaly scores will get a prediction of 1):

```
clf_name = 'KNN'
clf = KNN(contamination=0.1)
clf.fit(covidanalysisstand)
y_pred = clf.labels_
y_scores = clf.decision_scores_
```

4.  We can combine the two NumPy arrays with the predictions and anomaly scores – `y_pred` and `y_scores`, respectively – and convert them into the columns of a DataFrame. This makes it easier to view the range of anomaly scores and their associated predictions. 18 countries have been identified as outliers (this is a result of setting `contamination` to `0.1`). Outliers have anomaly scores of 1.77 to 9.34, while inliers have scores of 0.11 to 1.74:

```
pred = pd.DataFrame(zip(y_pred, y_scores),
  columns=['outlier','scores'],
  index=covidanalysis.index)

pred.outlier.value_counts()
0      156
1       18

pred.groupby(['outlier'])[['scores']].\
  agg(['min','median','max'])
```

|  | scores | | |
| --- | --- | --- | --- |
|  | min | median | max |
| outlier | | | |
| 0 | 0.11 | 0.84 | 1.74 |
| 1 | 1.77 | 2.48 | 9.34 |

5.  Let's take a closer look at the countries with the highest anomaly scores:

```
covidanalysis = covidanalysis.join(pred).\
  loc[:,analysisvars + ['scores']].\
  sort_values(['scores'], ascending=False)
covidanalysis.head(10)
```

| | location | total_cases_mill | total_deaths_mill |
|---|---|---|---|
| iso_code | | | |
| ... | | | |
| SGP | Singapore | 10,709.12 | 6.15 |
| HKG | Hong Kong | 1,593.31 | 28.28 |
| PER | Peru | 62,830.48 | 5,876.01 |
| QAT | Qatar | 77,373.61 | 206.87 |
| BHR | Bahrain | 156,793.41 | 803.37 |
| LUX | Luxembourg | 114,617.81 | 1,308.36 |
| BRN | Brunei | 608.02 | 6.86 |
| KWT | Kuwait | 86,458.62 | 482.14 |
| MDV | Maldives | 138,239.03 | 394.05 |
| ARE | United Arab Emirates | 65,125.17 | 186.75 |

| | aged_65_older | gdp_per_capita | scores |
|---|---|---|---|
| iso_code | | | |
| SGP | 12.92 | 85,535.38 | 9.34 |
| HKG | 16.30 | 56,054.92 | 8.03 |
| PER | 7.15 | 12,236.71 | 4.37 |
| QAT | 1.31 | 116,935.60 | 4.23 |
| BHR | 2.37 | 43,290.71 | 3.51 |
| LUX | 14.31 | 94,277.96 | 2.73 |
| BRN | 4.59 | 71,809.25 | 2.60 |
| KWT | 2.35 | 65,530.54 | 2.52 |
| MDV | 4.12 | 15,183.62 | 2.51 |
| ARE | 1.14 | 67,293.48 | 2.45 |

Several of the locations we identified as having high influence in the previous section have high anomaly scores, including Singapore, Hong Kong, Bahrain, and Maldives. This is more evidence that we need to take a closer look at the data for these countries. Perhaps there is invalid data or there are theoretical reasons why they are very different than the rest of the data.

Unlike the linear model in the previous section, there is no defined target. We include both total cases per million and total deaths per million in this case. Peru has been identified as an outlier here, though it was not with the linear model. This is partly because of Peru's very high deaths per million, which is the highest in the dataset (we did not use deaths per million in our linear regression model).

6.  Notice that Japan is not on this list of outliers. Let's take a look at its anomaly score:

```
covidanalysis.loc['JPN','scores']
2.03
```

The anomaly score is the 15th highest in the dataset. Compare this with the 4th highest Cook's D score for Japan from the previous section.

It is interesting to compare these results with a similar analysis we could conduct with Isolation Forest. We will do that in the next section.

> **Note**
> This has been a very simplified example of the approach we would take with a typical machine learning project. The most important omission here is that we are conducting our analysis on the full dataset. For reasons we will discuss at the beginning of *Chapter 4, Encoding, Transforming, and Scaling Features*, we want to split our data into training and testing datasets very early in the process. We will learn how to incorporate outlier detection in a machine learning pipeline in the remaining chapters of this book.

# Using Isolation Forest to find outliers

**Isolation Forest** is a relatively new machine learning technique for identifying anomalies. It has quickly become popular, partly because its algorithm is optimized to find outliers, rather than normal values. It finds outliers by successively partitioning the data until a data point has been isolated. Points that require fewer partitions to be isolated receive higher anomaly scores. This process turns out to be fairly easy on system resources. In this section, we will learn how to use it to detect outlier COVID-19 cases and deaths.

Isolation Forest is a good alternative to KNN, particularly when we're working with large datasets. The efficiency of the algorithm allows it to handle large samples and a high number of features. Let's get started:

1.  We can do an analysis similar to the one in the previous section with Isolation Forest rather than KNN. Let's start by loading scikit-learn's `StandardScaler` and `IsolationForest` modules, as well as the COVID-19 data:

    ```
    import pandas as pd
    import matplotlib.pyplot as plt
    from sklearn.preprocessing import StandardScaler
    from sklearn.ensemble import IsolationForest
    covidtotals = pd.read_csv("data/covidtotals.csv")
    covidtotals.set_index("iso_code", inplace=True)
    ```

2.  Next, we must standardize the data:

    ```
    analysisvars = ['location','total_cases_mill','total_
    deaths_mill',
      'population_density','aged_65_older','gdp_per_capita']
    standardizer = StandardScaler()
    covidanalysis = covidtotals.loc[:, analysisvars].dropna()
    covidanalysisstand =
      standardizer.fit_transform(covidanalysis.iloc[:, 1:])
    ```

3.  Now, we are ready to run our anomaly detection model. The `n_estimators` parameter indicates how many trees to build. Setting `max_features` to `1.0` will use all of our features. The `predict` method gives us the anomaly prediction, which is `-1` for an anomaly. This is based on the anomaly score, which we can get using `decision_function`:

    ```
    clf=IsolationForest(n_estimators=50,
      max_samples='auto', contamination=.1,
      max_features=1.0)
    clf.fit(covidanalysisstand)
    covidanalysis['anomaly'] =
      clf.predict(covidanalysisstand)
    covidanalysis['scores'] =
      clf.decision_function(covidanalysisstand)
    covidanalysis.anomaly.value_counts()
    ```

```
1      156
-1      18
Name: anomaly, dtype: int64
```

4.  Let's take a closer look at the outliers (we will also create a DataFrame of the inliers
    to use in a later step). We sort by anomaly score and show the countries with the
    highest (most negative) score. Singapore, Hong Kong, Bahrain, Qatar, and Peru are,
    again, the most anomalous:

```
inlier, outlier =
  covidanalysis.loc[covidanalysis.anomaly==1],\
  covidanalysis.loc[covidanalysis.anomaly==-1]
outlier[['location','total_cases_mill',
  'total_deaths_mill',
  'scores']].sort_values(['scores']).head(10)
         location    total_cases_mill    total_deaths_
mill     scores
iso_code
SGP    Singapore      10,709.12              6.15      -0.20
HKG    Hong Kong       1,593.31             28.28      -0.16
BHR    Bahrain       156,793.41            803.37      -0.14
QAT    Qatar          77,373.61            206.87      -0.13
PER    Peru           62,830.48          5,876.01      -0.12
LUX    Luxembourg    114,617.81          1,308.36      -0.09
JPN    Japan           6,420.87            117.40      -0.08
MDV    Maldives      138,239.03            394.05      -0.07
CZE    Czechia       155,782.97          2,830.43      -0.06
MNE    Montenegro    159,844.09          2,577.77      -0.03
```

5.  It's helpful to look at a visualization of the outliers and inliers:

```
fig = plt.figure()
ax = plt.axes(projection='3d')
ax.set_title('Isolation Forest Anomaly Detection')
ax.set_zlabel("Cases Per Million (thous.)")
ax.set_xlabel("GDP Per Capita (thous.)")
ax.set_ylabel("Aged 65 Plus %")
ax.scatter3D(inlier.gdp_per_capita/1000,
  inlier.aged_65_older, inlier.total_cases_mill/1000,
```

```
        label="inliers", c="blue")
    ax.scatter3D(outlier.gdp_per_capita/1000,
        outlier.aged_65_older,
        outlier.total_cases_mill/1000, label="outliers",
        c="red")
    ax.legend()
    plt.show()
```

This produces the following plot:

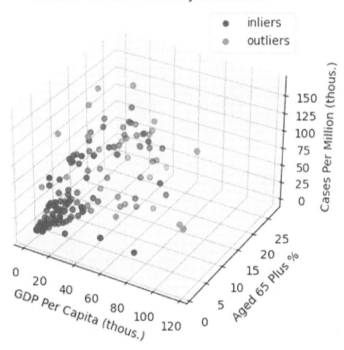

Figure 2.10 – Isolation Forest Anomaly Detection – GDP Per Capita and Cases Per Million

Although we are only able to see three dimensions with this visualization, the plot does illustrate some of what makes an outlier an outlier. We expect cases to increase as the GDP per capita and the age 65 plus percentage increase. We can see that the outliers deviate from the expected pattern, having cases per million noticeably above or below countries with similar GDPs and age 65 plus values.

# Summary

In this chapter, we used bivariate and multivariate statistical techniques and visualizations to get a better sense of bivariate relationships among features. We looked at common statistics, such as the Pearson correlation. We also examined bivariate relationships through visualizations, with scatter plots when both features are continuous, and with grouped boxplots when one feature is categorical. The last three sections of this chapter explored multivariate techniques for examining relationships and identifying outliers, including machine learning algorithms such as KNN and Isolation Forest.

Now that we have a good sense of the distribution of our data, we are ready to start engineering our features, including imputing missing values and encoding, transforming, and scaling our variables. This will be our focus for the next two chapters.

# 3
# Identifying and Fixing Missing Values

I think I speak for many data scientists when I say that rarely is there something so seemingly small and trivial that is as of much consequence as the missing value. We spend a good deal of our time worrying about missing values because they can have a dramatic, and surprising, effect on our analysis. This is most likely to happen when missing values are not random – that is, when they are correlated with a feature or target. For example, let's say we are doing a longitudinal study of earnings, but individuals with lower education are more likely to skip the earnings question each year. There is a decent chance that this will bias our parameter estimate for education.

Of course, identifying missing values is not even half of the battle. We then need to decide how to handle them. Do we remove any observation with a missing value for one or more features? Do we impute a value based on a sample-wide statistic such as the mean? Or do we assign a value based on a more targeted statistic, such as the mean for those in a certain class? Do we think of this differently for time series or longitudinal data where the nearest temporal value may make the most sense? Or should we use a more complex multivariate technique for imputing values, perhaps based on linear regression or **k-nearest neighbors (KNN)**?

The answer to all of these questions is *yes*. At some point, we will want to use each of these techniques. We will want to be able to answer why or why not to all of these possibilities when making a final choice about missing value imputation. Each will make sense, depending on the situation.

In this chapter, we'll look at techniques for identifying missing values for each feature or target, and for observations where values for a large number of the features are absent. Then, we will explore strategies for imputing values, such as setting values to the overall mean, to the mean for a given category, or forward filling. We will also examine multivariate techniques for imputing values for missing values and discuss when they are appropriate.

Specifically, in this chapter, we will cover the following topics:

- Identifying missing values
- Cleaning missing values
- Imputing values with regression
- Using KNN imputation
- Using random forest for imputation

# Technical requirements

This chapter will rely heavily on the pandas and NumPy libraries, but you don't require any prior knowledge of these. If you have installed Python from a scientific distribution, such as Anaconda or WinPython, these libraries are probably already installed. We will also be using the `statsmodels` library for linear regression, and machine learning algorithms from `sklearn` and `missingpy`. If you need to install any of these packages, you can do so by running `pip install [package name]` from a terminal window or Windows PowerShell.

# Identifying missing values

Since identifying missing values is such an important part of an analyst's workflow, any tool we use needs to make it easy to regularly check for such values. Fortunately, pandas makes it quite simple to identify missing values.

We will be working with the **National Longitudinal Survey** (**NLS**) in this chapter. The NLS has one observation per survey respondent. Data for employment, earnings, and college enrollment for each year are stored in columns with suffixes representing the year, such as `weeksworked16` and `weeksworked17` for weeks worked in 2016 and 2017, respectively.

> **Note**
>
> We will also work with the COVID-19 data again. This dataset has one observation for each country that specifies the total COVID-19 cases and deaths, as well as some demographic data for each country.

Follow these steps to identify our missing values:

1.  Let's start by loading the NLS and COVID-19 data:

    ```
    import pandas as pd
    import numpy as np
    nls97 = pd.read_csv("data/nls97b.csv")
    nls97.set_index("personid", inplace=True)
    covidtotals = pd.read_csv("data/covidtotals.csv")
    covidtotals.set_index("iso_code", inplace=True)
    ```

2.  Next, we count the number of missing values for columns that we may use as features. We can use the `isnull` method to test whether each feature value is missing. It will return `True` if the value is missing and `False` if not. Then, we can use `sum` to count the number of `True` values since `sum` will treat each `True` value as 1 and each `False` value as 0. We use `axis=0` to sum over the rows for each column:

    ```
    covidtotals.shape
    (221, 16)


    demovars = ['population_density','aged_65_older',
      'gdp_per_capita', 'life_expectancy',
      'diabetes_prevalence']
    covidtotals[demovars].isnull().sum(axis=0)
    population_density       15
    aged_65_older            33
    gdp_per_capita           28
    ```

```
life_expectancy              4
diabetes_prevalence         21
```

As we can see, 33 of the 221 countries have null values for `aged_65_older`. We have `life_expectancy` for almost all countries.

3.  If we want the number of missing values for each row, we can specify `axis=1` when summing. The following code creates a Series, `demovarsmisscnt`, with the number of missing values for the demographic features for each country. 181 countries have values for all of the features, 11 are missing values for four of the five features, and three are missing values for all of the features:

```
demovarsmisscnt = covidtotals[demovars].isnull().
sum(axis=1)
demovarsmisscnt.value_counts().sort_index()
0        181
1         15
2          6
3          5
4         11
5          3
dtype: int64
```

4.  Let's take a look at a few of the countries with four or more missing values. There is very little demographic data available for these countries:

```
covidtotals.loc[demovarsmisscnt > = 4, ['location'] +
    demovars].sample(6, random_state=1).T
```

| iso_code | FLK | NIU | MSR\ |
|---|---|---|---|
| location | Falkland Islands | Niue | Montserrat |
| population_density | NaN | NaN | NaN |
| aged_65_older | NaN | NaN | NaN |
| gdp_per_capita | NaN | NaN | NaN |
| life_expectancy | 81 | 74 | 74 |
| diabetes_prevalence | NaN | NaN | NaN |
| | | | |
| iso_code | COK | SYR | GGY |
| location | Cook Islands | Syria | Guernsey |
| population_density | NaN | NaN | NaN |

| | | | | | |
|---|---|---|---|---|---|
| aged_65_older | | NaN | NaN | | NaN |
| gdp_per_capita | | NaN | NaN | | NaN |
| life_expectancy | | 76 | 7 | | NaN |
| diabetes_prevalence | | NaN | NaN | | NaN |

5.  Let's also check missing values for total cases and deaths. 29 countries have missing values for cases per million in population, and 36 have missing deaths per million:

```
totvars =
  ['location','total_cases_mill','total_deaths_mill']
covidtotals[totvars].isnull().sum(axis=0)
location                0
total_cases_mill       29
total_deaths_mill      36
dtype: int64
```

6.  We should also get a sense of which countries are missing both. 29 countries are missing both cases and deaths, and we only have both for 185 countries:

```
totvarsmisscnt =
  covidtotals[totvars].isnull().sum(axis=1)
totvarsmisscnt.value_counts().sort_index()
0        185
1          7
2         29
dtype: int64
```

Sometimes, we have logical missing values that we need to transform into actual missing values. This happens when the dataset designers use valid values as codes for missing values. These are often values such as 9, 99, or 999, based on the allowable number of digits for the variable. Or it might be a more complicated coding scheme where there are codes for different reasons for there being missings. For example, in the NLS dataset, the codes reveal why the respondent did not provide an answer for a question: -3 is an invalid skip, -4 is a valid skip, and -5 is a non-interview.

7.  The last four columns in the NLS DataFrame have data at the highest grade completed for the respondent's mother and father, parental income, and mother's age when the respondent was born. Let's examine logical missings for those columns, starting with the highest grade that was completed for the respondent's mother:

```
nlsparents = nls97.iloc[:,-4:]
nlsparents.shape
(8984, 4)
nlsparents.loc[nlsparents.motherhighgrade.between(-5,
    -1), 'motherhighgrade'].value_counts()
-3          523
-4          165
Name: motherhighgrade, dtype: int64
```

8.  There are 523 invalid skips and 165 valid skips. Let's look at a few individuals that have at least one of these non-response values for these four features:

```
nlsparents.loc[nlsparents.apply(lambda x: x.between(
    -5,-1)).any(axis=1)]
```

| personid | motherage | parentincome | fatherhighgrade | motherhighgrade |
|---|---|---|---|---|
| 100284 | 22 | 50000 | 12 | -3 |
| 100931 | 23 | 60200 | -3 | 13 |
| 101122 | 25 | -4 | -3 | -3 |
| 101414 | 27 | 24656 | 10 | -3 |
| 101526 | -3 | 79500 | -4 | -4 |
| ... | ... | ... | ... | ... |
| 999087 | -3 | 121000 | -4 | 16 |
| 999103 | -3 | 73180 | 12 | -4 |
| 999406 | 19 | -4 | 17 | 15 |
| 999698 | -3 | 13000 | -4 | -4 |
| 999963 | 29 | -4 | 12 | 13 |

```
[3831 rows x 4 columns]
```

9.  For our analysis, the reason why there is a non-response is not important. Let's just count the number of non-responses for each of the features, regardless of the reason for the non-response:

```
nlsparents.apply(lambda x: x.between(-5,-1).sum())
motherage             608
parentincome         2396
fatherhighgrade      1856
motherhighgrade       688
dtype: int64
```

10. We should set these values to `missing` before using these features in our analysis. We can use `replace` to set all the values between -5 and -1 to `missing`. When we check for actual missings, we get the expected counts:

```
nlsparents.replace(list(range(-5,0)),
    np.nan, inplace=True)
nlsparents.isnull().sum()
motherage             608
parentincome         2396
fatherhighgrade      1856
motherhighgrade       688
dtype: int64
```

This section demonstrated some very handy pandas techniques for identifying the number of missing values for each feature, as well as observations with a large number of missing values. We also learned how to find logical missing values and convert them into actual missings. Next, we'll take our first look at cleaning missing values.

# Cleaning missing values

In this section, we'll go over some of the most straightforward approaches for handling missing values. This includes dropping observations where there are missing values; assigning a sample-wide summary statistic, such as the mean, to the missing values; and assigning values based on the mean value for an appropriate subset of the data:

1.  Let's load the NLS data and select some of the educational data:

    ```
    import pandas as pd
    nls97 = pd.read_csv("data/nls97b.csv")
    nls97.set_index("personid", inplace=True)
    schoolrecordlist =
       ['satverbal','satmath','gpaoverall','gpaenglish',
       'gpamath','gpascience','highestdegree',
       'highestgradecompleted']
    schoolrecord = nls97[schoolrecordlist]
    schoolrecord.shape
    (8984, 8)
    ```

2.  We can use the techniques we explored in the previous section to identify missing values. `schoolrecord.isnull().sum(axis=0)` gives us the number of missing values for each feature. The overwhelming majority of observations have missing values for `satverbal`, with 7,578 out of 8,984. Only 31 observations have missing values for `highestdegree`:

    ```
    schoolrecord.isnull().sum(axis=0)
    satverbal                7578
    satmath                  7577
    gpaoverall               2980
    gpaenglish               3186
    gpamath                  3218
    gpascience               3300
    highestdegree              31
    highestgradecompleted    2321
    dtype: int64
    ```

3.  We can create a Series, misscnt, that specifies the number of missing features for each observation with misscnt = schoolrecord.isnull(). sum(axis=1). 946 observations have seven missing values for the educational data, while 11 are missing values for all eight features:

```
misscnt = schoolrecord.isnull().sum(axis=1)
misscnt.value_counts().sort_index()
0            1087
1             312
2            3210
3            1102
4             176
5             101
6            2039
7             946
8              11
dtype: int64
```

4.  Let's also take a look at a few observations with seven or more missing values. It looks like highestdegree is often the one feature that is present, which is not surprising, given that we have already discovered that highestdegree is rarely missing:

```
schoolrecord.loc[misscnt>=7].head(4).T
personid              101705   102061   102648   104627
satverbal               NaN      NaN      NaN      NaN
satmath                 NaN      NaN      NaN      NaN
gpaoverall              NaN      NaN      NaN      NaN
gpaenglish              NaN      NaN      NaN      NaN
gpamath                 NaN      NaN      NaN      NaN
gpascience              NaN      NaN      NaN      NaN
highestdegree         1.GED   0.None   1.GED   0.None
highestgradecompleted   NaN      NaN      NaN      NaN
```

5.  Let's drop observations that have missing values for seven or more features out of eight. We can accomplish this by setting the `thresh` parameter of `dropna` to 2. This will drop observations that have fewer than two non-missing values; that is, 0 or 1 non-missing values. We get the expected number of observations after using `dropna`; that is, 8,984 - 946 - 11 = 8,027:

```
schoolrecord = schoolrecord.dropna(thresh=2)
schoolrecord.shape
(8027, 8)
schoolrecord.isnull().sum(axis=1).value_counts().sort_
index()
0       1087
1        312
2       3210
3       1102
4        176
5        101
6       2039
dtype: int64
```

There are a fair number of missing values for gpaoverall – that is, 2,980 – though we have valid values for two-thirds of observations ((8,984 – 2,980)/8,984). We might be able to salvage this as a feature if we do a good job of imputing the missing values. This is likely more desirable than just removing these observations. We do not want to lose that data if we can avoid it, particularly if individuals with missing gpaoverall are different from others in ways that will matter for our predictions.

6.  The most straightforward approach is to assign the overall mean for gpaoverall to the missing values. The following code uses the pandas Series `fillna` method to assign all missing values of gpaoverall to the mean value of the Series. The first argument to `fillna` is the value you want for all missing values – in this case, `schoolrecord.gpaoverall.mean()`. Note that we need to remember to set the `inplace` parameter to `True` to overwrite the existing values:

```
schoolrecord.gpaoverall.agg(['mean','std','count'])
mean         281.84
std           61.64
count      6,004.00
Name: gpaoverall, dtype: float64
```

```
schoolrecord.gpaoverall.fillna(
  schoolrecord.gpaoverall.mean(), inplace=True)
schoolrecord.gpaoverall.isnull().sum()
0
schoolrecord.gpaoverall.agg(['mean','std','count'])
mean          281.84
std            53.30
count       8,027.00
Name: gpaoverall, dtype: float64
```

The mean has not changed. However, there is a substantial reduction in the standard deviation, from 61.6 to 53.3. This is one of the disadvantages of using the dataset's mean for all missing values.

7. The NLS data also has a fair number of missing values for wageincome. The following code shows that 3,893 observations have missing values:

```
wageincome = nls97.wageincome.copy(deep=True)
wageincome.isnull().sum()
3893
wageincome.agg(['mean','std','count'])
mean         49,477.02
std          40,677.70
count         5,091.00
Name: wageincome, dtype: float64
wageincome.head().T
personid
100061        12,500
100139       120,000
100284        58,000
100292           NaN
100583        30,000
Name: wageincome, dtype: float64
```

> **Note**
>
> Here, we made a deep copy with the `copy` method, setting `deep` to `True`.
> We wouldn't normally do this but, in this case, we don't want to change the
> values of `wageincome` in the underlying DataFrame. We have avoided this
> here because we will demonstrate a different method of imputing values in the
> next couple of code blocks.

8. Rather than assigning the mean value of `wageincome` to the missings, we could
   use another common technique for imputing values: we could assign the nearest
   non-missing value from a preceding observation. The `ffill` option of `fillna`
   will do this for us:

```
wageincome.fillna(method='ffill', inplace=True)
wageincome.head().T
personid
100061        12,500
100139       120,000
100284        58,000
100292        58,000
100583        30,000
Name: wageincome, dtype: float64
wageincome.isnull().sum()
0
wageincome.agg(['mean','std','count'])
mean        49,549.33
std         40,014.34
count        8,984.00
Name: wageincome, dtype: float64
```

9. We could have done a backward fill instead by setting the `method` parameter of
   `fillna` to `bfill`. This sets missing values to the nearest following value. This
   produces the following output:

```
wageincome = nls97.wageincome.copy(deep=True)
wageincome.std()
40677.69679818673
wageincome.fillna(method='bfill', inplace=True)
wageincome.head().T
personid
```

```
100061          12,500
100139         120,000
100284          58,000
100292          30,000
100583          30,000
Name: wageincome, dtype: float64
wageincome.agg(['mean','std','count'])
mean      49,419.05
std       41,111.54
count      8,984.00
Name: wageincome, dtype: float64
```

If missing values are randomly distributed, then forward or backward filling has one advantage over using the mean: it is more likely to approximate the distribution of the non-missing values for the feature. Notice that the standard deviation did not drop substantially.

There are times when it makes sense to base our imputation of values on the mean or median value for similar observations; say, those that have the same value for a related feature. If we are imputing values for feature X1, and X1 is correlated with X2, we can use the relationship between X1 and X2 to impute a value for X1 that may make more sense than the dataset's mean. This is pretty straightforward when X2 is categorical. In this case, we can impute the mean value of X1 for the associated value of X2.

10. In the NLS DataFrame, weeks worked in 2017 correlates with the highest degree earned. The following code shows how the mean value of weeks worked changes with degree attainment. The mean for weeks worked is 39, but it is much lower for those without a degree (28.72) and much higher for those with a professional degree (47.20). In this case, it may be a better choice to assign 28.72 to the missing values for weeks worked for individuals who have not attained a degree, rather than 39:

```
nls97.weeksworked17.mean()
39.01664167916042
nls97.groupby(['highestdegree'])['weeksworked17'
    ].mean()
highestdegree
0. None                    28.72
1. GED                     34.59
2. High School             38.15
```

```
3. Associates          40.44
4. Bachelors           43.57
5. Masters             45.14
6. PhD                 44.31
7. Professional        47.20
Name: weeksworked17, dtype: float64
```

11. The following code assigns the mean value of weeks worked across observations with the same degree attainment level, for those observations missing weeks worked. We do this by using `groupby` to create a groupby DataFrame, `groupby(['highestdegree'])['weeksworked17']`. Then, we use `fillna` within `apply` to fill those missing values with the mean for the highest degree group. Notice that we make sure to only do this imputation for observations where the highest degree is not missing, `~nls97.highestdegree.isnull()`. We will still have missing values for observations that are missing both the highest degree and weeks worked:

```
nls97.loc[~nls97.highestdegree.isnull(),
  'weeksworked17imp'] =
  nls97.loc[ ~nls97.highestdegree.isnull() ].
  groupby(['highestdegree'])['weeksworked17'].
  apply(lambda group: group.fillna(np.mean(group)))


nls97[['weeksworked17imp','weeksworked17',
  'highestdegree']].head(10)
          weeksworked17imp  weeksworked17    highestdegree
personid
100061            48.00           48.00    2. High School
100139            52.00           52.00    2. High School
100284             0.00            0.00          0. None
100292            43.57             NaN      4. Bachelors
100583            52.00           52.00    2. High School
100833            47.00           47.00    2. High School
100931            52.00           52.00     3. Associates
101089            52.00           52.00    2. High School
101122            38.15             NaN    2. High School
101132            44.00           44.00          0. None
```

```
nls97[['weeksworked17imp','weeksworked17']].\
    agg(['mean','count'])
            weeksworked17imp    weeksworked17
mean                 38.52            39.02
count             8,953.00         6,670.00
```

These imputation strategies – removing observations with missing values, assigning a dataset's mean or median, using forward or backward filling, or using a group mean for a correlated feature – are fine for many predictive analytics projects. They work best when the missing values are not correlated with the target. When that is true, imputing values allows us to retain the other information from those observations without biasing our estimates.

Sometimes, however, that is not the case and more complicated imputation strategies are required. The next few sections will explore multivariate techniques for cleaning missing data.

# Imputing values with regression

We ended the previous section by assigning a group mean to the missing values rather than the overall sample mean. As we discussed, this is useful when the feature that determines the groups is correlated with the feature that has the missing values. Using regression to impute values is conceptually similar to this, but we typically use it when the imputation will be based on two or more features.

Regression imputation replaces a feature's missing values with values predicted by a regression model of correlated features. This particular kind of imputation is known as deterministic regression imputation since the imputed values all lie on the regression line, and no error or randomness is introduced.

One potential drawback of this approach is that it can substantially reduce the variance of the feature with missing values. We can use stochastic regression imputation to address this drawback. We will explore both approaches in this section.

The wageincome feature in the NLS dataset has several missing values. We can use linear regression to impute values. The wage income value is the reported earnings for 2016:

1. Let's start by loading the NLS data again and checking for missing values for wageincome and features that might be correlated with wageincome. We also load the statsmodels library.

   The info method tells us that we are missing values for wageincome for nearly 3,000 observations. There are fewer missing values for the other features:

```
import pandas as pd
import numpy as np
import statsmodels.api as sm
nls97 = pd.read_csv("data/nls97b.csv")
nls97.set_index("personid", inplace=True)
nls97[['wageincome','highestdegree','weeksworked16',
    'parentincome']].info()
<class 'pandas.core.frame.DataFrame'>
Int64Index: 8984 entries, 100061 to 999963
Data columns (total 4 columns):
 #   Column          Non-Null Count        Dtype
--   -------         --------------        -----
 0   wageincome      5091 non-null         float64
 1   highestdegree   8953 non-null         object
 2   weeksworked16   7068 non-null         float64
 3   parentincome    8984 non-null         int64
dtypes: float64(2), int64(1), object(1)
memory usage: 350.9+ KB
```

2. Let's convert the highestdegree feature into a numeric value. This will make the analysis we'll be doing in the rest of this section easier:

```
nls97['hdegnum'] =
    nls97.highestdegree.str[0:1].astype('float')
nls97.groupby(['highestdegree','hdegnum']).size()
highestdegree      hdegnum
0. None               0           953
1. GED                1          1146
2. High School        2          3667
3. Associates         3           737
```

|   | | |
|---|---|---|
| 4. Bachelors | 4 | 1673 |
| 5. Masters | 5 | 603 |
| 6. PhD | 6 | 54 |
| 7. Professional | 7 | 120 |

3.  As we've already discovered, we need to replace logical missing values for parentincome with actual missings. After that, we can run some correlations. Each of the features has some correlation with wageincome, particularly hdegnum:

```
nls97.parentincome.replace(list(range(-5,0)), np.nan,
    inplace=True)
nls97[['wageincome','hdegnum','weeksworked16',
    'parentincome']].corr()
```

|  | wageincome | hdegnum | weeksworked16 | parentin-come |
|---|---|---|---|---|
| wageincome | 1.00 | 0.40 | 0.18 | 0.27 |
| hdegnum | 0.40 | 1.00 | 0.24 | 0.33 |
| weeksworked16 | 0.18 | 0.24 | 1.00 | 0.10 |
| parentincome | 0.27 | 0.33 | 0.10 | 1.00 |

4.  We should check whether observations with missing values for wage income are different in some important way from those with non-missing values. The following code shows that these observations have significantly lower degree attainment levels, parental income, and weeks worked. This is a clear case where assigning the overall mean would not be the best choice:

```
nls97['missingwageincome'] =
    np.where(nls97.wageincome.isnull(),1,0)
nls97.groupby(['missingwageincome'])[['hdegnum',
    'parentincome', 'weeksworked16']].agg(['mean',
    'count'])
```

|  | hdegnum | | parentincome | | weeksworked16 | |
|---|---|---|---|---|---|---|
|  | mean | count | mean | count | mean | count |
| missingwageincome |  |  |  |  |  |  |
| 0 | 2.76 | 5072 | 48,409.13 | 3803 | 48.21 | 5052 |
| 1 | 1.95 | 3881 | 43,565.87 | 2785 | 16.36 | 2016 |

5.  Let's try regression imputation instead. Let's start by cleaning up the data a little bit more. We can replace the missing weeksworked16 and parentincome values with their means. We should also collapse hdegnum into those attaining less than a college degree, those with a college degree, and those with a post-graduate degree. We can set those up as dummy variables, with 0 or 1 values when they're False or True, respectively. This is a tried and true method for treating categorical data in regression analysis as it allows us to estimate different *y* intercepts based on group membership:

```
nls97.weeksworked16.fillna(nls97.weeksworked16.mean(),
    inplace=True)
nls97.parentincome.fillna(nls97.parentincome.mean(),
    inplace=True)
nls97['degltcol'] = np.where(nls97.hdegnum<=2,1,0)
nls97['degcol'] = np.where(nls97.hdegnum.between(3,4),
    1,0)
nls97['degadv'] = np.where(nls97.hdegnum>4,1,0)
```

> **Note**
>
> scikit-learn has preprocessing features that can help us with tasks like these. We will cover some of them in the next chapter.

6.  Next, we define a function, getlm, to run a linear model using the statsmodels module. This function has parameters for the name of the target or dependent variable, ycolname, and the names of the features or independent variables, xcolnames. Much of the work is done by the fit method of statsmodels; that is, OLS(y, X).fit():

```
def getlm(df, ycolname, xcolnames):
    df = df[[ycolname] + xcolnames].dropna()
    y = df[ycolname]
    X = df[xcolnames]
    X = sm.add_constant(X)
    lm = sm.OLS(y, X).fit()
    coefficients = pd.DataFrame(zip(['constant'] +
        xcolnames,lm.params, lm.pvalues), columns = [
        'features' , 'params','pvalues'])
    return coefficients, lm
```

7.  Now, we can use the `getlm` function to get the parameter estimates and the model summary. All of the coefficients are positive and significant at the 95% level since they have `pvalues` less than 0.05. As expected, wage income increases with the number of weeks worked and with parental income. Having a college degree gives a nearly $16K boost to earnings, compared with not having a college degree. A post-graduate degree bumps up the earnings prediction even more – almost $37K more than for those with less than a college degree:

> **Note**
>
> The coefficients of `degcol` and `degadv` are interpreted as relative to those without a college degree since that is the omitted dummy variable.

```
xvars = ['weeksworked16', 'parentincome', 'degcol',
    'degadv']
coefficients, lm = getlm(nls97, 'wageincome', xvars)
coefficients
```

|   | features | params | pvalues |
|---|----------|--------|---------|
| 0 | constant | 7,389.37 | 0.00 |
| 1 | weeksworked16 | 494.07 | 0.00 |
| 2 | parentincome | 0.18 | 0.00 |
| 3 | degcol | 15,770.07 | 0.00 |
| 4 | degadv | 36,737.84 | 0.00 |

8.  We can use this model to impute values for wage income where they are missing. We need to add a constant for the predictions since our model included a constant. We can convert the predictions into a DataFrame and then join it with the rest of the NLS data. Then, we can create a new wage income feature, `wageincomeimp`, that gets the predicted value when wage income is missing, and the original wage income value otherwise. Let's also take a look at some of the predictions to see whether they make sense:

```
pred = lm.predict(sm.add_constant(nls97[xvars])).
    to_frame().rename(columns= {0: 'pred'})
nls97 = nls97.join(pred)
nls97['wageincomeimp'] =
    np.where(nls97.wageincome.isnull(),
    nls97.pred, nls97.wageincome)

pd.options.display.float_format = '{:,.0f}'.format
```

```
nls97[['wageincomeimp','wageincome'] + xvars].head(10)
```

| personid | wageincomeimp | wageincome | weeksworked16 | parentincome | degcol | degadv |
|---|---|---|---|---|---|---|
| 100061 | 12,500 | 12,500 | 48 | 7,400 | 0 | 0 |
| 100139 | 120,000 | 120,000 | 53 | 57,000 | 0 | 0 |
| 100284 | 58,000 | 58,000 | 47 | 50,000 | 0 | 0 |
| 100292 | 36,547 | NaN | 4 | 62,760 | 1 | 0 |
| 100583 | 30,000 | 30,000 | 53 | 18,500 | 0 | 0 |
| 100833 | 39,000 | 39,000 | 45 | 37,000 | 0 | 0 |
| 100931 | 56,000 | 56,000 | 53 | 60,200 | 1 | 0 |
| 101089 | 36,000 | 36,000 | 53 | 32,307 | 0 | 0 |
| 101122 | 35,151 | NaN | 39 | 46,362 | 0 | 0 |
| 101132 | 0 | 0 | 22 | 2,470 | 0 | 0 |

9.  We should look at some summary statistics for our prediction and compare those with the actual wage income values. The mean for the imputed wage income feature is lower than the original wage income mean. This is not surprising since, as we have seen, individuals with missing wage income have lower values for positively correlated features. What is surprising is the sharp reduction in the standard deviation. This is one of the drawbacks of deterministic regression imputation:

```
nls97[['wageincomeimp','wageincome']].
   agg(['count','mean','std'])
```

|       | wageincomeimp | wageincome |
|-------|---------------|------------|
| count | 8,984         | 5,091      |
| mean  | 42,559        | 49,477     |
| std   | 33,406        | 40,678     |

10. Stochastic regression imputation adds a normally distributed error to the predictions based on the residuals from our model. We want this error to have a mean of 0 with the same standard deviation as our residuals. We can use NumPy's normal function for that with np.random.normal(0, lm.resid.std(), nls97.shape[0]). The lm.resid.std() parameter gets us the standard deviation of the residuals from our model. The final parameter value, nls97.shape[0], indicates how many values to create; in this case, we want a value for every row in our data.

We can join those values with our data and then add the error, `randomadd`, to our prediction:

```
randomadd = np.random.normal(0, lm.resid.std(),
    nls97.shape[0])
randomadddf = pd.DataFrame(randomadd,
    columns=['randomadd'], index=nls97.index)
nls97 = nls97.join(randomadddf)
nls97['stochasticpred'] = nls97.pred + nls97.randomadd
nls97['wageincomeimpstoc'] =
    np.where(nls97.wageincome.isnull(),
    nls97.stochasticpred, nls97.wageincome)
```

11. This should increase the variance but not have much of an effect on the mean. Let's confirm this:

```
nls97[['wageincomeimpstoc','wageincome']].agg([
    'count','mean','std'])
```

|  | wageincomeimpstoc | wageincome |
| --- | --- | --- |
| count | 8,984 | 5,091 |
| mean | 42,517 | 49,477 |
| std | 41,381 | 40,678 |

That seemed to have worked. Our stochastic prediction has pretty much the same standard deviation as the original wage income feature.

Regression imputation is a good way to take advantage of all the data we have to impute values for a feature. It is often superior to the imputation methods we examined in the previous section, particularly when missing values are not random. If we use stochastic regression imputation, we will not artificially reduce our variance.

Before we started using machine learning for this work, this was our go-to multivariate approach for imputation. We now have the option of using algorithms such as KNN for this task, which has advantages over regression imputation in some cases. KNN imputation, unlike regression imputation, does not assume a linear relationship between features, or that those features are normally distributed. We will explore KNN imputation in the next section.

# Using KNN imputation

KNN is a popular machine learning technique because it is intuitive, easy to run, and yields good results when there are not a large number of features and observations. For the same reasons, it is often used to impute missing values. As its name suggests, KNN identifies the k observations whose features are most similar to each observation. When it's used to impute missing values, KNN uses the nearest neighbors to determine what fill values to use.

We can use KNN imputation to do the same imputation we did in the previous section on regression imputation:

1.  Let's start by importing `KNNImputer` from scikit-learn and loading the NLS data again:

    ```
    import pandas as pd
    import numpy as np
    from sklearn.impute import KNNImputer
    nls97 = pd.read_csv("data/nls97b.csv")
    nls97.set_index("personid", inplace=True)
    ```

2.  Next, we must prepare the features. We collapse degree attainment into three categories – less than college, college, and post-college degree – with each category represented by a different dummy variable. We must also convert the logical missing values for parent income into actual missings:

    ```
    nls97['hdegnum'] =
        nls97.highestdegree.str[0:1].astype('float')
    nls97['degltcol'] = np.where(nls97.hdegnum<=2,1,0)
    nls97['degcol'] =
        np.where(nls97.hdegnum.between(3,4),1,0)
    nls97['degadv'] = np.where(nls97.hdegnum>4,1,0)
    nls97.parentincome.replace(list(range(-5,0)), np.nan,
        inplace=True)
    ```

3.  Let's create a DataFrame that contains just the wage income and a few correlated features:

    ```
    wagedatalist = ['wageincome','weeksworked16',
        'parentincome','degltcol','degcol','degadv']
    wagedata = nls97[wagedatalist]
    ```

4.  We are now ready to use the `fit_transform` method of the KNN imputer to get values for all the missing values in the passed DataFrame, `wagedata`. `fit_transform` returns a NumPy array that contains all the non-missing values from `wagedata`, plus the imputed ones. We can convert this array into a DataFrame using the same index as `wagedata`. This will make it easy to join the data in the next step.

> **Note**
>
> We will use this technique throughout this book when we're working with the NumPy arrays that are returned when we use scikit-learn's `transform` and `fit_transform` methods.

We need to specify the value to use for the number of nearest neighbors, for k. We use a general rule of thumb for determining k – the square root of the number of observations divided by 2 (*sqrt(N)/2*). That gives us 47 for k in this case:

```
impKNN = KNNImputer(n_neighbors=47)
newvalues = impKNN.fit_transform(wagedata)
wagedatalistimp = ['wageincomeimp','weeksworked16imp',
    'parentincomeimp','degltcol','degcol','degadv']
wagedataimp = pd.DataFrame(newvalues,
    columns=wagedatalistimp, index=wagedata.index)
```

5.  Now, we must join the imputed wage income and weeks worked columns with the original NLS wage data and make a few observations. Notice that, with KNN imputation, we did not need to do any pre-imputation for missing values of correlated features (with regression imputation, we set weeks worked and parent income to their dataset means). That does mean, however, that KNN imputation will return an imputation, even when there is not a lot of information, such as with `101122` for `personid` in the following code block:

```
wagedata = wagedata.join(wagedataimp[['wageincomeimp',
    'weeksworked16imp']])
wagedata[['wageincome','weeksworked16','parentincome',
    'degcol','degadv','wageincomeimp']].head(10)
```

| | wageincome | weeksworked16 | parentincome | degcol | degadv | wageincomeimp |
|---|---|---|---|---|---|---|
| personid | | | | | | |
| 100061 | 12,500 | 48 | 7,400 | 0 | 0 | 12,500 |
| 100139 | 120,000 | 53 | 57,000 | 0 | 0 | 120,000 |

| 100284 | 58,000 | 47 | 50,000 | 0 | 0 | 58,000 |
| 100292 | NaN | 4 | 62,760 | 1 | 0 | 28,029 |
| 100583 | 30,000 | 53 | 18,500 | 0 | 0 | 30,000 |
| 100833 | 39,000 | 45 | 37,000 | 0 | 0 | 39,000 |
| 100931 | 56,000 | 53 | 60,200 | 1 | 0 | 56,000 |
| 101089 | 36,000 | 53 | 32,307 | 0 | 0 | 36,000 |
| 101122 | NaN | NaN | NaN | 0 | 0 | 33,977 |
| 101132 | 0 | 22 | 2,470 | 0 | 0 | 0 |

6.  Let's take a look at the summary statistics for the original and imputed features. Not surprisingly, the imputed wage income's mean is lower than the original mean. As we discovered in the previous section, observations with missing wage incomes have lower degree attainment, weeks worked, and parental income. We also lose some of the variance in wage income:

```
wagedata[['wageincome','wageincomeimp']].agg(['count',
    'mean','std'])
           wageincome  wageincomeimp
count          5,091          8,984
mean          49,477         44,781
std           40,678         32,034
```

KNN does imputations without making any assumptions about the distribution of the underlying data. With regression imputation, the standard assumptions for linear regression apply – that is, that there is a linear relationship between features and that they are distributed normally. If this is not the case, KNN is likely a better approach for imputation.

Despite these advantages, KNN imputation does have limitations. First, we must tune the model with an initial assumption about a good value for k, sometimes informed by little more than our knowledge of the size of the dataset. KNN is also computationally expensive and may be impractical for very large datasets. Finally, KNN imputation may not perform well when the correlation is weak between the feature to be imputed and the predictor features. An alternative to KNN for imputation, random forest imputation, can help us avoid the disadvantages of both KNN and regression imputation. We will explore random forest imputation in the next section.

# Using random forest for imputation

Random forest is an ensemble learning method. It uses bootstrap aggregating, also known as bagging, to improve model accuracy. It makes predictions by repeatedly taking the mean of multiple trees, yielding progressively better estimates. We will use the `MissForest` algorithm in this section, which is an application of the random forest algorithm to find missing value imputation.

`MissForest` starts by filling in the median or mode (for continuous or categorical features, respectively) for missing values, then uses random forest to predict values. Using this transformed dataset, with missing values replaced with initial predictions, `MissForest` generates new predictions, perhaps replacing the initial prediction with a better one. `MissForest` will typically go through at least four iterations of this process.

Running `MissForest` is even easier than using the KNN imputer, which we used in the previous section. We will impute values for the same wage income data that we worked with previously:

1.  Let's start by importing the `MissForest` module and loading the NLS data:

    ```
    import pandas as pd
    import numpy as np
    import sys
    import sklearn.neighbors._base
    sys.modules['sklearn.neighbors.base'] =
       sklearn.neighbors._base
    from missingpy import MissForest
    nls97 = pd.read_csv("data/nls97b.csv")
    nls97.set_index("personid", inplace=True)
    ```

    > **Note**
    >
    > We need to address a conflict in the name of `sklearn.neighbors._base`, which can be either `sklearn.neighbors._base` or `sklearn.neighbors.base`, depending on your version of scikit-learn. At the time of writing, `MissForest` uses the older name.

2.  Let's do the same data cleaning that we did in the previous section:

    ```
    nls97['hdegnum'] =
       nls97.highestdegree.str[0:1].astype('float')
    nls97.parentincome.replace(list(range(-5,0)), np.nan,
    ```

```
        inplace=True)
nls97['degltcol'] = np.where(nls97.hdegnum<=2,1,0)
nls97['degcol'] = np.where(nls97.hdegnum.between(3,4),
    1,0)
nls97['degadv'] = np.where(nls97.hdegnum>4,1,0)
wagedatalist = ['wageincome','weeksworked16',
    'parentincome','degltcol','degcol','degadv']
wagedata = nls97[wagedatalist]
```

3.  Now, we are ready to run `MissForest`. Notice that this process is quite similar to our process of using the KNN imputer:

```
imputer = MissForest()
newvalues = imputer.fit_transform(wagedata)
wagedatalistimp = ['wageincomeimp','weeksworked16imp',
    'parentincomeimp','degltcol','degcol','degadv']
wagedataimp = pd.DataFrame(newvalues,
    columns=wagedatalistimp , index=wagedata.index)
Iteration: 0
Iteration: 1
Iteration: 2
Iteration: 3
```

4.  Let's take a look at a few of our imputed values and some summary statistics. The imputed values have a lower mean. This is not surprising, given that we have already learned that the missing values are not distributed randomly, that individuals with lower degree attainment and weeks worked are more likely to have missing values for wage income:

```
wagedata = wagedata.join(wagedataimp[['wageincomeimp',
    'weeksworked16imp']])
wagedata[['wageincome','weeksworked16','parentincome',
    'degcol','degadv','wageincomeimp']].head(10)
       wageincome  weeksworked16  parentin-
come   degcol  degadv  wageincomeimp
personid
100061     12,500       48     7,400    0    0      12,500
100139    120,000       53    57,000    0    0     120,000
100284     58,000       47    50,000    0    0      58,000
```

| 100292 | NaN | 4 | 62,760 | 1 | 0 | 42,065 |
| 100583 | 30,000 | 53 | 18,500 | 0 | 0 | 30,000 |
| 100833 | 39,000 | 45 | 37,000 | 0 | 0 | 39,000 |
| 100931 | 56,000 | 53 | 60,200 | 1 | 0 | 56,000 |
| 101089 | 36,000 | 5 | 32,307 | 0 | 0 | 36,000 |
| 101122 | NaN | NaN | NaN | 0 | 0 | 32,384 |
| 101132 | 0 | 22 | 2,470 | 0 | 0 | 0 |

```
wagedata[['wageincome','wageincomeimp',
  'weeksworked16','weeksworked16imp']].agg(['count',
  'mean','std'])
```

| | wageincome | wageincomeimp | weeksworked16 | weeksworked-16imp |
| --- | --- | --- | --- | --- |
| count | 5,091 | 8,984 | 7,068 | 8,984 |
| mean | 49,477 | 43,140 | 39 | 37 |
| std | 40,678 | 34,725 | 21 | 21 |

MissForest uses the random forest algorithm to generate highly accurate predictions. Unlike KNN, it doesn't need to be tuned with an initial value for k. It also is computationally less expensive than KNN. Perhaps most importantly, random forest imputation is less sensitive to low or very high correlation among features, though that was not an issue in this example.

# Summary

In this chapter, we explored the most popular approaches for missing value imputation and discussed the advantages and disadvantages of each approach. Assigning an overall sample mean is not usually a good approach, particularly when observations with missing values are different from other observations in important ways. We can also substantially reduce our variance. Forward or backward filling allows us to maintain the variance in our data, but it works best when the proximity of observations is meaningful, such as with time series or longitudinal data. In most non-trivial cases, we will want to use a multivariate technique, such as regression, KNN, or random forest imputation.

So far, we haven't touched on the important issue of data leakage and how to create separate training and testing datasets. To avoid data leakage, we need to work with training data independently of the testing data as soon as we begin our feature engineering. We will look at feature engineering in more detail in the next chapter. There, we will encode, transform, and scale features, while also being careful to separate the training and testing data.

# Section 2 – Preprocessing, Feature Selection, and Sampling

Anyone who has done a log transformation of a target, or scaled a feature, appreciates just how critical to our analysis preprocessing can be. Raise your hand if you were ever confident that your model approximated truth, but then tried a fairly obvious transformation and appreciated just how far from truth your original model was. Encoding, transforming, and scaling data is not a gimmick, though sometimes people have that impression. We apply that preprocessing because 1) it gets us closer to capturing a real-world process and 2) because many machine learning algorithms just work better with scaled data.

Feature selection is equally important. A good adage is never build a model with N features, when N - 1 features will do just as nicely. It is worth remembering that this is more complicated than having too many features. There are times when having 3 features is too many and others when having 103 features is perfectly fine. The issue is really whether the features are too correlated for their independent impact on the target to be disentangled. When that is not the case, the risks of overfitting and of unstable results increase substantially. We pay careful attention to feature selection in this part of the book and subsequently.

The chapter on model evaluation in this part prepares us for the work we will be doing in the rest of the book. We explore in detail the evaluation of both regression and classification models, those with continuous and categorical targets respectively. We also learn how to construct pipelines and do cross-validation. Most importantly, we learn about data leakage and how to avoid it.

This section comprises the following chapters:

# 4

# Encoding, Transforming, and Scaling Features

The first three chapters of this book focused on data cleaning, exploration, and how to identify missing values and outliers. The next few chapters will delve heavily into feature engineering, starting, in this chapter, with techniques to encode, transform, and scale data to improve the performance of machine learning models.

Typically, machine learning algorithms require some form of encoding of variables. Additionally, our models often perform better with scaling so that features with higher variability do not overwhelm the optimization. We will show you how to use different scaling techniques when your features have dramatically different ranges.

Specifically, in this chapter, we will explore the following main topics:

- Creating training datasets and avoiding data leakage
- Identifying irrelevant or redundant observations to be removed
- Encoding categorical features

- Encoding features with medium or high cardinality
- Transforming features
- Binning features
- Scaling features

# Technical requirements

In this chapter, we will work extensively with the `feature-engine` and `category_ encoders` packages alongside the `sklearn` library. You can use `pip` to install these packages with `pip install feature-engine`, `pip install category_ encoders`, and `pip install scikit-learn`. The code in this chapter uses version 0.24.2 of `sklearn`, version 1.1.2 of `feature-engine`, and version 2.2.2 of `category_encoders`. Note that either `pip install feature-engine` or `pip install feature_engine` will work.

All of the code for this chapter can be found on GitHub at `https://github.com/ PacktPublishing/Data-Cleaning-and-Exploration-with-Machine- Learning/tree/main/4.%20PruningEncodingandRescalingFeatures`.

# Creating training datasets and avoiding data leakage

One of the biggest threats to the performance of our models is data leakage. **Data leakage** occurs whenever our models are informed by data that is not in the training dataset. Sometimes, we inadvertently assist our model training with information that cannot be gleaned from the training data alone and end up with an overly rosy assessment of our model's accuracy.

Data scientists do not really intend for this to happen, hence the term *leakage*. This is not a *don't do it* kind of discussion. We all know not to do it. This is more of a *which steps should I take to avoid the problem?* discussion. It is actually quite easy to have some data leakage unless we develop routines to prevent it.

For example, if we have missing values for a feature, we might impute the mean across the whole dataset for those values. However, in order to validate our model, we subsequently split our data into training and testing datasets. We would then have accidentally introduced data leakage into our training dataset since the information from the full dataset (that is, the global mean) would have been used.

One of the practices that data scientists have adopted to avoid this is to establish separate training and testing datasets as close to the beginning of the analysis as possible. This can become a little more complicated with validation techniques such as cross-validation, but in the following chapters, we will go over how to avoid data leakage in a variety of situations.

We can use scikit-learn to create training and testing DataFrames for the National Longitudinal Survey of Youth data.

> **Note**
>
> The **National Longitudinal Survey (NLS)** of Youth is conducted by the United States Bureau of Labor Statistics. This survey started with a cohort of individuals in 1997 who were born between 1980 and 1985, with annual follow-ups each year through to 2017. For this section, I pulled 89 variables on grades, employment, income, and attitudes toward the government from the hundreds of data items within the survey. Separate files for SPSS, Stata, and SAS can be downloaded from the repository. The NLS data can be downloaded for public use from `https://www.nlsinfo.org/investigator/pages/search`.

Let's start creating the DataFrame:

1. First, we import the `train_test_split` module from `sklearn` and load the NLS data:

   ```
   import pandas as pd
   from sklearn.model_selection import train_test_split
   nls97 = pd.read_csv("data/nls97b.csv")
   nls97.set_index("personid", inplace=True)
   ```

2. Then, we can create training and testing DataFrames for the features (`X_train` and `X_test`) and the targets (`y_train` and `y_test`). In this example, `wageincome` is the target variable. We set the `test_size` parameter to `0.3` to leave 30% of the observations for testing. Note that we will only work with the **Scholastic Assessment Test (SAT)** and **grade point average (GPA)** data from the NLS:

   ```
   feature_cols = ['satverbal','satmath','gpascience',
     'gpaenglish','gpamath','gpaoverall']

   X_train, X_test, y_train, y_test =  \
     train_test_split(nls97[feature_cols],\
   ```

```
      nls97[['wageincome']], test_size=0.3, \
      random_state=0)
```

3.  Let's take a look at the training DataFrames created with `train_test_split`. We get the expected number of observations, 6,288, which is 70% of the total number of observations in the NLS DataFrame of 8,984:

```
nls97.shape[0]
8984


X_train.info()
<class 'pandas.core.frame.DataFrame'>
Int64Index: 6288 entries, 574974 to 370933
Data columns (total 6 columns):
 #   Column        Non-Null Count     Dtype
---  ------        --------------     -------
 0   satverbal      1001 non-null     float64
 1   satmath        1001 non-null    float64
 2   gpascience     3998 non-null     float64
 3   gpaenglish     4078 non-null     float64
 4   gpamath        4056 non-null     float64
 5   gpaoverall     4223 non-null     float64
dtypes: float64(6)
memory usage: 343.9 KB


y_train.info()
<class 'pandas.core.frame.DataFrame'>
Int64Index: 6288 entries, 574974 to 370933
Data columns (total 1 columns):
 #   Column        Non-Null Count     Dtype
---  ------        --------------     -------
 0   wageincome     3599 non-null     float64
dtypes: float64(1)
memory usage: 98.2 KB
```

4.  Additionally, let's look at the testing DataFrames. We get 30% of the total number of observations, as expected:

```
X_test.info()
<class 'pandas.core.frame.DataFrame'>
Int64Index: 2696 entries, 363170 to 629736
Data columns (total 6 columns):
 #   Column        Non-Null Count    Dtype
---  ------        --------------    -------
 0   satverbal      405 non-null     float64
 1   satmath        406 non-null     float64
 2   gpascience    1686 non-null     float64
 3   gpaenglish    1720 non-null     float64
 4   gpamath       1710 non-null     float64
 5   gpaoverall    1781 non-null     float64
dtypes: float64(6)
memory usage: 147.4 KB


y_test.info()
<class 'pandas.core.frame.DataFrame'>
Int64Index: 2696 entries, 363170 to 629736
Data columns (total 1 columns):
 #   Column        Non-Null Count    Dtype
---  ------        --------------    -------
 0   wageincome    1492 non-null     float64
dtypes: float64(1)
memory usage: 42.1 KB
```

We will use scikit-learn's `test_train_split` to create separate training and testing DataFrames in the rest of this chapter. We will introduce more complicated strategies for constructing testing datasets for validation in *Chapter 6, Preparing for Model Evaluation*.

Next, we begin our feature engineering work by removing features that are obviously unhelpful. This is because they have the same data as another feature or there is no variation in the responses.

# Removing redundant or unhelpful features

During the process of data cleaning and manipulation, we often end up with data that is no longer meaningful. Perhaps we subsetted data based on a single feature value, and we have retained that feature even though it now has the same value for all observations. Or, for the subset of the data that we are using, two features have the same value. Ideally, we catch those redundancies during our data cleaning. However, if we do not catch them during that process, we can use the open source `feature-engine` package to help us.

Additionally, there might be features that are so highly correlated that it is very unlikely that we could build a model that could use all of them effectively. `feature-engine` has a method, `DropCorrelatedFeatures`, that makes it easy to remove a feature when it is highly correlated with another feature.

In this section, we will work with land temperature data, along with the NLS data. Note that we will only load temperature data for Poland here.

> **Data Note**
>
> The land temperature dataset contains the average temperature readings (in Celsius) in 2019 from over 12,000 stations across the world, though the majority of the stations are in the United States. The raw data was retrieved from the Global Historical Climatology Network integrated database. It has been made available for public use by the United States National Oceanic and Atmospheric Administration at `https://www.ncdc.noaa.gov/data-access/land-based-station-data/land-based-datasets/global-historical-climatology-network-monthly-version-4`.

Let's start removing redundant and unhelpful features:

1.  Let's import the modules we need from `feature_engine` and `sklearn`, and load the NLS data and temperature data for Poland. The data from Poland was pulled from a larger dataset of 12,000 weather stations across the world. We use `dropna` to drop observations with any missing data:

    ```
    import pandas as pd
    import feature_engine.selection as fesel
    from sklearn.model_selection import train_test_split
    ```

```
nls97 = pd.read_csv("data/nls97b.csv")
nls97.set_index("personid", inplace=True)
ltpoland = pd.read_csv("data/ltpoland.csv")
ltpoland.set_index("station", inplace=True)
ltpoland.dropna(inplace=True)
```

2.  Next, we create training and testing DataFrames, as we did in the previous section:

```
feature_cols = ['satverbal','satmath','gpascience',
  'gpaenglish','gpamath','gpaoverall']

X_train, X_test, y_train, y_test =  \
  train_test_split(nls97[feature_cols],\
  nls97[['wageincome']], test_size=0.3, \
  random_state=0)
```

3.  We can use the pandas `corr` method to see how these features are correlated:

```
X_train.corr()
            satverbal   satmath   gpascience   gpaenglish \
satverbal     1.000      0.729      0.439        0.444
satmath       0.729      1.000      0.480        0.430
gpascience    0.439      0.480      1.000        0.672
gpaenglish    0.444      0.430      0.672        1.000
gpamath       0.375      0.518      0.606        0.600
gpaoverall    0.421      0.485      0.793        0.844


            gpamath   gpaoverall
satverbal     0.375      0.421
satmath       0.518      0.485
gpascience    0.606      0.793
gpaenglish    0.600      0.844
gpamath       1.000      0.750
gpaoverall    0.750      1.000
```

Here, gpaoverall is highly correlated with gpascience, gpaenglish, and gpamath. The corr method returns the Pearson coefficients by default. This is fine when we can assume a linear relationship between the features. However, when this assumption does not make sense, we should consider requesting Spearman coefficients instead. We can do that by passing spearman to the method parameter of corr.

4.  Let's drop features that have a correlation higher than 0.75 with another feature. We pass 0.75 to the threshold parameter of DropCorrelatedFeatures, indicating that we want to use Pearson coefficients and that we want to evaluate all the features by setting the variables to None. We use the fit method on the training data and then transform both the training and testing data. The info method shows that the resulting training DataFrame (X_train_tr) has all of the features except gpaoverall, which has correlations of 0.793 and 0.844 with gpascience and gpaenglish, respectively (DropCorrelatedFeatures will evaluate from left to right, so if gpamath and gpaoverall are highly correlated, it will drop gpaoverall. If gpaoverall had been to the left of gpamath, it would have dropped gpamath):

```
tr = fesel.DropCorrelatedFeatures(variables=None,
method='pearson', threshold=0.75)
tr.fit(X_train)
X_train_tr = tr.transform(X_train)
X_test_tr = tr.transform(X_test)
X_train_tr.info()
<class 'pandas.core.frame.DataFrame'>
Int64Index: 6288 entries, 574974 to 370933
Data columns (total 5 columns):
 #    Column        Non-Null Count      Dtype
---   ------        --------------      -------
 0    satverbal     1001 non-null       float64
 1    satmath       1001 non-null       float64
 2    gpascience    3998 non-null       float64
 3    gpaenglish    4078 non-null       float64
 4    gpamath       4056 non-null       float64
dtypes: float64(5)
memory usage: 294.8 KB
```

Typically, we would evaluate a feature more carefully before deciding to drop it. However, there are times when feature selection is part of a pipeline, and we need to automate the process. This can be done with `DropCorrelatedFeatures` since all of the `feature_engine` methods can be brought into a scikit-learn pipeline.

5.  Now, let's create training and testing DataFrames from the land temperature data for Poland. The value of `year` is the same for all observations, as is the value for `country`. Additionally, the value for `latabs` is the same as it is for `latitude` for each observation:

```
feature_cols = ['year','month','latabs',
  'latitude','elevation', 'longitude','country']


X_train, X_test, y_train, y_test =  \
  train_test_split(ltpoland[feature_cols],\
  ltpoland[['temperature']], test_size=0.3, \
  random_state=0)


X_train.sample(5, random_state=99)
          year  month  latabs  latitude  elevation  longi-
tude country
station
SIEDLCE   2019   11     52      52       152       22     Poland
OKECIE    2019    6     52      52       110       21     Poland
BALICE    2019    1     50      50       241       20     Poland
BALICE    2019    7     50      50       241       20     Poland
BIALYSTOK 2019   11     53      53       151       23     Poland


X_train.year.value_counts()
2019    84
Name: year, dtype: int64


X_train.country.value_counts()
Poland    84
Name: country, dtype: int64


(X_train.latitude!=X_train.latabs).sum()
0
```

6.  Let's drop features with the same values throughout the training dataset. Notice that year and country are removed after the transform:

```
tr = fesel.DropConstantFeatures()
tr.fit(X_train)
X_train_tr = tr.transform(X_train)
X_test_tr = tr.transform(X_test)
X_train_tr.head()
```

| station | month | latabs | latitude | elevation | longitude |
|---|---|---|---|---|---|
| OKECIE | 1 | 52 | 52 | 110 | 21 |
| LAWICA | 8 | 52 | 52 | 94 | 17 |
| LEBA | 11 | 55 | 55 | 2 | 18 |
| SIEDLCE | 10 | 52 | 52 | 152 | 22 |
| BIALYSTOK | 11 | 53 | 53 | 151 | 23 |

7.  Let's drop features that have the same values as other features. In this case, the transform drops latitude, which has the same values as latabs:

```
tr = fesel.DropDuplicateFeatures()
tr.fit(X_train_tr)
X_train_tr = tr.transform(X_train_tr)
X_train_tr.head()
```

| station | month | latabs | elevation | longitude |
|---|---|---|---|---|
| OKECIE | 1 | 52 | 110 | 21 |
| LAWICA | 8 | 52 | 94 | 17 |
| LEBA | 11 | 55 | 2 | 18 |
| SIEDLCE | 10 | 52 | 152 | 22 |
| BIALYSTOK | 11 | 53 | 151 | 23 |

This fixes some obvious problems with our features in the NLS data and the land temperature data for Poland. We dropped gpaoverall from a DataFrame that has the other GPA features because it is highly correlated with them. Additionally, we removed redundant data, dropping features with the same value throughout the DataFrame and features that duplicate the values of another feature.

The rest of this chapter explores somewhat messier feature engineering challenges: encoding, transforming, binning, and scaling.

# Encoding categorical features

There are several reasons why we might need to encode features before using them in most machine learning algorithms. First, these algorithms typically require numeric data. Second, when a categorical feature *is* represented with numbers, for example, 1 for female and 2 for male, we need to encode the values so that they are recognized as categorical. Third, the feature might actually be ordinal, with a discrete number of values that represent some meaningful ranking. Our models need to capture that ranking. Finally, a categorical feature might have a large number of values (known as high cardinality), and we might want our encoding to collapse categories.

We can handle the encoding of features with a limited number of values, say 15 or less, with one-hot encoding. In this section, we will, first, go over one-hot encoding and then discuss ordinal encoding. We will look at strategies for handling categorical features with high cardinality in the next section.

## One-hot encoding

One-hot encoding a feature creates a binary vector for each value of that feature. So, if a feature, called *letter*, has three unique values, A, B, and C, one-hot encoding creates three binary vectors to represent those values. The first binary vector, which we can call *letter_A*, has 1 whenever *letter* has a value of A, and 0 when it is B or C. *letter_B* and *letter_C* would be coded similarly. The transformed features, *letter_A*, *letter_B*, and *letter_C*, are often referred to as **dummy variables**. *Figure 4.1* illustrates one-hot encoding:

| letter | letter_A | letter_B | letter_C |
|--------|----------|----------|----------|
| A | 1 | 0 | 0 |
| B | 0 | 1 | 0 |
| C | 0 | 0 | 1 |

Figure 4.1 – The one-hot encoding of a categorical feature

A number of features from the NLS data are appropriate for one-hot encoding. In the following code blocks, we encode some of those features:

1. Let's start by importing the OneHotEncoder module from feature_engine and loading the data. Additionally, we import the OrdinalEncoder module from scikit-learn since we will use it later:

```
import pandas as pd
from feature_engine.encoding import OneHotEncoder
from sklearn.preprocessing import OrdinalEncoder
from sklearn.model_selection import train_test_split
```

```
nls97 = pd.read_csv("data/nls97b.csv")
nls97.set_index("personid", inplace=True)
```

2.  Next, we create training and testing DataFrames for the NLS data:

```
feature_cols =['gender','maritalstatus','colenroct99']
nls97demo = nls97[['wageincome'] + feature_cols].dropna()

X_demo_train, X_demo_test, y_demo_train, y_demo_test=\
    train_test_split(nls97demo[feature_cols],\
    nls97demo[['wageincome']], test_size=0.3, \
    random_state=0)
```

3.  One option we have for the encoding is the pandas get_dummies method. We can use it to indicate that we want to convert the gender and maritalstatus features. get_dummies gives us a dummy variable for each value of gender and maritalstatus. For example, gender has the values of Female and Male. get_dummies creates a feature, gender_Female, which is 1 when gender is Female and 0 when gender is Male. When gender is Male, gender_Male is 1 and gender_Female is 0. This is a tried-and-true method of doing this type of encoding and has served statisticians well for many years:

```
pd.get_dummies(X_demo_train, \
    columns=['gender','maritalstatus']).head(2).T
personid                        736081          832734
colenroct99                     1.Not enrolled  1.Not
enrolled
gender_Female                   1               0
gender_Male                     0               1
maritalstatus_Divorced          0               0
maritalstatus_Married           1               0
maritalstatus_Never-married 0                   1
maritalstatus_Separated         0               0
maritalstatus_Widowed           0               0
```

We are not saving the DataFrame created by get_dummies because, later in this section, we will be using a different technique to do the encoding.

Typically, we create *k-1* dummy variables for *k* unique values for a feature. So, if `gender` has two values in our data, we only need to create one dummy variable. If we know the value for `gender_Female`, we also know the value of `gender_Male`; therefore, the latter variable is redundant. Similarly, we know the value of `maritalstatus_Divorced` if we know the values of the other `maritalstatus` dummies. Creating a redundancy in this way is inelegantly referred to as the **dummy variable trap**. To avoid this problem, we drop one dummy from each group.

> **Note**
>
> For some machine learning algorithms, such as linear regression, dropping one dummy variable is actually required. In estimating the parameters of a linear model, the matrix is inverted. If our model has an intercept, and all dummy variables are included, the matrix cannot be inverted.

4.  We can set the `get_dummies` `drop_first` parameter to `True` to drop the first dummy from each group:

```
pd.get_dummies(X_demo_train, \
  columns=['gender','maritalstatus'],
  drop_first=True).head(2).T
personid                         736081         832734
colenroct99                      1. Not enrolled 1. Not
enrolled
gender_Male                      0              1
maritalstatus_Married            1              0
maritalstatus_Never-married      0              1
maritalstatus_Separated          0              0
maritalstatus_Widowed            0              0
```

An alternative to `get_dummies` is the one-hot encoder in either `sklearn` or `feature_engine`. These one-hot encoders have the advantage that they can be easily brought into a machine learning pipeline, and they can persist information gathered from the training dataset to the testing dataset.

5.  Let's use the `OneHotEncoder` module from `feature_engine` to do the encoding. We set `drop_last` to `True` to drop one of the dummies from each group. We fit the encoding to the training data and then transform both the training and testing data:

```
ohe = OneHotEncoder(drop_last=True,
  variables=['gender','maritalstatus'])
ohe.fit(X_demo_train)
X_demo_train_ohe = ohe.transform(X_demo_train)
X_demo_test_ohe = ohe.transform(X_demo_test)
X_demo_train_ohe.filter(regex='gen|mar', axis="columns").
head(2).T
```

| personid | 736081 | 832734 |
|---|---|---|
| gender_Female | 1 | 0 |
| maritalstatus_Married | 1 | 0 |
| maritalstatus_Never-married | 0 | 1 |
| maritalstatus_Divorced | 0 | 0 |
| maritalstatus_Separated | 0 | 0 |

This demonstrates that one-hot encoding is a fairly straightforward way to prepare nominal data for a machine learning algorithm. But what if our categorical features are ordinal, rather than nominal? In that case, we need to use ordinal encoding.

# Ordinal encoding

Categorical features can be either nominal or ordinal, as discussed in *Chapter 1, Examining the Distribution of Features and Targets*. Gender and marital status are nominal. Their values do not imply order. For example, "never married" is not a higher value than "divorced."

However, when a categorical feature is ordinal, we want the encoding to capture the ranking of the values. For example, if we have a feature that has the values of low, medium, and high, one-hot encoding would lose this ordering. Instead, a transformed feature with the values of 1, 2, and 3 for low, medium, and high, respectively, would be better. We can accomplish this with ordinal encoding.

The college enrollment feature on the NLS dataset can be considered an ordinal feature. The values range from *1. Not enrolled* to *3. 4-year college*. We should use ordinal encoding to prepare it for modeling. We will do that next:

1.  We can use the `OrdinalEncoder` module of `sklearn` to encode the college enrollment for 1999 feature. First, let's take a look at the values of `colenroct99` prior to encoding. The values are strings, but there is an implied order:

    ```
    X_demo_train.colenroct99.unique()
    array(['1. Not enrolled', '2. 2-year college ',
           '3. 4-year college'], dtype=object)
    ```

    ```
    X_demo_train.head()
                    gender    maritalstatus    colenroct99
    personid
    736081          Female    Married          1. Not enrolled
    832734          Male      Never-married    1. Not enrolled
    453537          Male      Married          1. Not enrolled
    322059          Female    Divorced         1. Not enrolled
    324323          Female    Married          2. 2-year college
    ```

2.  We can tell the `OrdinalEncoder` module to rank the values in the same order by passing the preceding array into the `categories` parameter. Then, we can use `fit_transform` to transform the college enrollment field, `colenroct99`. (The `fit_transform` method of the `sklearn` `OrdinalEncoder` module returns a NumPy array, so we need to use the pandas DataFrame method to create a DataFrame.) Finally, we join the encoded features with the other features from the training data:

    ```
    oe = OrdinalEncoder(categories=\
      [X_demo_train.colenroct99.unique()])
    colenr_enc = \
      pd.DataFrame(oe.fit_transform(X_demo_
    train[['colenroct99']]),
        columns=['colenroct99'], index=X_demo_train.index)
    X_demo_train_enc = \
      X_demo_train[['gender','maritalstatus']].\
      join(colenr_enc)
    ```

3. Let's take a look at a few observations of the resulting DataFrame. Additionally, we should compare the counts of the original college enrollment feature to the transformed feature:

```
X_demo_train_enc.head()
                    gender        maritalstatus    colenroct99

personid
736081              Female        Married          0
832734              Male          Never-married    0
453537              Male          Married          0
322059              Female        Divorced         0
324323              Female        Married          1


X_demo_train.colenroct99.value_counts().sort_index()
1. Not enrolled            3050
2. 2-year college          142
3. 4-year college          350
Name: colenroct99, dtype: int64


X_demo_train_enc.colenroct99.value_counts().sort_index()
0          3050
1          142
2          350
Name: colenroct99, dtype: int64
```

The ordinal encoding replaces the initial values for colenroct99 with numbers from 0 to 2. It is now in a form that is consumable by many machine learning models, and we have retained the meaningful ranking information.

> **Note**
> Ordinal encoding is appropriate for non-linear models such as decision trees. It might not make sense in a linear regression model because that would assume that the distance between values was equally meaningful across the whole distribution. In this example, that would assume that the increase from 0 to 1 (that is, from no enrollment to 2-year enrollment) is the same thing as the increase from 1 to 2 (that is, from 2-year enrollment to 4-year enrollment).

One-hot encoding and ordinal encoding are relatively straightforward approaches to engineering categorical features. It can be more complicated to deal with categorical features when there are many more unique values. In the next section, we will go over a couple of techniques for handling those features.

# Encoding categorical features with medium or high cardinality

When we are working with a categorical feature that has many unique values, say 10 or more, it can be impractical to create a dummy variable for each value. When there is high cardinality, that is, a very large number of unique values, there might be too few observations with certain values to provide much information for our models. At the extreme, with an ID variable, there is just one observation for each value.

There are a couple of ways in which to handle medium or high cardinality. One way is to create dummies for the top *k* categories and group the remaining values into an *other* category. Another way is to use feature hashing, also known as the hashing trick. In this section, we will explore both strategies. We will be using the COVID-19 dataset for this example:

1.  Let's create training and testing DataFrames from COVID-19 data, and import the feature_engine and category_encoders libraries:

    ```
    import pandas as pd
    from feature_engine.encoding import OneHotEncoder
    from category_encoders.hashing import HashingEncoder
    from sklearn.model_selection import train_test_split

    covidtotals = pd.read_csv("data/covidtotals.csv")
    feature_cols = ['location','population',
        'aged_65_older','diabetes_prevalence','region']
    covidtotals = covidtotals[['total_cases'] + feature_
    cols].dropna()

    X_train, X_test, y_train, y_test =  \
      train_test_split(covidtotals[feature_cols],\
      covidtotals[['total_cases']], test_size=0.3,
      random_state=0)
    ```

The feature region has 16 unique values, the first 6 of which have counts of 10 or more:

```
X_train.region.value_counts()
Eastern Europe   16
East Asia   12
Western Europe   12
West Africa   11
West Asia   10
East Africa   10
South America   7
South Asia   7
Central Africa   7
Southern Africa   7
Oceania / Aus   6
Caribbean   6
Central Asia   5
North Africa   4
North America   3
Central America   3
Name: region, dtype: int64
```

2. We can use the `OneHotEncoder` module from `feature_engine` again to encode the `region` feature. This time, we use the `top_categories` parameter to indicate that we only want to create dummies for the top six category values. Any values that do not fall into the top six will have a 0 for all of the dummies:

```
ohe = OneHotEncoder(top_categories=6,
variables=['region'])
covidtotals_ohe = ohe.fit_transform(covidtotals)
covidtotals_ohe.filter(regex='location|region',
  axis="columns").sample(5, random_state=99).T
```

|  | 97 | 173 | 92 | 187 | 104 |
|---|---|---|---|---|---|
| Location | Israel | Senegal | Indonesia | Sri Lanka | Kenya |
| region_Eastern Europe | 0 | 0 | 0 | 0 | 0 |
| region_Western Europe | 0 | 0 | 0 | 0 | 0 |
| region_West Africa | 0 | 1 | 0 | 0 | 0 |
| region_East Asia | 0 | 0 | 1 | 0 | 0 |

```
        region_West Asia        1      0      0      0      0
        region_East Africa      0      0      0      0      1
```

An alternative approach to one-hot encoding, when a categorical feature has many unique values, is to use **feature hashing**.

# Feature hashing

Feature hashing maps a large number of unique feature values to a smaller number of dummy variables. We can specify the number of dummy variables to create. However, collisions are possible; that is, some feature values might map to the same dummy variable combination. The number of collisions increases as we decrease the number of requested dummy variables.

We can use `HashingEncoder` from `category_encoders` to do feature hashing. We use `n_components` to indicate that we want six dummy variables (we copy the `region` feature before we do the transform so that we can compare the original values to the new dummies):

```
X_train['region2'] = X_train.region
he = HashingEncoder(cols=['region'], n_components=6)
X_train_enc = he.fit_transform(X_train)
X_train_enc.\
  groupby(['col_0','col_1','col_2','col_3','col_4',
    'col_5','region2']).\
    size().reset_index().rename(columns={0:'count'})
     col_0 col_1 col_2 col_3 col_4 col_5 region2            count
0      0     0     0     0     0     1    Caribbean         6
1      0     0     0     0     0     1    Central Africa    7
2      0     0     0     0     0     1    East Africa       10
3      0     0     0     0     0     1    North Africa      4
4      0     0     0     0     1     0    Central America   3
5      0     0     0     0     1     0    Eastern Europe    16
6      0     0     0     0     1     0    North America     3
7      0     0     0     0     1     0    Oceania / Aus     6
8      0     0     0     0     1     0    Southern Africa   7
9      0     0     0     0     1     0    West Asia         10
10     0     0     0     0     1     0    Western Europe    12
11     0     0     0     1     0     0    Central Asia      5
```

| 12 | 0 | 0 | 0 | 1 | 0 | 0 | East Asia | 12 |
| 13 | 0 | 0 | 0 | 1 | 0 | 0 | South Asia | 7 |
| 14 | 0 | 0 | 1 | 0 | 0 | 0 | West Africa | 11 |
| 15 | 1 | 0 | 0 | 0 | 0 | 0 | South America | 7 |

Unfortunately, this gives us a large number of collisions. For example, Caribbean, Central Africa, East Africa, and North Africa all get the same dummy variable values. In this case at least, using one-hot encoding and specifying the number of categories, as we did in the last section, was a better solution.

In the previous two sections, we covered common encoding strategies: one-hot encoding, ordinal encoding, and feature hashing. Almost all of our categorical features will require some kind of encoding before we can use them in a model. However, sometimes, we need to alter our features in other ways, including with transformations, binning, and scaling. In the next three sections, we will consider the reasons why we might need to alter our features in these ways and explore tools for doing that.

# Using mathematical transformations

Sometimes, we want to use features that do not have a Gaussian distribution with a machine learning algorithm that assumes our features are distributed in that way. When that happens, we either need to change our minds about which algorithm to use (for example, we could choose KNN rather than linear regression) or transform our features so that they approximate a Gaussian distribution. In this section, we will go over a couple of strategies for doing the latter:

1.  We start by importing the transformation module from `feature_engine`, `train_test_split` from `sklearn`, and `stats` from `scipy`. Additionally, we create training and testing DataFrames with the COVID-19 data:

```
import pandas as pd
from feature_engine import transformation as vt
from sklearn.model_selection import train_test_split
import matplotlib.pyplot as plt
from scipy import stats

covidtotals = pd.read_csv("data/covidtotals.csv")
feature_cols = ['location','population',
    'aged_65_older','diabetes_prevalence','region']
covidtotals = covidtotals[['total_cases'] + feature_
```

```
cols].dropna()

X_train, X_test, y_train, y_test =  \
   train_test_split(covidtotals[feature_cols],\
   covidtotals[['total_cases']], test_size=0.3, \
   random_state=0)
```

2.  Let's take a look at how the total number of cases by country is distributed. We should also calculate the skew:

```
y_train.total_cases.skew()
6.313169268923333
```

```
plt.hist(y_train.total_cases)
plt.title("Total COVID Cases  (in millions)")
plt.xlabel('Cases')
plt.ylabel("Number of Countries")
plt.show()
```

This produces the following histogram:

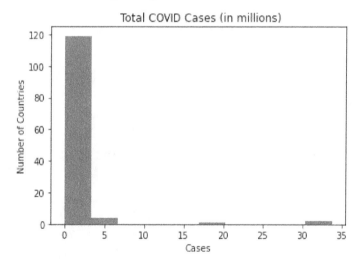

Figure 4.2 – A histogram of the total number of COVID cases

This illustrates the very high skew for the total number of cases. In fact, it looks log-normal, which is not surprising given the large number of very low values and several very high values.

> **Note**
>
> For more information about the measures of skew and kurtosis, please refer to *Chapter 1*, *Examining the Distribution of Features and Targets*.

3.  Let's try a log transformation. All we need to do to get `feature_engine` to do the transformation is call `LogTranformer` and pass the feature or features that we would like to transform:

```
tf = vt.LogTransformer(variables = ['total_cases'])
y_train_tf = tf.fit_transform(y_train)


y_train_tf.total_cases.skew()
-1.3872728024141519


plt.hist(y_train_tf.total_cases)
plt.title("Total COVID Cases (log transformation)")
plt.xlabel('Cases')
plt.ylabel("Number of Countries")
plt.show()
```

This produces the following histogram:

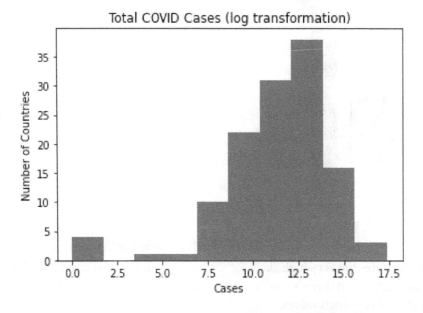

Figure 4.3 – A histogram of the total number of COVID cases with log transformation

Effectively, log transformations increase variability at the lower end of the distribution and decrease variability at the upper end. This produces a more symmetrical distribution. This is because the slope of the logarithmic function is steeper for smaller values than for larger ones.

4.  This is definitely a big improvement, but there is now some negative skew. Perhaps a Box-Cox transformation will yield better results. Let's try that:

```
tf = vt.BoxCoxTransformer(variables = ['total_cases'])
y_train_tf = tf.fit_transform(y_train)


y_train_tf.total_cases.skew()
0.07333475786753735


plt.hist(y_train_tf.total_cases)
plt.title("Total COVID Cases (Box-Cox transformation)")
plt.xlabel('Cases')
plt.ylabel("Number of Countries")
plt.show()
```

This produces the following plot:

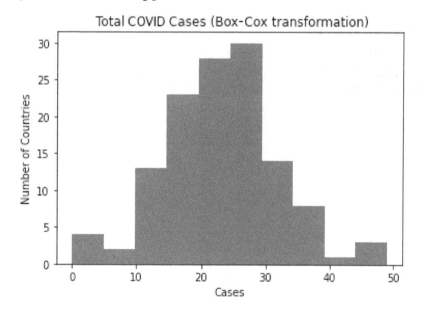

Figure 4.4 – A histogram of the total number of COVID cases with a Box-Cox transformation

Box-Cox transformations identify a value for lambda between -5 and 5 that generates a distribution that is closest to normal. It uses the following equation for the transformation:

$$y(\lambda) = (y\lambda - 1)/\lambda, if\ \lambda \neq 0$$

or

$$y(\lambda) = log\ y, if\ \lambda = 0$$

Here, $y(\lambda)$ is our transformed feature. Just for fun, let's see the value of the lambda that was used to transform `total_cases`:

```
stats.boxcox(y_train.total_cases)[1]
0.10435377585681517
```

The lambda for the Box-Cox transformation is `0.104`. For comparison, the lambda for a feature with a Gaussian distribution would be 1.000, meaning that no transformation would be necessary.

Now that our transformed total cases feature looks good, we can build a model with it as the target. Additionally, we can set up our pipeline to restore values to their original scaling when we make predictions. `feature_engine` has a number of other transformations that are implemented similarly to the log and Box-Cox transformations.

# Feature binning

Sometimes, we will want to convert a continuous feature into a categorical feature. The process of creating *k* equally spaced intervals from the minimum to the maximum value of a distribution is called **binning** or, the somewhat less-friendly term, **discretization**. Binning can address several important issues with a feature: skew, excessive kurtosis, and the presence of outliers.

# Equal-width and equal-frequency binning

Binning might be a good choice with the COVID case data. Let's try that (this might also be useful with other variables in the dataset, including total deaths and population, but we will only work with total cases for now. `total_cases` is the target variable in the following code, so it is a column – the only column – on the `y_train` DataFrame):

1. First, we need to import `EqualFrequencyDiscretiser` and `EqualWidthDiscretiser` from `feature_engine`. Additionally, we need to create training and testing DataFrames from the COVID data:

```
import pandas as pd
from feature_engine.discretisation import
EqualFrequencyDiscretiser as efd
from feature_engine.discretisation import
EqualWidthDiscretiser as ewd
from sklearn.preprocessing import KBinsDiscretizer
from sklearn.model_selection import train_test_split

covidtotals = pd.read_csv("data/covidtotals.csv")

feature_cols = ['location','population',
    'aged_65_older','diabetes_prevalence','region']
covidtotals = covidtotals[['total_cases'] + feature_
cols].dropna()

X_train, X_test, y_train, y_test = \
  train_test_split(covidtotals[feature_cols],\
  covidtotals[['total_cases']], test_size=0.3, random_
state=0)
```

2. We can use the pandas qcut method, and its q parameter, to create 10 bins of relatively equal frequency:

```
y_train['total_cases_group'] = pd.qcut(y_train.total_
cases, q=10, labels=[0,1,2,3,4,5,6,7,8,9])
y_train.total_cases_group.value_counts().sort_index()
0    13
1    13
2    12
3    13
```

```
4    12
5    13
6    12
7    13
8    12
9    13
Name: total_cases_group, dtype: int64
```

3. We can accomplish the same thing with `EqualFrequencyDiscretiser`. First, we define a function to run the transformation. The function takes a feature_ engine transformation and the training and testing DataFrames. It returns the transformed DataFrames (it is not necessary to define a function, but it makes sense here since we will repeat these steps later):

```
def runtransform(bt, dftrain, dftest):
    bt.fit(dftrain)
    train_bins = bt.transform(dftrain)
    test_bins = bt.transform(dftest)
    return train_bins, test_bins
```

4. Next, we create an `EqualFrequencyDiscretiser` transformer and call the `runtransform` function that we just created:

```
y_train.drop(['total_cases_group'], axis=1, inplace=True)
bintransformer = efd(q=10, variables=['total_cases'])
y_train_bins, y_test_bins = runtransform(bintransformer,
y_train, y_test)
y_train_bins.total_cases.value_counts().sort_index()
0    13
1    13
2    12
3    13
4    12
5    13
6    12
7    13
8    12
9    13
Name: total_cases, dtype: int64
```

This gives us the same results as qcut, but it has the advantage of being easier to bring into a machine learning pipeline since we are using `feature_engine` to produce it. The equal-frequency binning addresses both the skew and outlier problems.

> **Note**
>
> We will explore machine learning pipelines in detail in this book, starting with *Chapter 6, Preparing for Model Evaluation*. Here, the key point is that feature engine transformers can be a part of a pipeline that includes other `sklearn`-compatible transformers, even ones we construct ourselves.

5. `EqualWidthDiscretiser` works similarly:

   ```
   bintransformer = ewd(bins=10, variables=['total_cases'])
   ```

   ```
   y_train_bins, y_test_bins = runtransform(bintransformer,
   y_train, y_test)
   y_train_bins.total_cases.value_counts().sort_index()
   0   119
   1   4
   5   1
   9   2
   Name: total_cases, dtype: int64
   ```

   This is a far less successful transformation. Almost all of the values are at the bottom of the distribution in the data prior to the binning, so it is not surprising that equal-width binning would have the same problem. It results in only 4 bins, even though we requested 10.

6. Let's examine the range of each bin. Here, we can see that the equal-width binner is not even able to construct equal-width bins because of the small number of observations at the top of the distribution:

   ```
   pd.options.display.float_format = '{:,.0f}'.format
   ```

   ```
   y_train_bins = y_train_bins.\
      rename(columns={'total_cases':'total_cases_group'}).\
      join(y_train)
   ```

   ```
   y_train_bins.groupby("total_cases_group")["total_cases"].
   ```

```
agg(['min','max'])
   min   max
total_cases_group
0   1            3,304,135
1   3,740,567    5,856,682
5   18,909,037   18,909,037
9   30,709,557   33,770,444
```

Although in this case, equal-width binning was a bad choice, there are many times when it makes sense. It can be useful when data is more uniformly distributed or when the equal widths make sense substantively.

# K-means binning

Another option is to use *k*-means clustering to determine the bins. The *k*-means algorithm randomly selects *k* data points as centers of clusters and then assigns the other data points to the closest cluster. The mean of each cluster is computed, and the data points are reassigned to the nearest new cluster. This process is repeated until the optimal centers are found.

When *k*-means is used for binning, all data points in the same cluster will have the same ordinal value:

1.  We can use scikit-learn's KBinsDiscretizer to create bins with the COVID cases data:

```
kbins = KBinsDiscretizer(n_bins=10, encode='ordinal',
  strategy='kmeans')
y_train_bins = \
  pd.DataFrame(kbins.fit_transform(y_train),
  columns=['total_cases'])
y_train_bins.total_cases.value_counts().sort_index()
0   49
1   24
2   23
3   11
4   6
5   6
6   4
7   1
```

```
8   1
9   1
Name: total_cases, dtype: int64
```

2.  Let's compare the skew and kurtosis of the original total cases variable to that of the binned variable. Recall that we would expect a skew of 0 and a kurtosis near 3 for a variable with a Gaussian distribution. The distribution of the binned variable is much closer to Gaussian:

```
y_train.total_cases.agg(['skew','kurtosis'])
skew            6.313
kurtosis       41.553
Name: total_cases, dtype: float64


y_train_bins.total_cases.agg(['skew','kurtosis'])
skew            1.439
kurtosis        1.923
Name: total_cases, dtype: float64
```

Binning can help us to address skew, kurtosis, and outliers in our data. However, it does mask much of the variation in the feature and reduces its explanatory potential. Often, some form of scaling, such as min-max or z-score, is a better option. Let's examine feature scaling next.

# Feature scaling

Often, the features we want to use in our model are on very different scales. Put simply, the distance between the minimum and maximum values, or the range, varies substantially across possible features. For example, in the COVID-19 data, the total cases feature goes from 1 to almost 34 million, while aged 65 or older goes from 9 to 27 (the number represents the percentage of the population).

Having features on very different scales impacts many machine learning algorithms. For example, KNN models often use Euclidean distance, and features with greater ranges will have a greater influence on the model. Scaling can address this problem.

In this section, we will go over two popular approaches to scaling: **min-max scaling** and **standard** (or **z-score**) scaling. Min-max scaling replaces each value with its location in the range. More precisely, the following happens:

$$z_{ij} = (x_{ij} - min_j)/(max_j - min_j)$$

Here, $z_{ij}$ is the min-max score, $x_{ij}$ is the value for the $i^{th}$ observation of the $j^{th}$ feature, and $min_j$ and $max_j$ are the minimum and maximum values of the $j^{th}$ feature.

Standard scaling normalizes the feature values around a mean of 0. Those who studied undergraduate statistics will recognize it as the z-score. Specifically, it is as follows:

$$z_{ij} = (x_{ij} - u_j)/s_j$$

Here, $x_{ij}$ is the value for the $i^{th}$ observation of the $j^{th}$ feature, $u_j$ is the mean for feature $j$, and $s_j$ is the standard deviation for that feature.

We can use scikit-learn's preprocessing module to get the min-max and standard scalers:

1.  We start by importing the preprocessing module and creating training and testing DataFrames from the COVID-19 data:

```
import pandas as pd
from sklearn.model_selection import train_test_split
from sklearn.preprocessing import MinMaxScaler,
StandardScaler, RobustScaler


covidtotals = pd.read_csv("data/covidtotals.csv")
feature_cols = ['population','total_deaths',
    'aged_65_older','diabetes_prevalence']
covidtotals = covidtotals[['total_cases'] + feature_
cols].dropna()


X_train, X_test, y_train, y_test =  \
  train_test_split(covidtotals[feature_cols],\
  covidtotals[['total_cases']], test_size=0.3, random_
state=0)
```

2.  Now, we can run the min-max scaler. The `fit_transform` method for `sklearn` will return a numpy array. We convert it into a pandas DataFrame using the columns and index from the training DataFrame. Notice how all features now have values between 0 and 1:

```
scaler = MinMaxScaler()
X_train_mms = pd.DataFrame(scaler.fit_transform(X_train),
    columns=X_train.columns, index=X_train.index)
X_train_mms.describe()
        population total_deaths aged_65_older diabetes_
```

```
prevalence
count   123.00          123.00          123.00          123.00
mean    0.04            0.04            0.30            0.41
std     0.13            0.14            0.24            0.23
min     0.00            0.00            0.00            0.00
25%     0.00            0.00            0.10            0.26
50%     0.01            0.00            0.22            0.37
75%     0.02            0.02            0.51            0.54
max     1.00            1.00            1.00            1.00
```

3.  We run the standard scaler in the same manner:

```
scaler = StandardScaler()
X_train_ss = pd.DataFrame(scaler.fit_transform(X_train),
   columns=X_train.columns, index=X_train.index)
X_train_ss.describe()
        population   total_deaths   aged_65_older   diabetes_
prevalence
count   123.00        123.00         123.00          123.00
mean   -0.00         -0.00          -0.00           -0.00
std     1.00          1.00           1.00            1.00
min    -0.29         -0.32          -1.24           -1.84
25%    -0.27         -0.31          -0.84           -0.69
50%    -0.24         -0.29          -0.34           -0.18
75%    -0.11         -0.18           0.87            0.59
max     7.58          6.75           2.93            2.63
```

If we have outliers in our data, robust scaling might be a good option. Robust scaling subtracts the median from each value of a variable and divides that value by the interquartile range. So, each value is as follows:

$$z_{ij} = (x_{ij} - median_j)/(3rd\ quantile_j - 1st\ quantile_j)$$

Here, $x_{ij}$ is the value of the $j^{th}$ feature, and $median_j$, $3rd\ quantile_j$, and $1st\ quantile_j$ are the median, third, and first quantiles of the $j^{th}$ feature. Robust scaling is less sensitive to extreme values since it does not use the mean or variance.

4.  We can use scikit-learn's `RobustScaler` module to do robust scaling:

```
scaler = RobustScaler()
X_train_rs = pd.DataFrame(
  scaler.fit_transform(X_train),
  columns=X_train.columns, index=X_train.index)


X_train_rs.describe()
```

|       | population | total_deaths | aged_65_older | diabetes_prevalence |
|-------|-----------|--------------|---------------|---------------------|
| count | 123.00 | 123.00 | 123.00 | 123.00 |
| mean | 1.47 | 2.22 | 0.20 | 0.14 |
| std | 6.24 | 7.65 | 0.59 | 0.79 |
| min | -0.35 | -0.19 | -0.53 | -1.30 |
| 25% | -0.24 | -0.15 | -0.30 | -0.40 |
| 50% | 0.00 | 0.00 | 0.00 | 0.00 |
| 75% | 0.76 | 0.85 | 0.70 | 0.60 |
| max | 48.59 | 53.64 | 1.91 | 2.20 |

We use feature scaling with most machine learning algorithms. Although it is not often required, it yields noticeably better results. Min-max scaling and standard scaling are popular scaling techniques, but there are times when robust scaling might be the better option.

# Summary

In this chapter, we covered a wide range of feature engineering techniques. We used tools to drop redundant or highly correlated features. We explored the most common kinds of encoding – one-hot encoding, ordinal encoding, and hashing encoding. Following this, we used transformations to improve the distribution of our features. Finally, we used common binning and scaling approaches to address skew, kurtosis, and outliers, and to adjust for features with widely different ranges.

Some of the techniques we discussed in this chapter are required for most machine learning models. We almost always need to encode our features for algorithms in order to understand them correctly. For example, most algorithms cannot make sense of *female* or *male* values or know not to treat ZIP codes as ordinal. Although not typically necessary, scaling is often a very good idea when we have features with vastly different ranges. When we are using algorithms that assume a Gaussian distribution of our features, some form of transformation might be required for our features to be consistent with that assumption.

We now have a good sense of how our features are distributed, have imputed missing values, and have done some feature engineering where necessary. We are now prepared to begin perhaps the most interesting and meaningful part of the model building process – feature selection.

In the next chapter, we will examine key feature selection tasks, building on the feature cleaning, exploration, and engineering work that we have done so far.

# 5
# Feature Selection

Depending on how you began your data analytic work and your own intellectual interests, you might have a different perspective on the topic of **feature selection**. You might think, *yeah, yeah, it is an important topic, but I really want to get to the model building.* Or, at the other extreme, you might view feature selection as at the core of model building and believe that you are 90% of the way toward having your model once you have chosen your features. For now, let's just agree that we should spend a good chunk of time understanding the relationships between features – and their relationship to a target if we are building a supervised model – before we do any serious model specification.

It is helpful to approach our feature selection work with the attitude that less is more. If we can reach nearly the same degree of accuracy or explain as much of the variance with fewer features, we should select the simpler model. Sometimes, we can actually get better accuracy with fewer features. This can be hard to wrap our brains around, and even be a tad disappointing for those of us who cut our teeth on building models that told rich and complicated stories.

But we are less concerned with parameter estimates than with the accuracy of our predictions when fitting machine learning models. Unnecessary features can contribute to overfitting and tax hardware resources.

We can sometimes spend months specifying the features of our model, even when there is a limited number of columns in the data. Bivariate correlations, such as those created in *Chapter 2, Examining Bivariate and Multivariate Relationships between Features and Targets*, give us some sense of what to expect, but the importance of a feature can vary significantly once other potentially explanatory features are introduced. The feature may no longer be significant, or, conversely, may only be significant when other features are included. Two features might be so highly correlated that including both of them offers very little additional information than including just one.

This chapter takes a close look at feature selection techniques applicable to a variety of predictive modeling tasks. Specifically, we will explore the following topics:

- Selecting features for classification models
- Selecting features for regression models
- Using forward and backward feature selection
- Using exhaustive feature selection
- Eliminating features recursively in a regression model
- Eliminating features recursively in a classification model
- Using Boruta for feature selection
- Using regularization and other embedded methods
- Using principal component analysis

# Technical requirements

We will work with the `feature_engine`, `mlxtend`, and `boruta` packages in this chapter, in addition to the `scikit-learn` library. You can use `pip` to install these packages. I have chosen a dataset with a small number of observations for our work in this chapter, so the code should work fine even on suboptimal workstations.

> **Note**
> We will work exclusively in this chapter with data from The National Longitudinal Survey of Youth, conducted by the United States Bureau of Labor Statistics. This survey started with a cohort of individuals in 1997 who were born between 1980 and 1985, with annual follow-ups each year through 2017. We will work with educational attainment, household demographic, weeks worked, and wage income data. The wage income column represents wages earned in 2016. The NLS dataset can be downloaded for public use at `https://www.nlsinfo.org/investigator/pages/search`.

# Selecting features for classification models

The most straightforward feature selection methods are based on each feature's relationship with a target variable. The next two sections examine techniques for determining the $k$ best features based on their linear or non-linear relationship with the target. These are known as filter methods. They are also sometimes called univariate methods since they evaluate the relationship between the feature and the target independent of the impact of other features.

We use somewhat different strategies when the target is categorical than when it is continuous. We'll go over the former in this section and the latter in the next.

## Mutual information classification for feature selection with a categorical target

We can use **mutual information** classification or **analysis of variance** (**ANOVA**) tests to select features when we have a categorical target. We will try mutual information classification first, and then ANOVA for comparison.

Mutual information is a measure of how much information about a variable is provided by knowing the value of another variable. At the extreme, when features are completely independent, the mutual information score is 0.

We can use `scikit-learn`'s `SelectKBest` class to select the $k$ features that have the highest predictive strength based on mutual information classification or some other appropriate measure. We can use hyperparameter tuning to select the value of $k$. We can also examine the scores of all features, whether they were identified as one of the $k$ best or not, as we will see in this section.

Let's first try mutual information classification to identify features that are related to completing a bachelor's degree. Later, we will compare that with using ANOVA F-values as the basis for selection:

1.  We start by importing `OneHotEncoder` from `feature_engine` to encode some of the data, and `train_test_split` from `scikit-learn` to create training and testing data. We will also need `scikit-learn`'s `SelectKBest`, `mutual_info_classif`, and `f_classif` modules for our feature selection:

    ```
    import pandas as pd
    from feature_engine.encoding import OneHotEncoder
    from sklearn.model_selection import train_test_split
    from sklearn.preprocessing import StandardScaler
    from sklearn.feature_selection import SelectKBest,\
      mutual_info_classif, f_classif
    ```

2. We load NLS data that has a binary variable for having completed a bachelor's degree and features possibly related to degree attainment: **Scholastic Assessment Test (SAT)** score, high school GPA, parental educational attainment and income, and gender. Observations with missing values for any feature have been removed. We then create training and testing DataFrames, encode the gender feature, and scale the other data:

```
nls97compba = pd.read_csv("data/nls97compba.csv")

feature_cols = ['gender','satverbal','satmath',
  'gpascience', 'gpaenglish','gpamath','gpaoverall',
  'motherhighgrade','fatherhighgrade','parentincome']

X_train, X_test, y_train, y_test =  \
  train_test_split(nls97compba[feature_cols],\
  nls97compba[['completedba']], test_size=0.3, random_
state=0)

ohe = OneHotEncoder(drop_last=True, variables=['gender'])
X_train_enc = ohe.fit_transform(X_train)

scaler = StandardScaler()
standcols = X_train_enc.iloc[:,:-1].columns
X_train_enc = \
  pd.DataFrame(scaler.fit_transform(X_train_
enc[standcols]),
    columns=standcols, index=X_train_enc.index).\
  join(X_train_enc[['gender_Female']])
```

> **Note**
>
> We will do a complete case analysis of the NLS data throughout this chapter; that is, we will remove all observations that have missing values for any of the features. This is not usually a good approach and is particularly problematic when data is not missing at random or when there is a large number of missing values for one or more features. In such cases, it would be better to use some of the approaches that we used in *Chapter 3, Identifying and Fixing Missing Values*. We will do a complete case analysis in this chapter to keep the examples as straightforward as possible.

3.  Now we are ready to select features for our model of bachelor's degree completion. One approach is to use mutual information classification. To do that, we set the `score_func` value of `SelectKBest` to `mutual_info_classif` and indicate that we want the five best features. Then, we call `fit` and use the `get_support` method to get the five best features:

```
ksel = SelectKBest(score_func=mutual_info_classif, k=5)
ksel.fit(X_train_enc, y_train.values.ravel())
selcols = X_train_enc.columns[ksel.get_support()]
selcols
Index(['satverbal', 'satmath', 'gpascience',
'gpaenglish', 'gpaoverall'], dtype='object')
```

4.  If we also want to see the score for each feature, we can use the `scores_` attribute, though we need to do a little work to associate the scores with a particular feature name and sort the scores in descending order:

```
pd.DataFrame({'score': ksel.scores_,
   'feature': X_train_enc.columns},
   columns=['feature','score']).\
   sort_values(['score'], ascending=False)
```

|   | feature | score |
|---|---|---|
| 5 | gpaoverall | 0.108 |
| 1 | satmath | 0.074 |
| 3 | gpaenglish | 0.072 |
| 0 | satverbal | 0.069 |
| 2 | gpascience | 0.047 |
| 4 | gpamath | 0.038 |
| 8 | parentincome | 0.024 |
| 7 | fatherhighgrade | 0.022 |
| 6 | motherhighgrade | 0.022 |
| 9 | gender_Female | 0.015 |

> **Note**
>
> This is a stochastic process, so we will get different results each time we run it.

To get the same results each time, you can pass a partial function to `score_func`:

```
from functools import partial
SelectKBest(score_func=partial(mutual_info_classif,
                               random_state=0), k=5)
```

5.   We can create a DataFrame with just the important features using the `selcols` array we created using `get_support`. (We could have used the `transform` method of `SelectKBest` instead. This would have returned the values of the selected features as a NumPy array.)

```
X_train_analysis = X_train_enc[selcols]
X_train_analysis.dtypes
satverbal        float64
satmath          float64
gpascience       float64
gpaenglish       float64
gpaoverall       float64
dtype: object
```

That is all we need to do to select the *k* best features for our model using mutual information.

## ANOVA F-value for feature selection with a categorical target

Alternatively, we can use ANOVA instead of mutual information. ANOVA evaluates how different the mean for a feature is for each target class. This is a good metric for univariate feature selection when we can assume a linear relationship between features and the target and our features are normally distributed. If those assumptions do not hold, mutual information classification is a better choice.

Let's try using ANOVA for our feature selection. We can set the `score_func` parameter of `SelectKBest` to `f_classif` to select based on ANOVA:

```
ksel = SelectKBest(score_func=f_classif, k=5)
ksel.fit(X_train_enc, y_train.values.ravel())
selcols = X_train_enc.columns[ksel.get_support()]
selcols
Index(['satverbal', 'satmath', 'gpascience', 'gpaenglish',
'gpaoverall'], dtype='object')
```

```
pd.DataFrame({'score': ksel.scores_,
  'feature': X_train_enc.columns},
  columns=['feature','score']).\
  sort_values(['score'], ascending=False)
```

|   | feature | score |
|---|---------|-------|
| 5 | gpaoverall | 119.471 |
| 3 | gpaenglish | 108.006 |
| 2 | gpascience | 96.824 |
| 1 | satmath | 84.901 |
| 0 | satverbal | 77.363 |
| 4 | gpamath | 60.930 |
| 7 | fatherhighgrade | 37.481 |
| 6 | motherhighgrade | 29.377 |
| 8 | parentincome | 22.266 |
| 9 | gender_Female | 15.098 |

This selected the same features as were selected when we used mutual information. Showing the scores gives us some indication of whether the selected value for $k$ makes sense. For example, there is a greater drop in score from the fifth- to the sixth-best feature (77-61) than from the fourth to the fifth (85-77). There is an even bigger decline from the sixth to the seventh, however (61-37), suggesting that we should at least consider a value for $k$ of 6.

ANOVA tests, and the mutual information classification we did earlier, do not take into account features that are only important in multivariate analysis. For example, fatherhighgrade might matter among individuals with similar GPA or SAT scores. We use multivariate feature selection methods later in this chapter. We do more univariate feature selection in the next section where we explore selection techniques appropriate for continuous targets.

# Selecting features for regression models

**Regression models** have a continuous target. The statistical techniques we used in the previous section are not appropriate for such targets. Fortunately, scikit-learn's selection module provides several options for selecting features when building regression models. (By regression models here, I do not mean linear regression models. I am only referring to *models with continuous targets*.) Two good options are selection based on F-tests and selection based on mutual information for regression. Let's start with F-tests.

# F-tests for feature selection with a continuous target

The F-statistic is a measure of the strength of the linear correlation between a target and a single regressor. `Scikit-learn` has an `f_regression` scoring function, which returns F-statistics. We can use it with `SelectKBest` to select features based on that statistic.

Let's use F-statistics to select features for a model of wages. We use mutual information for regression in the next section to select features for the same target:

1.  We start by importing the one-hot encoder from `feature_engine` and `train_test_split` and `SelectKBest` from `scikit-learn`. We also import `f_regression` to get F-statistics later:

    ```
    import pandas as pd
    import numpy as np
    from feature_engine.encoding import OneHotEncoder
    from sklearn.model_selection import train_test_split
    from sklearn.preprocessing import StandardScaler
    from sklearn.feature_selection import SelectKBest, f_
    regression
    ```

2.  Next, we load the NLS data, including educational attainment, parental income, and wage income data:

    ```
    nls97wages = pd.read_csv("data/nls97wages.csv")
    feature_cols = ['satverbal','satmath','gpascience',
        'gpaenglish','gpamath','gpaoverall','gender',
        'motherhighgrade','fatherhighgrade','parentincome',
        'completedba']
    ```

3.  Then, we create training and testing DataFrames, encode the `gender` feature, and scale the training data. We need to scale the target in this case since it is continuous:

    ```
    X_train, X_test, y_train, y_test =  \
      train_test_split(nls97wages[feature_cols],\
      nls97wages[['wageincome']], test_size=0.3, random_
    state=0)

    ohe = OneHotEncoder(drop_last=True, variables=['gender'])
    X_train_enc = ohe.fit_transform(X_train)

    scaler = StandardScaler()
    ```

```
standcols = X_train_enc.iloc[:,:-1].columns
X_train_enc = \
   pd.DataFrame(scaler.fit_transform(X_train_
enc[standcols]),
   columns=standcols, index=X_train_enc.index).\
   join(X_train_enc[['gender_Male']])

y_train = \
   pd.DataFrame(scaler.fit_transform(y_train),
   columns=['wageincome'], index=y_train.index)
```

> **Note**
> You may have noticed that we are not encoding or scaling the testing data.
> We will need to do that eventually to validate our models. We will introduce
> validation later in this chapter and go over it in much more detail in the next
> chapter.

4.  Now, we are ready to select features. We set `score_func` of `SelectKBest` to `f_regression` and indicate that we want the five best features. The `get_support` method of `SelectKBest` returns `True` for each feature that was selected:

```
ksel = SelectKBest(score_func=f_regression, k=5)
ksel.fit(X_train_enc, y_train.values.ravel())
selcols = X_train_enc.columns[ksel.get_support()]
selcols
Index(['satmath', 'gpascience', 'parentincome',
  'completedba','gender_Male'],
      dtype='object')
```

5.  We can use the `scores_` attribute to see the score for each feature:

```
pd.DataFrame({'score': ksel.scores_,
   'feature': X_train_enc.columns},
   columns=['feature','score']).\
   sort_values(['score'], ascending=False)
```

|   | feature | score |
|---|---------|-------|
| 1 | satmath | 45 |
| 9 | completedba | 38 |

| 10 | gender_Male | 26 |
|----|-------------|----|
| 8  | parentincome | 24 |
| 2  | gpascience | 21 |
| 0  | satverbal | 19 |
| 5  | gpaoverall | 17 |
| 4  | gpamath | 13 |
| 3  | gpaenglish | 10 |
| 6  | motherhighgrade | 9 |
| 7  | fatherhighgrade | 8 |

The disadvantage of the F-statistic is that it assumes a linear relationship between each feature and the target. When that assumption does not make sense, we can use mutual information for regression instead.

## Mutual information for feature selection with a continuous target

We can also use `SelectKBest` to select features using mutual information for regression:

1.  We need to set the `score_func` parameter of `SelectKBest` to `mutual_info_regression`, but there is a small complication. To get the same results each time we run the feature selection, we need to set a `random_state` value. As we discussed in the previous section, we can use a partial function for that. We pass `partial(mutual_info_regression, random_state=0)` to the score function.

2.  We can then run the `fit` method and use `get_support` to get the selected features. We can use the `scores_` attribute to give us the score for each feature:

```
from functools import partial
ksel = SelectKBest(score_func=\
  partial(mutual_info_regression, random_state=0),
  k=5)
ksel.fit(X_train_enc, y_train.values.ravel())

selcols = X_train_enc.columns[ksel.get_support()]
selcols
Index(['satmath', 'gpascience', 'fatherhighgrade',
'completedba','gender_Male'],dtype='object')
pd.DataFrame({'score': ksel.scores_,
  'feature': X_train_enc.columns},
```

```
columns=['feature','score']).\
    sort_values(['score'], ascending=False)
```

|    | feature | score |
|----|---------|-------|
| 1  | satmath | 0.101 |
| 10 | gender_Male | 0.074 |
| 7  | fatherhighgrade | 0.047 |
| 2  | gpascience | 0.044 |
| 9  | completedba | 0.044 |
| 4  | gpamath | 0.016 |
| 8  | parentincome | 0.015 |
| 6  | motherhighgrade | 0.012 |
| 0  | satverbal | 0.000 |
| 3  | gpaenglish | 0.000 |
| 5  | gpaoverall | 0.000 |

We get fairly similar results with mutual information for regression as we did with
F-tests. parentincome was selected with F-tests and fatherhighgrade with mutual
information. Otherwise, the same features are selected.

A key advantage of mutual information for regression compared with F-tests is that it does
not assume a linear relationship between the feature and the target. If that assumption
turns out to be unwarranted, mutual information is a better approach. (Again, there is also
some randomness in the scoring and the score for each feature can bounce around within
a limited range.)

> **Note**
>
> Our choice of k=5 to get the five best features is quite arbitrary. We can make
> it much more scientific with some hyperparameter tuning. We will go over
> tuning in the next chapter.

The feature selection methods we have used so far are known as *filter methods*. They examine
the univariate relationship between each feature and the target. They are a good starting
point. Similar to our discussion in previous chapters of the usefulness of having correlations
handy before we start examining multivariate relationships, it is helpful to at least explore
filter methods. Often, though, our model fitting will require taking into account features
that are important, or not, when other features are also included. To do that, we need to use
wrapper or embedded methods for feature selection. We explore wrapper methods in the
next few sections, starting with forward and backward feature selection.

# Using forward and backward feature selection

Forward and backward feature selection, as their names suggest, select features by adding them one by one – or subtracting them for backward selection – and assessing the impact on model performance after each iteration. Since both methods assess that performance based on a given algorithm, they are considered **wrapper** selection methods.

Wrapper feature selection methods have two advantages over the filter methods we have explored so far. First, they evaluate the importance of features as other features are included. Second, since features are evaluated based on their contribution to the performance of a specific algorithm, we get a better sense of which features will ultimately matter. For example, `satmath` seemed to be an important feature based on our results from the previous section. But it is possible that `satmath` is only important when we use a particular model, say linear regression, and not an alternative such as decision tree regression. Wrapper selection methods can help us discover that.

The main disadvantage of wrapper methods is that they can be quite expensive computationally since they retrain the model after each iteration. We will look at both forward and backward feature selection in this section.

## Using forward feature selection

**Forward feature selection** starts by identifying a subset of features that individually have a significant relationship with a target, not unlike the filter methods. But it then evaluates all possible combinations of the selected features for the combination that performs best with the chosen algorithm.

We can use forward feature selection to develop a model of bachelor's degree completion. Since wrapper methods require us to choose an algorithm, and this is a binary target, let's use `scikit-learn`'s **random forest classifier**. We will also need the `feature_selection` module of `mlxtend` to do the iteration required to select features:

1.  We start by importing the necessary libraries:

    ```
    import pandas as pd
    from feature_engine.encoding import OneHotEncoder
    from sklearn.model_selection import train_test_split
    from sklearn.preprocessing import StandardScaler
    from sklearn.ensemble import RandomForestClassifier
    from mlxtend.feature_selection import
    SequentialFeatureSelector
    ```

2.  Then, we load the NLS data again. We also create a training DataFrame, encode the
    gender feature, and standardize the remaining features:

```
nls97compba = pd.read_csv("data/nls97compba.csv")

feature_cols = ['satverbal','satmath','gpascience',
   'gpaenglish','gpamath','gpaoverall','gender',
   'motherhighgrade','fatherhighgrade','parentincome']

X_train, X_test, y_train, y_test =  \
   train_test_split(nls97compba[feature_cols],\
   nls97compba[['completedba']], test_size=0.3, random_
state=0)

ohe = OneHotEncoder(drop_last=True, variables=['gender'])
X_train_enc = ohe.fit_transform(X_train)

scaler = StandardScaler()
standcols = X_train_enc.iloc[:,:-1].columns
X_train_enc = \
   pd.DataFrame(scaler.fit_transform(X_train_
enc[standcols]),
   columns=standcols, index=X_train_enc.index).\
   join(X_train_enc[['gender_Female']])
```

3.  We create a random forest classifier object and then pass that object to the feature
    selector of mlxtend. We indicate that we want it to select five features and that
    it should forward select. (We can also use the sequential feature selector to select
    backward.) After running fit, we can use the k_feature_idx_ attribute to get
    the list of selected features:

```
rfc = RandomForestClassifier(n_estimators=100, n_jobs=-1,
random_state=0)

sfs = SequentialFeatureSelector(rfc, k_features=5,
   forward=True, floating=False, verbose=2,
   scoring='accuracy', cv=5)

sfs.fit(X_train_enc, y_train.values.ravel())
```

```
selcols = X_train_enc.columns[list(sfs.k_feature_idx_)]
selcols
Index(['satverbal', 'satmath', 'gpaoverall',
'parentincome', 'gender_Female'], dtype='object')
```

You might recall from the first section of this chapter that our univariate feature selection for the completed bachelor's degree target gave us somewhat different results:

```
Index(['satverbal', 'satmath', 'gpascience',
  'gpaenglish', 'gpaoverall'], dtype='object')
```

Three of the features – satmath, satverbal, and gpaoverall – are the same. But our forward feature selection has identified parentincome and gender_Female as more important than gpascience and gpaenglish, which were selected in the univariate analysis. Indeed, gender_Female had among the lowest scores in the earlier analysis. These differences likely reflect the advantages of wrapper feature selection methods. We can identify features that are not important unless other features are included, and we are evaluating the effect on the performance of a particular algorithm, in this case, random forest classification.

One disadvantage of forward selection is that *once a feature is selected, it is not removed, even though it may decline in importance as additional features are added.* (Recall that forward feature selection adds features iteratively based on the contribution of that feature to the model.)

Let's see whether our results vary with backward feature selection.

## Using backward feature selection

Backward feature selection starts with all features and eliminates the least important. It then repeats this process with the remaining features. We can use mlxtend's SequentialFeatureSelector for backward selection in pretty much the same way we used it for forward selection.

We instantiate a RandomForestClassifier object from the scikit-learn library and then pass it to mlxtend's sequential feature selector:

```
rfc = RandomForestClassifier(n_estimators=100, n_jobs=-1,
random_state=0)


sfs = SequentialFeatureSelector(rfc, k_features=5,
  forward=False, floating=False, verbose=2,
  scoring='accuracy', cv=5)
```

```
sfs.fit(X_train_enc, y_train.values.ravel())

selcols = X_train_enc.columns[list(sfs.k_feature_idx_)]
selcols
Index(['satverbal', 'gpascience', 'gpaenglish',
  'gpaoverall', 'gender_Female'], dtype='object')
```

Perhaps unsurprisingly, we get different results for our feature selection. satmath and parentincome are no longer selected, and gpascience and gpaenglish are.

Backward feature selection has the opposite drawback to forward feature selection. *Once a feature has been removed, it is not re-evaluated, even though its importance may change with different feature mixtures.* Let's try exhaustive feature selection instead.

# Using exhaustive feature selection

If your results from forward and backward selection are unpersuasive, and you do not mind running a model while you go out for coffee or lunch, you can try exhaustive feature selection. **Exhaustive feature selection** trains a given model on all possible combinations of features and selects the best subset of features. But it does this at a price. As the name suggests, this procedure might exhaust both system resources and your patience.

Let's use exhaustive feature selection for our model of bachelor's degree completion:

1.  We start by loading the required libraries, including the RandomForestClassifier and LogisticRegression modules from scikit-learn and ExhaustiveFeatureSelector from mlxtend. We also import the accuracy_score module so that we can evaluate a model with the selected features:

    ```
    import pandas as pd
    from feature_engine.encoding import OneHotEncoder
    from sklearn.model_selection import train_test_split
    from sklearn.preprocessing import StandardScaler
    from sklearn.ensemble import RandomForestClassifier
    from sklearn.linear_model import LogisticRegression
    from mlxtend.feature_selection import
    ExhaustiveFeatureSelector
    from sklearn.metrics import accuracy_score
    ```

2. Next, we load the NLS educational attainment data and create training and testing DataFrames:

```
nls97compba = pd.read_csv("data/nls97compba.csv")

feature_cols = ['satverbal','satmath','gpascience',
    'gpaenglish','gpamath','gpaoverall','gender',
    'motherhighgrade','fatherhighgrade','parentincome']

X_train, X_test, y_train, y_test =  \
    train_test_split(nls97compba[feature_cols],\
    nls97compba[['completedba']], test_size=0.3, random_
state=0)
```

3. Then, we encode and scale the training and testing data:

```
ohe = OneHotEncoder(drop_last=True, variables=['gender'])
ohe.fit(X_train)
X_train_enc, X_test_enc = \
    ohe.transform(X_train), ohe.transform(X_test)

scaler = StandardScaler()
standcols = X_train_enc.iloc[:,:-1].columns
scaler.fit(X_train_enc[standcols])
X_train_enc = \
    pd.DataFrame(scaler.transform(X_train_enc[standcols]),
    columns=standcols, index=X_train_enc.index).\
    join(X_train_enc[['gender_Female']])
X_test_enc = \
    pd.DataFrame(scaler.transform(X_test_enc[standcols]),
    columns=standcols, index=X_test_enc.index).\
    join(X_test_enc[['gender_Female']])
```

4.  We create a random forest classifier object and pass it to `mlxtend`'s
    `ExhaustiveFeatureSelector`. We tell the feature selector to evaluate all
    combinations of one to five features and return the combination with the highest
    accuracy in predicting degree attainment. After running `fit`, we can use the
    `best_feature_names_` attribute to get the selected features:

```
rfc = RandomForestClassifier(n_estimators=100, max_
depth=2,n_jobs=-1, random_state=0)

efs = ExhaustiveFeatureSelector(rfc, max_features=5,
   min_features=1, scoring='accuracy',
   print_progress=True, cv=5)

efs.fit(X_train_enc, y_train.values.ravel())
efs.best_feature_names_
('satverbal', 'gpascience', 'gpamath', 'gender_Female')
```

5.  Let's evaluate the accuracy of this model. We first need to transform the training
    and testing data to include only the four selected features. Then, we can fit the
    random forest classifier again with just those features and generate the predicted
    values for bachelor's degree completion. We can then calculate the percentage of the
    time we predicted the target correctly, which is 67%:

```
X_train_efs = efs.transform(X_train)
X_test_efs = efs.transform(X_test)
rfc.fit(X_train_efs, y_train.values.ravel())
y_pred = rfc.predict(X_test_efs)

confusion = pd.DataFrame(y_pred, columns=['pred'],
   index=y_test.index).\
   join(y_test)

confusion.loc[confusion.pred==confusion.completedba].
shape[0]\
   /confusion.shape[0]
0.6703296703296703
```

6.  We get the same answer if we just use scikit-learn's `accuracy score` instead. (We calculate it in the previous step because it is pretty straightforward and it gives us a better sense of what is meant by accuracy in this case.)

```
accuracy_score(y_test, y_pred)
0.6703296703296703
```

> **Note**
>
> The accuracy score is often used to assess the performance of a classification model. We will lean on it in this chapter, but other measures might be equally or more important depending on the purposes of your model. For example, we are sometimes more concerned with sensitivity, the ratio of our correct positive predictions to the number of actual positives. We examine the evaluation of classification models in detail in *Chapter 6*, *Preparing for Model Evaluation*.

7.  Let's now try exhaustive feature selection with a logistic model:

```
lr = LogisticRegression(solver='liblinear')
efs = ExhaustiveFeatureSelector(lr, max_features=5,
    min_features=1, scoring='accuracy',
    print_progress=True, cv=5)

efs.fit(X_train_enc, y_train.values.ravel())
efs.best_feature_names_
('satmath', 'gpascience', 'gpaenglish',
'motherhighgrade', 'gender_Female')
```

8.  Let's look at the accuracy of the logistic model. We get a fairly similar accuracy score:

```
X_train_efs = efs.transform(X_train_enc)
X_test_efs = efs.transform(X_test_enc)
lr.fit(X_train_efs, y_train.values.ravel())
y_pred = lr.predict(X_test_efs)
accuracy_score(y_test, y_pred)
0.6923076923076923
```

9.  One key advantage of the logistic model is that it is much faster to train, which really makes a difference with exhaustive feature selection. If we time the training for each model (probably not a good idea to do that on your computer unless it's a pretty high-end machine or you don't mind walking away from your computer for a while), we see a substantial difference in average training time – from an amazing 5 minutes for the random forest to 4 seconds for the logistic regression. (Of course, the absolute numbers are machine-dependent.)

```
rfc = RandomForestClassifier(n_estimators=100, max_
depth=2,
  n_jobs=-1, random_state=0)
efs = ExhaustiveFeatureSelector(rfc, max_features=5,
  min_features=1, scoring='accuracy',
  print_progress=True, cv=5)


%timeit efs.fit(X_train_enc, y_train.values.ravel())
5min 8s ± 3 s per loop (mean ± std. dev. of 7 runs, 1
loop each)



lr = LogisticRegression(solver='liblinear')
efs = ExhaustiveFeatureSelector(lr, max_features=5,
  min_features=1, scoring='accuracy',
  print_progress=True, cv=5)


%timeit efs.fit(X_train_enc, y_train.values.ravel())
4.29 s ± 45.5 ms per loop (mean ± std. dev. of 7 runs, 1
loop each)
```

Exhaustive feature selection can provide very clear guidance about the features to select, as I have mentioned, but that may come at too high a price for many projects. It may actually be better suited for *diagnostic work* than for use in a machine learning pipeline. If a linear model is appropriate, it can lower the computational costs considerably.

Wrapper methods, such as forward, backward, and exhaustive feature selection, tax system resources because they need to be trained with each iteration, and the more difficult the chosen algorithm is to implement, the more this is an issue. **Recursive feature elimination** (**RFE**) is something of a compromise between the simplicity of filter methods and the better information provided by wrapper methods. It is similar to backward feature selection, except it simplifies the removal of a feature at each iteration by basing it on the model's overall performance rather than re-evaluating each feature. We explore recursive feature selection in the next two sections.

# Eliminating features recursively in a regression model

A popular wrapper method is RFE. This method starts with all features, removes the lowest weighted one (based on a coefficient or feature importance measure), and repeats the process until the best-fitting model has been identified. When a feature is removed, it is given a ranking reflecting the point at which it was removed.

RFE can be used for both regression models and classification models. We will start by using it in a regression model:

1.  We import the necessary libraries, three of which we have not used yet: the `RFE`, `RandomForestRegressor`, and `LinearRegression` modules from `scikit-learn`:

    ```
    import pandas as pd
    from feature_engine.encoding import OneHotEncoder
    from sklearn.model_selection import train_test_split
    from sklearn.preprocessing import StandardScaler
    from sklearn.feature_selection import RFE
    from sklearn.ensemble import RandomForestRegressor
    from sklearn.linear_model import LinearRegression
    ```

2.  Next, we load the NLS data on wages and create training and testing DataFrames:

    ```
    nls97wages = pd.read_csv("data/nls97wages.csv")

    feature_cols = ['satverbal','satmath','gpascience',
        'gpaenglish','gpamath','gpaoverall','motherhighgrade',
        'fatherhighgrade','parentincome','gender','complet-
    edba']
    ```

```
X_train, X_test, y_train, y_test =   \
  train_test_split(nls97wages[feature_cols],\
  nls97wages[['weeklywage']], test_size=0.3, random_
state=0)
```

3.  We need to encode the `gender` feature and standardize the other features and the target (`wageincome`). We do not do any encoding or scaling of `completedba`, which is a binary feature:

```
ohe = OneHotEncoder(drop_last=True, variables=['gender'])
ohe.fit(X_train)
X_train_enc, X_test_enc = \
  ohe.transform(X_train), ohe.transform(X_test)


scaler = StandardScaler()
standcols = feature_cols[:-2]
scaler.fit(X_train_enc[standcols])
X_train_enc = \
  pd.DataFrame(scaler.transform(X_train_enc[standcols]),
  columns=standcols, index=X_train_enc.index).\
  join(X_train_enc[['gender_Male','completedba']])
X_test_enc = \
  pd.DataFrame(scaler.transform(X_test_enc[standcols]),
  columns=standcols, index=X_test_enc.index).\
  join(X_test_enc[['gender_Male','completedba']])


scaler.fit(y_train)
y_train, y_test = \
  pd.DataFrame(scaler.transform(y_train),
  columns=['weeklywage'], index=y_train.index),\
  pd.DataFrame(scaler.transform(y_test),
  columns=['weeklywage'], index=y_test.index)
```

Now, we are ready to do some recursive feature selection. Since RFE is a wrapper method, we need to choose an algorithm around which the selection will be *wrapped*. Random forests for regression make sense in this case. We are modeling a continuous target and do not want to assume a linear relationship between the features and the target.

4.  RFE is fairly easy to implement with `scikit-learn`. We instantiate an RFE object, telling it what estimator we want in the process. We indicate `RandomForestRegressor`. We then fit the model and use `get_support` to get the selected features. We limit `max_depth` to 2 to avoid overfitting:

```
rfr = RandomForestRegressor(max_depth=2)
treesel = RFE(estimator=rfr, n_features_to_select=5)
treesel.fit(X_train_enc, y_train.values.ravel())
selcols = X_train_enc.columns[treesel.get_support()]
selcols
  Index(['satmath', 'gpaoverall', 'parentincome', 'gender_
Male', 'completedba'], dtype='object')
```

Note that this gives us a somewhat different list of features than using a filter method (with F-tests) for the wage income target. `gpaoverall` and `motherhighgrade` are selected here and not the `gender` flag or `gpascience`.

5.  We can use the `ranking_` attribute to see when each of the eliminated features was removed:

```
pd.DataFrame({'ranking': treesel.ranking_,
  'feature': X_train_enc.columns},
  columns=['feature','ranking']).\
  sort_values(['ranking'], ascending=True)
```

|    | feature | ranking |
|----|---------|---------|
| 1  | satmath | 1 |
| 5  | gpaoverall | 1 |
| 8  | parentincome | 1 |
| 9  | gender_Male | 1 |
| 10 | completedba | 1 |
| 6  | motherhighgrade | 2 |
| 2  | gpascience | 3 |
| 0  | satverbal | 4 |

| 3 | **gpaenglish** | 5 |
| 4 | **gpamath** | 6 |
| 7 | **fatherhighgrade** | 7 |

`fatherhighgrade` was removed after the first interaction and `gpamath` after the second.

6.  Let's run some test statistics. We fit only the selected features on a random forest regressor model. The `transform` method of the RFE selector gives us just the selected features with `treesel.transform(X_train_enc)`. We can use the `score` method to get the r-squared value, also known as the coefficient of determination. R-squared is a measure of the percentage of total variation explained by our model. We get a very low score, indicating that our model explains only a little of the variation. (Note that this is a stochastic process, so we will likely get different results each time we fit the model.)

    ```
    rfr.fit(treesel.transform(X_train_enc), y_train.values.
    ravel())
    ```

    ```
    rfr.score(treesel.transform(X_test_enc), y_test)
    0.13612629794428466
    ```

7.  Let's see whether we get any better results using RFE with a linear regression model. This model returns the same features as the random forest regressor:

    ```
    lr = LinearRegression()
    lrsel = RFE(estimator=lr, n_features_to_select=5)
    lrsel.fit(X_train_enc, y_train)
    selcols = X_train_enc.columns[lrsel.get_support()]
    selcols
    Index(['satmath', 'gpaoverall', 'parentincome', 'gender_
    Male', 'completedba'], dtype='object')
    ```

8.  Let's evaluate the linear model:

    ```
    lr.fit(lrsel.transform(X_train_enc), y_train)
    ```

    ```
    lr.score(lrsel.transform(X_test_enc), y_test)
    0.17773742846314056
    ```

The linear model is not really much better than the random forest model. This is likely a sign that, collectively, the features available to us only capture a small part of the variation in wages per week. This is an important reminder that we can identify several significant features and still have a model with limited explanatory power. (Perhaps it is also good news that our scores on standardized tests, and even our degree attainment, are important but not determinative of our wages many years later.)

Let's try RFE with a classification model.

# Eliminating features recursively in a classification model

RFE can also be a good choice for classification problems. We can use RFE to select features for a model of bachelor's degree completion. You may recall that we used exhaustive feature selection to select features for that model earlier in this chapter. Let's see whether we get better accuracy or an easier-to-train model with RFE:

1. We import the same libraries we have been working with so far in this chapter:

```
import pandas as pd
from feature_engine.encoding import OneHotEncoder
from sklearn.model_selection import train_test_split
from sklearn.preprocessing import StandardScaler
from sklearn.ensemble import RandomForestClassifier
from sklearn.feature_selection import RFE
from sklearn.metrics import accuracy_score
```

2. Next, we create training and testing data from the NLS educational attainment data:

```
nls97compba = pd.read_csv("data/nls97compba.csv")
feature_cols = ['satverbal','satmath','gpascience',
    'gpaenglish','gpamath','gpaoverall','gender',
    'motherhighgrade','fatherhighgrade','parentincome']

X_train, X_test, y_train, y_test =  \
    train_test_split(nls97compba[feature_cols],\
    nls97compba[['completedba']], test_size=0.3,
    random_state=0)
```

3.  Then, we encode and scale the training and testing data:

```
ohe = OneHotEncoder(drop_last=True, variables=['gender'])
ohe.fit(X_train)
X_train_enc, X_test_enc = \
  ohe.transform(X_train), ohe.transform(X_test)


scaler = StandardScaler()
standcols = X_train_enc.iloc[:,:-1].columns
scaler.fit(X_train_enc[standcols])
X_train_enc = \
  pd.DataFrame(scaler.transform(X_train_enc[standcols]),
  columns=standcols, index=X_train_enc.index).\
  join(X_train_enc[['gender_Female']])
X_test_enc = \
  pd.DataFrame(scaler.transform(X_test_enc[standcols]),
  columns=standcols, index=X_test_enc.index).\
  join(X_test_enc[['gender_Female']])
```

4.  We instantiate a random forest classifier and pass it to the RFE selection method.
    We can then fit the model and get the selected features.

```
rfc = RandomForestClassifier(n_estimators=100, max_
depth=2,
  n_jobs=-1, random_state=0)
treesel = RFE(estimator=rfc, n_features_to_select=5)
treesel.fit(X_train_enc, y_train.values.ravel())
selcols = X_train_enc.columns[treesel.get_support()]
selcols
Index(['satverbal', 'satmath', 'gpascience',
'gpaenglish', 'gpaoverall'], dtype='object')
```

5.  We can also show how the features are ranked by using the RFE `ranking_` attribute:

```
pd.DataFrame({'ranking': treesel.ranking_,
  'feature': X_train_enc.columns},
  columns=['feature','ranking']).\
  sort_values(['ranking'], ascending=True)
```

|   | feature | ranking |
|---|---------|---------|
| 0 | satverbal | 1 |
| 1 | satmath | 1 |
| 2 | gpascience | 1 |
| 3 | gpaenglish | 1 |
| 5 | gpaoverall | 1 |
| 4 | gpamath | 2 |
| 8 | parentincome | 3 |
| 7 | fatherhighgrade | 4 |
| 6 | motherhighgrade | 5 |
| 9 | gender_Female | 6 |

6.  Let's look at the accuracy of a model with the selected features using the same random forest classifier we used for our baseline model:

```
rfc.fit(treesel.transform(X_train_enc), y_train.values.
ravel())


y_pred = rfc.predict(treesel.transform(X_test_enc))


accuracy_score(y_test, y_pred)
0.684981684981685
```

Recall that we had 67% accuracy with the exhaustive feature selection. We get about the same accuracy here. The benefit of RFE though is that it can be significantly easier to train than exhaustive feature selection.

Another option among wrapper and wrapper-like feature selection methods is the **Boruta** library. Originally developed as an R package, it can now be used with any `scikit-learn` ensemble method. We use it with `scikit-learn`'s random forest classifier in the next section.

# Using Boruta for feature selection

The Boruta package takes a unique approach to feature selection, though it has some similarities with wrapper methods. For each feature, Boruta creates a shadow feature, one with the same range of values as the original feature but with shuffled values. It then evaluates whether the original feature offers more information than the shadow feature, gradually removing features providing the least information. Boruta outputs confirmed, tentative, and rejected features with each iteration.

Let's use the Boruta package to select features for a classification model of bachelor's degree completion (you can install the Boruta package with `pip` if you have not yet installed it):

1.  We start by loading the necessary libraries:

    ```
    import pandas as pd
    from feature_engine.encoding import OneHotEncoder
    from sklearn.model_selection import train_test_split
    from sklearn.preprocessing import StandardScaler
    from sklearn.ensemble import RandomForestClassifier
    from boruta import BorutaPy
    from sklearn.metrics import accuracy_score
    ```

2.  We load the NLS educational attainment data again and create the training and test DataFrames:

    ```
    nls97compba = pd.read_csv("data/nls97compba.csv")

    feature_cols = ['satverbal','satmath','gpascience',
        'gpaenglish','gpamath','gpaoverall','gender',
        'motherhighgrade','fatherhighgrade','parentincome']

    X_train, X_test, y_train, y_test =  \
        train_test_split(nls97compba[feature_cols],\
        nls97compba[['completedba']], test_size=0.3, random_
    state=0)
    ```

3.  Next, we encode and scale the training and test data:

    ```
    ohe = OneHotEncoder(drop_last=True, variables=['gender'])
    ohe.fit(X_train)
    X_train_enc, X_test_enc = \
        ohe.transform(X_train), ohe.transform(X_test)

    scaler = StandardScaler()
    standcols = X_train_enc.iloc[:,:-1].columns
    scaler.fit(X_train_enc[standcols])
    X_train_enc = \
        pd.DataFrame(scaler.transform(X_train_enc[standcols]),
    ```

```
    columns=standcols, index=X_train_enc.index).\
    join(X_train_enc[['gender_Female']])
X_test_enc = \
    pd.DataFrame(scaler.transform(X_test_enc[standcols]),
    columns=standcols, index=X_test_enc.index).\
    join(X_test_enc[['gender_Female']])
```

4. We run Boruta feature selection in much the same way that we ran RFE feature selection. We use random forest as our baseline method again. We instantiate a random forest classifier and pass it to Boruta's feature selector. We then fit the model, which stops at 100 iterations, identifying 9 features that provide information:

```
rfc = RandomForestClassifier(n_estimators=100,
    max_depth=2, n_jobs=-1, random_state=0)
borsel = BorutaPy(rfc, random_state=0, verbose=2)
borsel.fit(X_train_enc.values, y_train.values.ravel())
BorutaPy finished running.

Iteration:             100 / 100
Confirmed:             9
Tentative:             1
Rejected:              0


selcols = X_train_enc.columns[borsel.support_]
selcols
Index(['satverbal', 'satmath', 'gpascience',
'gpaenglish', 'gpamath', 'gpaoverall', 'motherhighgrade',
'fatherhighgrade', 'parentincome', 'gender_Female'],
dtype='object')
```

5. We can use the ranking_ property to view the rankings of the features:

```
pd.DataFrame({'ranking': borsel.ranking_,
    'feature': X_train_enc.columns},
    columns=['feature','ranking']).\
    sort_values(['ranking'], ascending=True)
```

|   | feature | ranking |
|---|---------|---------|
| 0 | satverbal | 1 |
| 1 | satmath | 1 |
| 2 | gpascience | 1 |

| 3 | gpaenglish | 1 |
| 4 | gpamath | 1 |
| 5 | gpaoverall | 1 |
| 6 | motherhighgrade | 1 |
| 7 | fatherhighgrade | 1 |
| 8 | parentincome | 1 |
| 9 | gender_Female | 2 |

6.  To evaluate the model's accuracy, we fit the random forest classifier model with just the selected features. We can then make predictions for the testing data and compute accuracy:

```
rfc.fit(borsel.transform(X_train_enc.values), y_train.
values.ravel())
y_pred = rfc.predict(borsel.transform(X_test_enc.values))
accuracy_score(y_test, y_pred)
0.684981684981685
```

Part of Boruta's appeal is the persuasiveness of its selection of each feature. If a feature has been selected, then it likely does provide information that is not captured by combinations of features that exclude it. However, it is quite computationally expensive, not unlike exhaustive feature selection. It can help us sort out which features matter, but it may not always be suitable for pipelines where training speed matters.

The last few sections have shown some of the advantages and some disadvantages of wrapper feature selection methods. We explore embedded selection methods in the next section. These methods provide more information than filter methods but without the computational costs of wrapper methods. They do this by embedding feature selection into the training process. We will explore embedded methods with the same data we have worked with so far.

# Using regularization and other embedded methods

**Regularization** methods are embedded methods. Like wrapper methods, embedded methods evaluate features relative to a given algorithm. But they are not as expensive computationally. That is because feature selection is built into the algorithm already and so happens as the model is being trained.

Embedded models use the following process:

1.  Train a model.

2.  Estimate each feature's importance to the model's predictions.

3.  Remove features with low importance.

Regularization accomplishes this by adding a penalty to any model to constrain the parameters. **L1 regularization**, also referred to as **lasso regularization**, shrinks some of the coefficients in a regression model to 0, effectively eliminating those features.

## Using L1 regularization

1.  We will use L1 regularization with logistic regression to select features for a bachelor's degree attainment model:We need to first import the required libraries, including a module we will be using for the first time, `SelectFromModel` from `scikit-learn`:

    ```
    import pandas as pd
    from feature_engine.encoding import OneHotEncoder
    from sklearn.model_selection import train_test_split
    from sklearn.preprocessing import StandardScaler
    from sklearn.linear_model import LogisticRegression
    from sklearn.ensemble import RandomForestClassifier
    from sklearn.feature_selection import SelectFromModel
    from sklearn.metrics import accuracy_score
    ```

2.  Next, we load NLS data on educational attainment:

    ```
    nls97compba = pd.read_csv("data/nls97compba.csv")

    feature_cols = ['satverbal','satmath','gpascience',
        'gpaenglish','gpamath','gpaoverall','gender',
        'motherhighgrade','fatherhighgrade','parentincome']

    X_train, X_test, y_train, y_test =  \
        train_test_split(nls97compba[feature_cols],\
        nls97compba[['completedba']], test_size=0.3,
        random_state=0)
    ```

3.   Then, we encode and scale the training and testing data:

```
ohe = OneHotEncoder(drop_last=True,
                         variables=['gender'])
ohe.fit(X_train)
X_train_enc, X_test_enc = \
  ohe.transform(X_train), ohe.transform(X_test)

scaler = StandardScaler()
standcols = X_train_enc.iloc[:,:-1].columns
scaler.fit(X_train_enc[standcols])
X_train_enc = \
  pd.DataFrame(scaler.transform(X_train_enc[standcols]),
  columns=standcols, index=X_train_enc.index).\
  join(X_train_enc[['gender_Female']])
X_test_enc = \
  pd.DataFrame(scaler.transform(X_test_enc[standcols]),
  columns=standcols, index=X_test_enc.index).\
  join(X_test_enc[['gender_Female']])
```

4.   Now we are ready to do feature selection based on logistic regression with an L1 penalty:

```
lr = LogisticRegression(C=1, penalty="l1",
                           solver='liblinear')
regsel = SelectFromModel(lr, max_features=5)
regsel.fit(X_train_enc, y_train.values.ravel())
selcols = X_train_enc.columns[regsel.get_support()]
selcols
Index(['satmath', 'gpascience', 'gpaoverall',
'fatherhighgrade', 'gender_Female'], dtype='object')
```

5.  Let's evaluate the accuracy of the model. We get an accuracy score of 0.68:

```
lr.fit(regsel.transform(X_train_enc),
        y_train.values.ravel())


y_pred = lr.predict(regsel.transform(X_test_enc))


accuracy_score(y_test, y_pred)
0.684981684981685
```

This gives us fairly similar results to that of the forward feature selection for bachelor's degree completion. We used a random forest classifier as a wrapper method in that example.

Lasso regularization is a good choice for feature selection in a case like this, particularly when performance is a key concern. It does, however, assume a linear relationship between the features and the target, which might not be appropriate. Fortunately, there are embedded feature selection methods that do not make that assumption. A good alternative to logistic regression for the embedded model is a random forest classifier. We try that next with the same data.

## Using a random forest classifier

In this section, let's use a random forest classifier:

1.  We can use SelectFromModel to use a random forest classifier rather than logistic regression:

```
rfc = RandomForestClassifier(n_estimators=100,
  max_depth=2, n_jobs=-1, random_state=0)
rfcsel = SelectFromModel(rfc, max_features=5)
rfcsel.fit(X_train_enc, y_train.values.ravel())
selcols = X_train_enc.columns[rfcsel.get_support()]
selcols
Index(['satverbal', 'gpascience', 'gpaenglish',
   'gpaoverall'], dtype='object')
```

This actually selects very different features from the lasso regression. satmath, fatherhighgrade, and gender_Female are no longer selected, while satverbal and gpaenglish are. This is likely partly due to the relaxation of the assumption of linearity.

2.  Let's evaluate the accuracy of the random forest classifier model. We get an accuracy score of **0.67**. This is pretty much the same score that we got with the lasso regression:

```
rfc.fit(rfcsel.transform(X_train_enc),
        y_train.values.ravel())


y_pred = rfc.predict(rfcsel.transform(X_test_enc))


accuracy_score(y_test, y_pred)
0.673992673992674
```

Embedded methods are generally less CPU-/GPU-intensive than wrapper methods but can nonetheless produce good results. With our models of bachelor's degree completion in this section, we get the same accuracy as we did with our models based on exhaustive feature selection.

Each of the methods we have discussed so far has important use cases, as we have discussed. However, we have not yet really discussed one very challenging feature selection problem. What do you do if you simply have too many features, many of which independently account for something important in your model? By too many, here I mean that there are so many features that the model cannot run efficiently, either for training or for predicting target values. How can we reduce the feature set without sacrificing some of the predictive power of our model? In that situation, **principal component analysis** (**PCA**) might be a good approach. We'll discuss PCA in the next section.

# Using principal component analysis

A very different approach to feature selection than any of the methods we have discussed so far is PCA. PCA allows us to replace the existing feature set with a limited number of components, each of which explains an important amount of the variance. It does this by finding a component that captures the largest amount of variance, followed by a second component that captures the largest amount of remaining variance, and then a third component, and so on. One key advantage of this approach is that these components, known as **principal components**, are uncorrelated. We discuss PCA in detail in *Chapter 15, Principal Component Analysis*.

Although I include PCA here as a feature selection approach, it is probably better to think of it as a tool for dimension reduction. We use it for feature selection when we need to limit the number of dimensions without sacrificing too much explanatory power.

Let's work with the NLS data again and use PCA to select features for a model of bachelor's degree completion:

1. We start by loading the necessary libraries. The only module we have not already used in this chapter is scikit-learn's PCA:

```
import pandas as pd
from feature_engine.encoding import OneHotEncoder
from sklearn.model_selection import train_test_split
from sklearn.preprocessing import StandardScaler
from sklearn.decomposition import PCA
from sklearn.ensemble import RandomForestClassifier
from sklearn.metrics import accuracy_score
```

2. Next, we create training and testing DataFrames once again:

```
nls97compba = pd.read_csv("data/nls97compba.csv")

feature_cols = ['satverbal','satmath','gpascience',
    'gpaenglish','gpamath','gpaoverall','gender',
    'motherhighgrade', 'fatherhighgrade','parentincome']

X_train, X_test, y_train, y_test =  \
    train_test_split(nls97compba[feature_cols],\
    nls97compba[['completedba']], test_size=0.3,
    random_state=0)
```

3. We need to scale and encode the data. Scaling is particularly important with PCA:

```
ohe = OneHotEncoder(drop_last=True,
                        variables=['gender'])
ohe.fit(X_train)
X_train_enc, X_test_enc = \
    ohe.transform(X_train), ohe.transform(X_test)

scaler = StandardScaler()
standcols = X_train_enc.iloc[:,:-1].columns
scaler.fit(X_train_enc[standcols])
X_train_enc = \
```

```
    pd.DataFrame(scaler.transform(X_train_enc[standcols]),
    columns=standcols, index=X_train_enc.index).\
    join(X_train_enc[['gender_Female']])
X_test_enc = \
    pd.DataFrame(scaler.transform(X_test_enc[standcols]),
    columns=standcols, index=X_test_enc.index).\
    join(X_test_enc[['gender_Female']])
```

4.  Now, we instantiate a PCA object and fit a model:

```
pca = PCA(n_components=5)
pca.fit(X_train_enc)
```

5.  The components_ attribute of the PCA object returns the scores of all 10 features
    on each of the 5 components. The features that drive the first component most
    are those with scores that have the highest absolute value. In this case, that is
    gpaoverall, gpaenglish, and gpascience. For the second component,
    the most important features are motherhighgrade, fatherhighgrade, and
    parentincome. satverbal and satmath drive the third component.

    In the following output, columns **0** through **4** are the five principal components:

```
    pd.DataFrame(pca.components_,
        columns=X_train_enc.columns).T
```

|                 | 0      | 1      | 2      | 3      | 4      |
|-----------------|--------|--------|--------|--------|--------|
| satverbal       | -0.34  | -0.16  | -0.61  | -0.02  | -0.19  |
| satmath         | -0.37  | -0.13  | -0.56  | 0.10   | 0.11   |
| gpascience      | -0.40  | 0.21   | 0.18   | 0.03   | 0.02   |
| gpaenglish      | -0.40  | 0.22   | 0.18   | 0.08   | -0.19  |
| gpamath         | -0.38  | 0.24   | 0.12   | 0.08   | 0.23   |
| gpaoverall      | -0.43  | 0.25   | 0.23   | -0.04  | -0.03  |
| motherhighgrade | -0.19  | -0.51  | 0.24   | -0.43  | -0.59  |
| fatherhighgrade | -0.20  | -0.51  | 0.18   | -0.35  | 0.70   |
| parentincome    | -0.16  | -0.46  | 0.28   | 0.82   | -0.08  |
| gender_Female   | -0.02  | 0.08   | 0.12   | -0.04  | -0.11  |

Another way to understand these scores is that they indicate how much each feature
contributes to the component. (Indeed, if for each component, you square each of
the 10 scores and then sum the squares, you get a total of 1.)

6.  Let's also examine how much of the variance in the features is explained by each component. The first component accounts for 46% of the variance alone, followed by an additional 19% for the second component. We can use NumPy's cumsum method to see how much of feature variance is explained by the five components cumulatively. We can explain 87% of the variance in the 10 features with 5 components:

```
pca.explained_variance_ratio_
array([0.46073387, 0.19036089, 0.09295703, 0.07163009,
0.05328056])


np.cumsum(pca.explained_variance_ratio_)
array([0.46073387, 0.65109476, 0.74405179, 0.81568188,
0.86896244])
```

7.  Let's transform our features in the testing data based on these five principal components. This returns a NumPy array with only the five principal components. We look at the first few rows. We also need to transform the testing DataFrame:

```
X_train_pca = pca.transform(X_train_enc)
X_train_pca.shape
(634, 5)


np.round(X_train_pca[0:6],2)
array([[ 2.79, -0.34,  0.41,  1.42, -0.11],
       [-1.29,  0.79,  1.79, -0.49, -0.01],
       [-1.04, -0.72, -0.62, -0.91,  0.27],
       [-0.22, -0.8 , -0.83, -0.75,  0.59],
       [ 0.11, -0.56,  1.4 ,  0.2 , -0.71],
       [ 0.93,  0.42, -0.68, -0.45, -0.89]])


X_test_pca = pca.transform(X_test_enc)
```

We can now fit a model of bachelor's degree completion using these principal components. Let's run a random forest classification.

8.  We first create a random forest classifier object. We then pass the training data with the principal components and the target values to its `fit` method. We pass the testing data with the components to the classifier's `predict` method and then get an accuracy score:

```
rfc = RandomForestClassifier(n_estimators=100,
   max_depth=2, n_jobs=-1, random_state=0)


rfc.fit(X_train_pca, y_train.values.ravel())


y_pred = rfc.predict(X_test_pca)


accuracy_score(y_test, y_pred)
0.7032967032967034
```

A dimension reduction technique such as PCA can be a good option when the feature selection challenge is that we have highly correlated features and we want to reduce the number of dimensions without substantially reducing the explained variance. In this example, the high school GPA features moved together, as did the parental education and income levels and the SAT features. They became the key features for our first three components. (An argument can be made that our model could have had just those three components since together they accounted for 74% of the variance of the features.)

There are several modifications to PCA that might be useful depending on your data and modeling objectives. This includes strategies to handle outliers and regularization. PCA can also be extended to situations where the components are not linearly separable by using kernels. We discuss PCA in detail in *Chapter 15, Principal Component Analysis*.

Let's summarize what we've learned in this chapter.

# Summary

In this chapter, we went over a range of feature selection methods, from filter to wrapper to embedded methods. We also saw how they work with categorical and continuous targets. For wrapper and embedded methods, we considered how they work with different algorithms.

Filter methods are very easy to run and interpret and are easy on system resources. However, they do not take other features into account when evaluating each feature. Nor do they tell us how that assessment might vary by the algorithm used. Wrapper methods do not have any of these limitations but they are computationally expensive. Embedded methods are often a good compromise, selecting features based on multivariate relationships and a given algorithm without taxing system resources as much as wrapper methods. We also explored how a dimension reduction method, PCA, could improve our feature selection.

You also probably noticed that I slipped in a little bit of model validation during this chapter. We will go over model validation in much more detail in the next chapter.

# 6
# Preparing for Model Evaluation

It is a good idea to think through how you will evaluate your model's performance before you begin to run it. A common technique is to separate data into training and testing datasets. We do this relatively early in the process to avoid what is known as data leakage; that is, conducting analyses based on data that is intended to be set aside for model evaluation. In this chapter, we will look at approaches for creating training datasets, including how to ensure that training data is representative. We will look into cross-validation strategies such as **K-fold**, which address some of the limitations of using static training/testing splits. We will also begin to look more closely at assessing the performance of models.

You might be wondering why we are discussing model evaluation before going over any algorithms in detail. This is because there is a practical consideration. We tend to use the same metrics and evaluation techniques across algorithms with similar purposes. We examine accuracy and sensitivity when evaluating classification models, and mean absolute error and R-squared when examining regression models. We do cross-validation with all supervised learning models. So, we will repeat the strategies introduced here several times in the following chapters. You may even find yourself coming back to these pages when the concepts are re-introduced later.

Beyond those practical considerations, our modeling work improves when we do not see data extraction, data cleaning, exploratory analysis, feature engineering and Preprocessing, model specification, and model evaluation as discrete, sequential tasks. If you have been building machine learning models for just 6 months or over 30 years, you probably appreciate that such rigid sequencing is inconsistent with our workflow as data scientists. We are always preparing for model validation, and always cleaning data. This is a good thing. We do better work when we integrate these tasks; when we continue to interrogate our data cleaning as we select features, and when we look back at bivariate correlations or scatter plots after calculating precision or root mean squared error.

We will also spend a fair bit of time constructing visualizations of these concepts. It is a good idea to get in the habit of looking at confusion matrices and cumulative accuracy profiles when working on classification problems, and plots of residuals when working with a continuous target. This, too, will serve us well in subsequent chapters.

Specifically, in this chapter, we will cover the following topics:

- Measuring accuracy, sensitivity, specificity, and precision for binary classification
- Examining CAP, ROC, and precision-sensitivity curves for binary classification
- Evaluating multiclass models
- Evaluating regression models
- Using K-fold cross-validation
- Preprocessing data with pipelines

# Technical requirements

In this chapter, we will work with the `feature_engine` and `matplotlib` libraries, in addition to the scikit-learn library. You can use `pip` to install these packages. The code files for this chapter can be found in this book's GitHub repository at `https://github.com/PacktPublishing/Data-Cleaning-and-Exploration-with-Machine-Learning`.

# Measuring accuracy, sensitivity, specificity, and precision for binary classification

When assessing a classification model, we typically want to know how often we are right. In the case of a binary target – one where the target has two possible categorical values – we calculate **accuracy** as the ratio of times we predict the correct classification against the total number of observations.

But, depending on the classification problem, accuracy may not be the most important performance measure. Perhaps we are willing to accept more false positives for a model that can identify more true positives, even if that means lower accuracy. This might be true for a model that would predict the likelihood of having breast cancer, a security breach, or structural damage in a bridge. In these cases, we may emphasize **sensitivity** (the propensity to identify positive cases) over accuracy.

On the other hand, we may want a model that could identify negative cases with high reliability, even if that meant it did not do as good a job of identifying positives. **Specificity** is a measure of the percentage of all negatives identified by the model.

**Precision**, the percentage of predicted positives that are actually positives, is another important measure. For some applications, it is important to limit false positives, even if we have to tolerate lower sensitivity. An apple grower, using image recognition to identify bad apples, may prefer a high-precision model to a more sensitive one, not wanting to discard apples unnecessarily.

This can be made clearer by looking at a confusion matrix:

| Confusion Matrix | | | |
|---|---|---|---|
| | | Predicted Value | |
| | | **Negative** | **Positive** |
| **Actual Value** | **Negative** | True Negative (TN) | False Positive (FP) |
| | **Positive** | False Negative (FN) | True Positive (TP) |

Figure 6.1 – Confusion matrix

The confusion matrix helps us conceptualize accuracy, sensitivity, specificity, and precision. Accuracy is the percentage of observations for which our prediction was correct. This can be stated more precisely as follows:

$$accuracy = \frac{\sum TP + \sum TN}{Number\ of\ observations}$$

Sensitivity is the number of times we predicted positives correctly divided by the number of positives. It might be helpful to glance again at the confusion matrix and confirm that actual positive values can either be **predicted positives** (**TP**) or **predicted negatives** (**FN**). Sensitivity is also referred to as **recall** or the **true positive rate**:

$$sensitivity = \frac{\sum TP}{\sum TP + \sum FN}$$

Specificity is the number of times we correctly predicted a **negative value** (**TN**) divided by the number of actual negative values (**TN + FP**). Specificity is also known as the **true negative rate**:

$$specificity = \frac{\sum TN}{\sum TN + \sum FP}$$

Precision is the number of times we correctly predicted a **positive value** (**TP**) divided by the number of positive values predicted:

$$precision = \frac{\sum TP}{\sum TP + \sum FP}$$

When there is class imbalance, measures such as accuracy and sensitivity can give us very different estimates of the performance of our model. An extreme example will illustrate this. Chimpanzees sometimes *termite fish*, putting a stick in a termite mound with the hopes of catching a few termites. This is only occasionally successful. I am no primatologist, but we can perhaps model a successful fishing attempt as a function of the size of the stick used, the time of year, and the age of the chimpanzee. In our testing data, fishing attempts are only successful 2% of the time. (This data has been made up for this demonstration.)

Let's also say that we build a classification model of successful termite fishing that has a sensitivity of 50%. So, if there are 100 fishing attempts in our testing data, we would predict only one of the two successful attempts correctly. There is also one false positive, where our model predicted successful fishing when the fishing failed. This gives us the following confusion matrix:

| Successful Termite Fishing Confusion Matrix | | | |
|---|---|---|---|
| | | Predicted Value | |
| | | Negative | Positive |
| Actual Value | Negative | True Negative (TN) = 97 | False Positive (FP) = 1 |
| | Positive | False Negative (FN) = 1 | True Positive (TP) = 1 |

Figure 6.2 – Successful termite fishing confusion matrix

Notice that we get a very high accuracy of 98% – that is, (97+1) / 100. We get high accuracy and low sensitivity because a large percentage of the fishing attempts are negative and that is easy to predict. A model that just predicts failure always would also have an accuracy of 98%.

Now, let's look at these model evaluation measures with real data. We can experiment with a **k-nearest neighbors (KNN)** model to predict bachelor's degree attainment and evaluate its accuracy, sensitivity, specificity, and precision:

1.  We will start by loading libraries for encoding and standardizing data, and for creating training and testing DataFrames. We will also load scikit-learn's KNN classifier and the metrics library:

```
import pandas as pd
import numpy as np
from feature_engine.encoding import OneHotEncoder
from sklearn.model_selection import train_test_split
from sklearn.preprocessing import StandardScaler
from sklearn.neighbors import KNeighborsClassifier
import sklearn.metrics as skmet
import matplotlib.pyplot as plt
```

2.  Now, we can create training and testing DataFrames and encode and scale the data:

```
nls97compba = pd.read_csv("data/nls97compba.csv")
feature_cols = ['satverbal','satmath','gpaoverall',
  'parentincome','gender']

X_train, X_test, y_train, y_test =  \
  train_test_split(nls97compba[feature_cols],\
  nls97compba[['completedba']], test_size=0.3, random_
```

```
                    state=0)

ohe = OneHotEncoder(drop_last=True, variables=['gender'])
ohe.fit(X_train)
X_train_enc, X_test_enc = \
  ohe.transform(X_train), ohe.transform(X_test)

scaler = StandardScaler()
standcols = X_train_enc.iloc[:,:-1].columns
scaler.fit(X_train_enc[standcols])
X_train_enc = \
  pd.DataFrame(scaler.transform(X_train_enc[standcols]),
  columns=standcols, index=X_train_enc.index).\
  join(X_train_enc[['gender_Female']])
X_test_enc = \
  pd.DataFrame(scaler.transform(X_test_enc[standcols]),
  columns=standcols, index=X_test_enc.index).\
  join(X_test_enc[['gender_Female']])
```

3.  Let's create a KNN classification model. We will not worry too much about how we specify it since we just want to focus on evaluation measures in this section. We will use all of the features listed in `feature_cols` We use the predict method of the KNN classifier to generate predictions from the testing data:

```
knn = KNeighborsClassifier(n_neighbors = 5)
knn.fit(X_train_enc, y_train.values.ravel())
pred = knn.predict(X_test_enc)
```

4.  We can use scikit-learn to plot a confusion matrix. We will pass the actual values in the testing data (`y_test`) and predicted values to the `confusion_matrix` method:

```
cm = skmet.confusion_matrix(y_test, pred, labels=knn.
classes_)
cmplot = skmet.ConfusionMatrixDisplay(
  confusion_matrix=cm,
  display_labels=['Negative', 'Positive'])
cmplot.plot()
cmplot.ax_.set(title='Confusion Matrix',
  xlabel='Predicted Value', ylabel='Actual Value')
```

This generates the following plot:

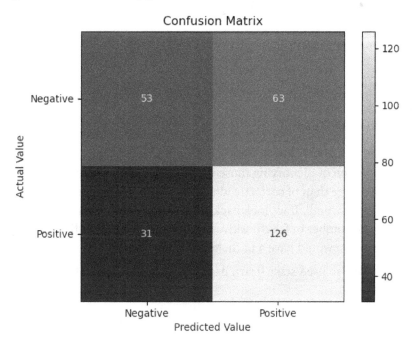

Figure 6.3 – Confusion matrix of actual and predicted values

5. We can also just return the true negative, false positive, false negative, and true positive counts:

```
tn, fp, fn, tp = skmet.confusion_matrix(
    y_test.values.ravel(), pred).ravel()

tn, fp, fn, tp
(53, 63, 31, 126)
```

6. We now have what we need to calculate accuracy, sensitivity, specificity, and precision:

```
accuracy = (tp + tn) / pred.shape[0]
accuracy
0.6556776556776557

sensitivity = tp / (tp + fn)
sensitivity
0.802547770700637
```

```
specificity = tn / (tn+fp)
specificity
0.45689655172413796


precision = tp / (tp + fp)
precision
0.6666666666666666
```

This model has relatively low accuracy, but somewhat better sensitivity; that is, it does a better job of identifying those in the testing data who have completed a bachelor's degree than of correctly identifying both degree completers and non-completers overall. If we look back at the confusion matrix, we will see that there are a fair number of false positives, as our model predicts that 63 individuals in the testing data would have a bachelor's degree who did not.

7.  We could have also used scikit-learn handy methods for generating these statistics directly:

```
skmet.accuracy_score(y_test.values.ravel(), pred)
0.6556776556776557


skmet.recall_score(y_test.values.ravel(), pred)
0.802547770700637


skmet.precision_score(y_test.values.ravel(), pred)
0.6666666666666666
```

Just for comparison, let's try a random forest classifier and see if we get any better results.

8.  Let's fit a random forest classifier to the same data and call `confusion_matrix` again:

```
rfc = RandomForestClassifier(n_estimators=100,
  max_depth=2, n_jobs=-1, random_state=0)
rfc.fit(X_train_enc, y_train.values.ravel())
pred = rfc.predict(X_test_enc)

tn, fp, fn, tp = skmet.confusion_matrix(
  y_test.values.ravel(), pred).ravel()
```

```
tn, fp, fn, tp
(49, 67, 17, 140)


accuracy = (tp + tn) / pred.shape[0]
accuracy
0.6923076923076923


sensitivity = tp / (tp + fn)
sensitivity
0.89171974522293


specificity = tn / (tn+fp)
specificity
0.4224137931034483


precision = tp / (tp + fp)
precision
0.6763285024154589
```

The second model gets us significantly fewer false negatives and more true positives than the first model. It is less likely to predict no bachelor's degree when individuals in the test data have completed a bachelor's degree, and more likely to predict a bachelor's degree when the person has completed one. The main impact of the lower FP and higher TP is a significantly higher sensitivity. The second model identifies actual positives 89% of the time, compared with 80% for the first model.

The measures we have discussed in this section – accuracy, sensitivity, specificity, and precision – are worth looking at whenever we are evaluating a classification model. But it can be hard to get a good sense of the tradeoffs we are sometimes confronted with, between precision and sensitivity, for example. Data scientists rely on several standard visualizations to improve our sense of these tradeoffs when building classification models. We will examine these visualizations in the next section.

# Examining CAP, ROC, and precision-sensitivity curves for binary classification

There are several ways to visualize the performance of a binary classification model. A relatively straightforward visualization is the **Cumulative Accuracy Profile (CAP)**, which shows the ability of our model to identify in-class, or positive, cases. It shows the cumulative cases on the $X$-axis and the cumulative positive outcomes on the $Y$-axis. A CAP curve is a good way to see how good a job our model does at discriminating in-class observations. (When discussing binary classification models, I will use the terms *in-class* and *positive* interchangeably.)

**Receiver operating characteristic (ROC)** curves illustrate the tradeoff between model sensitivity (being able to identify positive values) and the false positive rate as we adjust the threshold for classifying a positive value. Similarly, precision-sensitivity curves show the relationship between the reliability of our positive predictions (their precision) and sensitivity (our model's ability to identify positive actual values) as we adjust the threshold.

## Constructing CAP curves

Let's start with CAP curves for our bachelor's completion KNN model. Let's also compare that with a decision tree model. Again, we will not do much with feature selection here. The previous chapter went over feature selection in some detail.

In addition to curves for our models, CAP curves also have plots of a **random model** and a **perfect model** to view for comparison. The random model provides no information other than the overall distribution of positive values. The perfect model predicts positive values precisely. To illustrate how those plots are drawn, we will start with a hypothetical example. Imagine that you sample the first six cards of a nicely shuffled deck of playing cards. You create a table with the cumulative card total in one column and the number of red cards in the next column. It may look something like this:

| Sample of Playing Cards | | |
|---|---|---|
| Card | Cum Cards | Cum Red |
| Queen of Hearts | 1 | 1 |
| Four of Spades | 2 | 1 |
| Five of Diamonds | 3 | 2 |
| Ace of Hearts | 4 | 3 |
| Queen of Clubs | 5 | 3 |
| Seven of Spades | 6 | 3 |

Figure 6.4 – Sample of playing cards

We can plot a random model based on just our knowledge of the number of red cards. The random model has just two points, (0,0) and (6,3), but that is all we need.

The perfect model plot requires a bit more explanation. If our model predicted red cards perfectly and we sorted by the prediction in descending order, we would get *Figure 6.5*. The cumulative in-class count matches the number of cards until the red cards have been exhausted, which is 3 in this case. A plot of the cumulative in-class total with a perfect model would have two slopes; equal to 1 up until the in-class total was reached, and then 0 after that:

| Sample of Playing Cards | | | |
| :--- | :--- | :--- | :--- |
| Sorted and with Perfect Predictions | | | |
| Card | Cum Obs | Cum Red | Pred Red |
| Queen of Hearts | 1 | 1 | 1 |
| Five of Diamonds | 2 | 2 | 1 |
| Ace of Hearts | 3 | 3 | 1 |
| Four of Spades | 4 | 3 | 0 |
| Queen of Clubs | 5 | 3 | 0 |
| Seven of Spades | 6 | 3 | 0 |

Figure 6.5 – Sample of playing cards

We now know enough to plot both the random model and the perfect model. The perfect model will have three points: (0,0), (in-class count, in-class count), and (number of cards, in-class count). In this case, in-class count is 3 and the number of cards is 6:

```
numobs = 6
inclasscnt = 3
plt.yticks([1,2,3])
plt.plot([0, numobs], [0, inclasscnt], c = 'b', label = 'Random
Model')
plt.plot([0, inclasscnt, numobs], [0, inclasscnt, inclasscnt],
c = 'grey', linewidth = 2, label = 'Perfect Model')
plt.title("Cumulative Accuracy Profile")
plt.xlabel("Total Cards")
plt.ylabel("In-class (Red) Cards")
```

This produces the following plot:

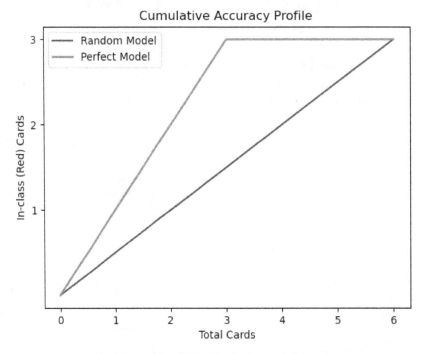

Figure 6.6 – CAP with playing card data

One way to understand the improvement of the perfect model over the random model is to consider how many red cards the random model would predict at the midpoint – that is, 3 cards. At that point, the random model would predict 1.5 red cards. However, the perfect model would predict 3. (Remember that we have sorted the cards by prediction in descending order.)

Having constructed plots for random and perfect models with made-up data, let's try it with our bachelor's degree completion data:

1.  First, we must import the same modules as in the previous section:

    ```
    import pandas as pd
    import numpy as np
    from feature_engine.encoding import OneHotEncoder
    from sklearn.model_selection import train_test_split
    from sklearn.preprocessing import StandardScaler
    from sklearn.neighbors import KNeighborsClassifier
    from sklearn.ensemble import RandomForestClassifier
    ```

```
import sklearn.metrics as skmet
import matplotlib.pyplot as plt
import seaborn as sb
```

2. Then, we load, encode, and scale the NLS bachelor's degree data:

```
nls97compba = pd.read_csv("data/nls97compba.csv")
feature_cols = ['satverbal','satmath','gpaoverall',
  'parentincome','gender']

X_train, X_test, y_train, y_test =  \
  train_test_split(nls97compba[feature_cols],\
  nls97compba[['completedba']], test_size=0.3, random_
state=0)

ohe = OneHotEncoder(drop_last=True, variables=['gender'])
ohe.fit(X_train)
X_train_enc, X_test_enc = \
  ohe.transform(X_train), ohe.transform(X_test)

scaler = StandardScaler()
standcols = X_train_enc.iloc[:,:-1].columns
scaler.fit(X_train_enc[standcols])
X_train_enc = \
  pd.DataFrame(scaler.transform(X_train_enc[standcols]),
  columns=standcols, index=X_train_enc.index).\
  join(X_train_enc[['gender_Female']])
X_test_enc = \
  pd.DataFrame(scaler.transform(X_test_enc[standcols]),
  columns=standcols, index=X_test_enc.index).\
  join(X_test_enc[['gender_Female']])
```

3. Next, we create KNeighborsClassifier and RandomForestClassifier instances:

```
knn = KNeighborsClassifier(n_neighbors = 5)
rfc = RandomForestClassifier(n_estimators=100, max_
depth=2,
  n_jobs=-1, random_state=0)
```

We are now ready to start plotting our CAP curves. We will start by drawing a random model and then a perfect model. These are models that use no information (other than the overall distribution of positive values) and that provide perfect information, respectively.

4.  We count the number of observations in the test data and the number of positive values. We will use (0,0) and (the number of observations, in-class count) to draw the random model line. For the perfect model, we will plot a line from (0,0) to (in-class count, in-class count) since that model can perfectly discriminate in-class values (it is never wrong). It is flat to the right of that point since there are no more positive values to find.

    We will also draw a vertical line at the midpoint and a horizontal line where that intersects the random model line. This will be more useful later:

    ```python
    numobs = y_test.shape[0]
    inclasscnt = y_test.iloc[:,0].sum()


    plt.plot([0, numobs], [0, inclasscnt], c = 'b', label =
    'Random Model')
    plt.plot([0, inclasscnt, numobs], [0, inclasscnt,
    inclasscnt], c = 'grey', linewidth = 2, label = 'Perfect
    Model')
    plt.axvline(numobs/2, color='black', linestyle='dashed',
    linewidth=1)
    plt.axhline(numobs/2, color='black', linestyle='dashed',
    linewidth=1)
    plt.title("Cumulative Accuracy Profile")
    plt.xlabel("Total Observations")
    plt.ylabel("In-class Observations")
    plt.legend()
    ```

    This produces the following plot:

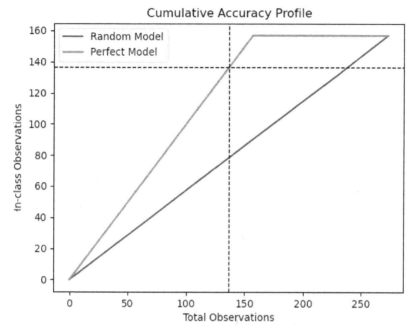

Figure 6.7 – CAP with just random and perfect models

5.   Next, we define a function to plot a CAP curve for a model we pass to it. We will use the `predict_proba` method to get an array with the probability that each observation in the test data is in-class (in this case, has completed a bachelor's degree). Then, we will create a DataFrame with those probabilities and the actual target value, sort it by probability in reverse order, and calculate a running total of positive actual target values.

We will also get the value of the running total at the middle observation and draw a horizontal line at that point. Finally, we will plot a line that has an array from 0 to the number of observations as *x* values, and the running in-class totals as *y* values:

```
def addplot (model, X, Xtest, y, modelname, linecolor):
    model.fit(X, y.values.ravel())
    probs = model.predict_proba(Xtest)[:, 1]

    probdf = pd.DataFrame(zip(probs, y_test.values.
ravel()),
        columns=(['prob','inclass']))
    probdf.loc[-1] = [0,0]
    probdf = probdf.sort_values(['prob','inclass'],
        ascending=False).\
```

```
    assign(inclasscum = lambda x: x.inclass.cumsum())
inclassmidpoint = \
    probdf.iloc[int(probdf.shape[0]/2)].inclasscum
plt.axhline(inclassmidpoint, color=linecolor,
    linestyle='dashed', linewidth=1)
plt.plot(np.arange(0, probdf.shape[0]),
    probdf.inclasscum, c = linecolor,
    label = modelname, linewidth = 4)
```

6.  Now, let's run the function for the KNN and random forest classifier models using the same data:

```
addplot(knn, X_train_enc, X_test_enc, y_train,
    'KNN', 'red')
addplot(rfc, X_train_enc, X_test_enc, y_train,
    'Random Forest', 'green')
plt.legend()
```

This updates our earlier plot:

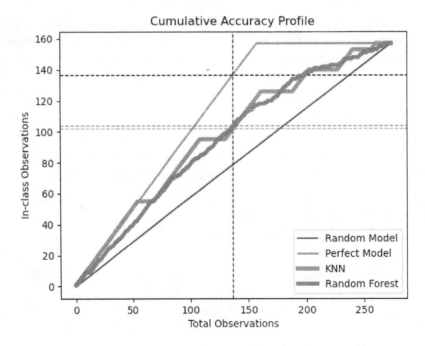

Figure 6.8 – CAP updated with KNN and random forest models

Not surprisingly, the CAP curves show that our KNN and random forest models are better than randomly guessing, but not as good as a perfect model. The question is, how much better and how much worse, respectively. The horizontal lines give us some idea. A perfect model would have correctly identified 138 positive values out of 138 observations. (Recall that the observations are sorted so that the observations with the highest likelihood of being positive are first.) The random model would have identified 70 (line not shown), while the KNN and random forest models would have identified 102 and 103, respectively. Our two models are 74% and 75% as good as a perfect model would have been at discriminating positive values. Anything between 70% and 80% is considered to be a good model; percentages above that are very good, while percentages below that are poor.

## Plotting a receiver operating characteristic (ROC) curve

ROC curves illustrate the tradeoff between the false positive rate and the true positive rate (also known as sensitivity) as we adjust the threshold. We should discuss the false positive rate before going further. It is the percentage of actual negatives (true negatives plus false positives) that our model falsely identifies as positive:

$$False\ positive\ rate = \frac{\sum FP}{\sum TN + \sum FP}$$

Here, you can see the relationship that the false positive rate has with specificity, which was discussed at the beginning of this chapter. The difference is the numerator. Specificity is the percentage of actual negatives that our model correctly identifies as negative:

$$specificity = \frac{\sum TN}{\sum TN + \sum FP}$$

We can also compare the false positive rate with sensitivity, which is the percentage of actual positives (true positives plus false negatives) that our model correctly identifies as positive:

$$sensitivity = \frac{\sum TP}{\sum TP + \sum FN}$$

We are typically confronted with a tradeoff between sensitivity and the false positive rate. We want our models to be able to identify a large percentage of the actual positives, but we do not want a problematically high false positive rate. What is *problematically high* depends on your context.

The tradeoff between sensitivity and the false positive rate is trickier the more difficult it is to discriminate between negative and positive cases. We can see this with our bachelor's degree completion model when we plot the predicted probabilities:

1.  First, let's fit our random forest classifier again and generate predictions and prediction probabilities. We will see that this model predicts that the person completes a bachelor's degree when the predicted probability is greater than $0.500$:

```
rfc.fit(X_train_enc, y_train.values.ravel())
pred = rfc.predict(X_test_enc)
pred_probs = rfc.predict_proba(X_test_enc)[:, 1]


probdf = pd.DataFrame(zip(
  pred_probs, pred, y_test.values.ravel()),
  columns=(['prob','pred','actual']))


probdf.groupby(['pred'])['prob'].agg(['min','max'])
```

```
             min              max
pred
0.000        0.305            0.500
1.000        0.502            0.883
```

2.  It is helpful to compare the distribution of these probabilities with the actual class values. We can do this with density plots:

```
sb.kdeplot(probdf.loc[probdf.actual==1].prob,
  shade=True, color='red',
  label="Completed BA")
sb.kdeplot(probdf.loc[probdf.actual==0].prob,
  shade=True, color='green',
  label="Did Not Complete")
plt.axvline(0.5, color='black', linestyle='dashed',
linewidth=1)
plt.axvline(0.65, color='black', linestyle='dashed',
linewidth=1)
plt.title("Predicted Probability Distribution")
plt.legend(loc="upper left")
```

This produces the following plot:

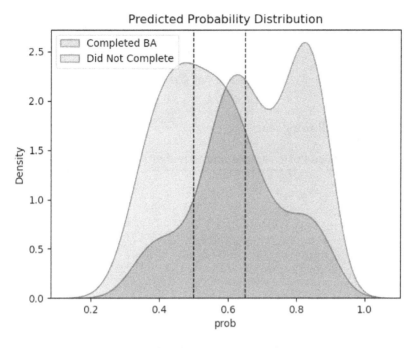

Figure 6.9 – Density plot of in-class and out-of-class observations

Here, we can see that our model has some trouble discriminating between actual positive and negative values since there is a fair bit of in-class and out-of-class overlap. A threshold of 0.500 (the left dotted line) gets us a lot of false positives since a good portion of the distribution of out-of-class observations (those not completing bachelor's degrees) have predicted probabilities greater than 0.500. If we move the threshold higher, say to 0.650, we get many more false negatives since many in-class observations have probabilities lower than 0.65.

3.  It is easy to construct a ROC curve based on the testing data and the random forest model. The `roc_curve` method returns both the false positive rate (`fpr`) and sensitivity (true positive rate, `tpr`) at different thresholds (`ths`).

First, let's draw separate false positive rate and sensitivity lines by threshold:

```
fpr, tpr, ths = skmet.roc_curve(y_test, pred_probs)
ths = ths[1:]
fpr = fpr[1:]
tpr = tpr[1:]
fig, ax = plt.subplots()
```

```
ax.plot(ths, fpr, label="False Positive Rate")
ax.plot(ths, tpr, label="Sensitivity")
ax.set_title('False Positive Rate and Sensitivity by
Threshold')
ax.set_xlabel('Threshold')
ax.set_ylabel('False Positive Rate and Sensitivity')
ax.legend()
```

This produces the following plot:

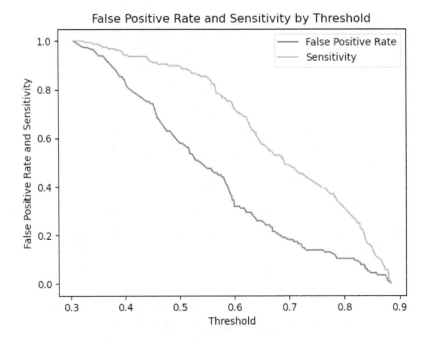

Figure 6.10 – False positive rate and sensitivity lines

Here, we can see that increasing the threshold will improve (reduce) our false positive rate, but also lower our sensitivity.

4.  Now, let's draw the associated ROC curve, which plots the false positive rate against sensitivity for each threshold:

```
fig, ax = plt.subplots()
ax.plot(fpr, tpr, linewidth=4, color="black")
ax.set_title('ROC curve')
ax.set_xlabel('False Positive Rate')
ax.set_ylabel('Sensitivity')
```

This produces the following plot:

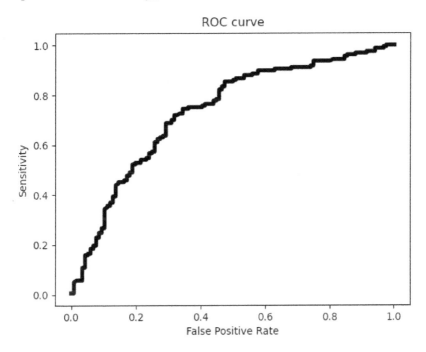

Figure 6.11 – ROC curve with false positive rate and sensitivity

The ROC curve indicates that the tradeoff between the false positive rate and sensitivity is pretty steep until the false positive rate is about 0.5 or higher. Let's see what that means for the threshold of 0.5 that was used for the random forest model predictions.

5.  Let's select an index from the threshold array that is near 0.5, and also one near 0.4 and 0.6 for comparison. Then, we will draw vertical lines for the false positive rate at those indexes, and horizontal lines for the sensitivity values at those indexes:

```
tholdind = np.where((ths>0.499) & (ths<0.501))[0][0]
tholdindlow = np.where((ths>0.397) & (ths<0.404))[0][0]
tholdindhigh = np.where((ths>0.599) & (ths<0.601))[0][0]
plt.vlines((fpr[tholdindlow],fpr[tholdind],
    fpr[tholdindhigh]), 0, 1, linestyles ="dashed",
    colors =["green","blue","purple"])
plt.hlines((tpr[tholdindlow],tpr[tholdind],
    tpr[tholdindhigh]), 0, 1, linestyles ="dashed",
    colors =["green","blue","purple"])
```

This updates our plot:

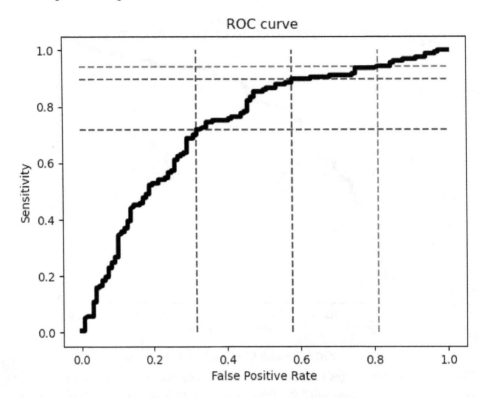

Figure 6.12 – ROC curve with lines for thresholds

This illustrates the tradeoff between the false positive rate and sensitivity at the 0.5 threshold (the blue dashed line) used for predictions. The ROC curve has very little slope with thresholds above 0.5, such as with the 0.6 threshold (the green dashed line). So, reducing the threshold from 0.6 to 0.5 results in a substantially lower false positive rate (from above 0.8 to below 0.6), but not much reduction in sensitivity. However, improving (reducing) the false positive rate by reducing the threshold from 0.5 to 0.4 (from the blue to the purple line) leads to significantly worse sensitivity. It drops from nearly 90% to just above 70%.

# Plotting precision-sensitivity curves

It is often helpful to examine the relationship between precision and sensitivity as the threshold is adjusted. Remember that precision tells us the percentage of the time we are correct when we predict a positive value:

$$precision = \frac{\sum TP}{\sum TP + \sum FP}$$

We can improve precision by increasing the threshold for classifying a value as positive. However, this will likely mean a reduction in sensitivity. As we improve how often we are correct when we predict a positive value (precision), we will decrease the number of positive values we are able to identify (sensitivity). Precision-sensitivity curves, often called precision-recall curves, illustrate this tradeoff.

Before drawing the precision-sensitivity curve, let's look at separate precision and sensitivity lines plotted against thresholds:

1.  We can get the points for the precision-sensitivity curves with the `precision_recall_curve` method. We remove some squirreliness at the highest threshold values, which can sometimes happen:

    ```
    prec, sens, ths = skmet.precision_recall_curve(y_test,
    pred_probs)


    prec = prec[1:-10]
    sens = sens[1:-10]
    ths  = ths[:-10]


    fig, ax = plt.subplots()
    ax.plot(ths, prec, label='Precision')
    ax.plot(ths, sens, label='Sensitivity')
    ax.set_title('Precision and Sensitivity by Threshold')
    ax.set_xlabel('Threshold')
    ax.set_ylabel('Precision and Sensitivity')
    ax.set_xlim(0.3,0.9)
    ax.legend()
    ```

This produces the following plot:

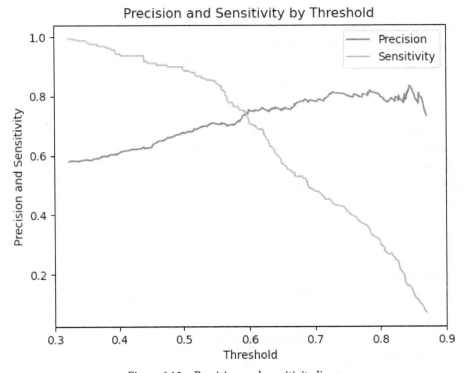

Figure 6.13 – Precision and sensitivity lines

Here, we can see that sensitivity declines more steeply with thresholds above 0.5. This decline does not buy us much improved precision beyond the 0.6 threshold.

2.  Now, let's plot sensitivity against precision to view the precision-sensitivity curve:

```
fig, ax = plt.subplots()
ax.plot(sens, prec)
ax.set_title('Precision-Sensitivity Curve')
ax.set_xlabel('Sensitivity')
ax.set_ylabel('Precision')
plt.yticks(np.arange(0.2, 0.9, 0.2))
```

This produces the following plot:

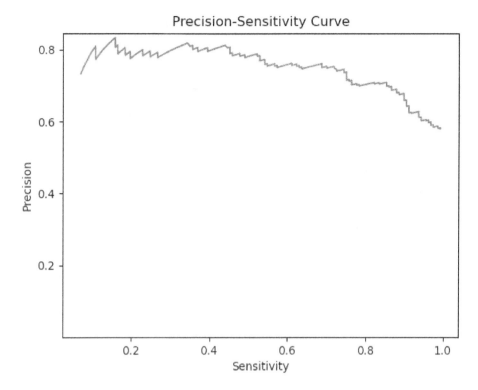

Figure 6.14 – Precision-sensitivity curve

The precision-sensitivity curve reflects the fact that sensitivity is much more responsive to threshold than is precision with this particular model. This means that we could decrease the threshold below 0.5 to get greater sensitivity, without a significant reduction in precision.

> **Note**
>
> The choice of threshold is partly a matter of judgment and domain knowledge, and is mostly an issue when we have significant class imbalance. However, in *Chapter 10, Logistic Regression* we will explore how to calculate an optimal threshold.

This section, and the previous one, demonstrated how to evaluate binary classification models. They showed that model evaluation is not just a thumbs up and thumbs down process. It is much more like tasting your batter as you make a cake. We make good initial assumptions about our model specification and use the model evaluation process to make improvements. This often involves tradeoffs between accuracy, sensitivity, specificity, and precision, and modeling decisions that resist one-size-fits-all recommendations. These decisions are very much domain-dependent and a matter of professional judgment.

The discussion in this section, and most of the techniques, apply as much to multiclass modeling. We discuss evaluating multiclass models in the next section.

# Evaluating multiclass models

All of the same principles that we used to evaluate binary classification models apply to multiclass model evaluation. Computing a confusion matrix is just as important, though a fair bit more difficult to interpret. We also still need to examine somewhat competing measures, such as precision and sensitivity. This, too, is messier than doing so with binary classification.

Once again, we will work with the NLS degree completion data. We will alter the target in this case, from bachelor's degree completion or not to high school completion, bachelor's degree completion, and post-graduate degree completion:

1. We will start by loading the necessary libraries. These are the same libraries we used in the previous two sections:

```
import pandas as pd
import numpy as np
from feature_engine.encoding import OneHotEncoder
from sklearn.model_selection import train_test_split
from sklearn.preprocessing import StandardScaler
from sklearn.neighbors import KNeighborsClassifier
import sklearn.metrics as skmet
import matplotlib.pyplot as plt
```

2. Next, we will load the NLS degree attainment data, create training and testing DataFrames, and encode and scale the data:

```
nls97degreelevel = pd.read_csv("data/nls97degreelevel.
csv")
feature_cols = ['satverbal','satmath','gpaoverall',
    'parentincome','gender']

X_train, X_test, y_train, y_test =  \
    train_test_split(nls97degreelevel[feature_cols],\
    nls97degreelevel[['degreelevel']], test_size=0.3,
random_state=0)
```

```
ohe = OneHotEncoder(drop_last=True, variables=['gender'])
ohe.fit(X_train)
X_train_enc, X_test_enc = \
  ohe.transform(X_train), ohe.transform(X_test)

scaler = StandardScaler()
standcols = X_train_enc.iloc[:,:-1].columns
scaler.fit(X_train_enc[standcols])
X_train_enc = \
  pd.DataFrame(scaler.transform(X_train_enc[standcols]),
  columns=standcols, index=X_train_enc.index).\
  join(X_train_enc[['gender_Female']])
X_test_enc = \
  pd.DataFrame(scaler.transform(X_test_enc[standcols]),
  columns=standcols, index=X_test_enc.index).\
  join(X_test_enc[['gender_Female']])
```

3. Now, we will run a KNN model and predict values for each degree level category:

```
knn = KNeighborsClassifier(n_neighbors = 5)
knn.fit(X_train_enc, y_train.values.ravel())
pred = knn.predict(X_test_enc)
pred_probs = knn.predict_proba(X_test_enc)[:, 1]
```

4. We can use those predictions to generate a confusion matrix:

```
cm = skmet.confusion_matrix(y_test, pred)
cmplot = skmet.ConfusionMatrixDisplay(confusion_
matrix=cm, display_labels=['High School',
'Bachelor','Post-Graduate'])
cmplot.plot()
cmplot.ax_.set(title='Confusion Matrix',
  xlabel='Predicted Value', ylabel='Actual Value')
```

This generates the following plot:

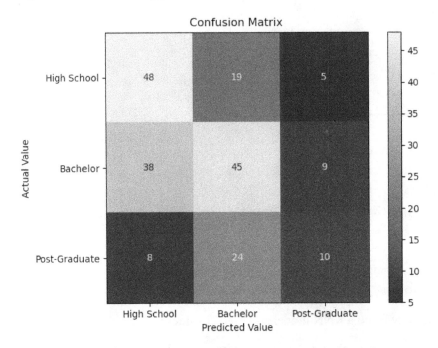

Figure 6.15 – Confusion matrix with a multiclass target

It is possible to calculate evaluation measures by hand. Precision is the percentage of our in-class predictions that are actually in-class. So, for our prediction of high school, it is 48 / (48 + 38 + 8) = 0.51. Sensitivity for the high school class – that is, the percentage of actual values of high school that our model predicts – is 48 / (48 + 19 + 5) = 0.67. However, this is fairly tedious. Fortunately, scikit-learn can do this for us.

5.  We can call the `classification_report` method to get these statistics, passing actual and predicted values (remember that recall and sensitivity are the same measure):

```
print(skmet.classification_report(y_test, pred,
    target_names=['High School', 'Bachelor', 'Post-
Graduate']))


                   precision    recall  f1-score   support

     High School        0.51      0.67      0.58        72
        Bachelor        0.51      0.49      0.50        92
   Post-Graduate        0.42      0.24      0.30        42
```

| | | | | |
|---|---|---|---|---|
| accuracy | | | 0.50 | 206 |
| macro avg | 0.48 | 0.46 | 0.46 | 206 |
| weighted avg | 0.49 | 0.50 | 0.49 | 206 |

In addition to precision and sensitivity rates by class, we get some other statistics. The F1-score is the harmonic mean of precision and sensitivity.

$$f1\text{-}score = 2\frac{p * s}{p + s}$$

Here, $p$ is precision and $s$ is sensitivity.

To get the average precision, sensitivity, and F1-score across classes, we can either use the simple average (macro average) or a weighted average that adjusts for class size. Using the weighted average, we get precision, sensitivity, and F1-score values of 0.49, 0.50, and 0.49, respectively. (Since the classes are relatively balanced here, there is not much difference between the macro average and the weighted average.)

This demonstrates how to extend the evaluation measures we discussed for binary classification models to multiclass evaluation. The same concepts and techniques apply, though they are more difficult to implement.

So far, we have focused on metrics and visualizations to help us evaluate classification models. We have not examined metrics for evaluating regression models yet. These metrics can be somewhat more straightforward than those for classification. We will discuss them in the next section.

# Evaluating regression models

Metrics for regression evaluation are typically based on the distance between the actual values for the target variable and a model's predicted values. The most common measures – mean squared error, root mean squared error, mean absolute error, and R-squared – all track how successfully our predictions capture variation in a target.

The distance between the actual value and our prediction is known as the residual, or error. The **mean squared error** (**MSE**) is the mean of the square of the residuals:

$$MSE = \frac{1}{N}\sum_{i=1}^{N}(y_i - \hat{y}_i)^2$$

Here, $y_i$ is the actual target variable value at the ith observation and $\hat{y}_i$ is our prediction for the target. The residuals are squared to handle negative values, where the predicted value is higher than the actual value. To return our measurement to a more meaningful scale, we often use the square root of MSE. That is known as **root mean squared error (RMSE)**.

Due to the squaring, MSE will penalize larger residuals much more than it will smaller residuals. For example, if we have predictions for five observations, with one having a residual of 25, and the other four having a residual of 0, we will get an MSE of $(0+0+0+0+625)/5 = 125$. However, if all five observations had residuals of 5, the MSE would be $(25+25+25+25+25)/5 = 25$.

A good alternative to squaring the residuals is to take their absolute value. This gives us the mean absolute error:

$$MAE = \frac{1}{N} \sum_{i=1}^{N} |y_i - \hat{y}_i|$$

R-squared, also known as the coefficient of determination, is an estimate of the proportion of the variation in the target variable captured by our model. We square the residuals, as we do when calculating MSE, and divide that by the deviation of each actual target value from its sample mean. This gives us the still unexplained variation, which we subtract from 1 to get the explained variation:

$$R\text{-}squared = 1 - \frac{sum\ of\ the\ residuals\ squared}{total\ sum\ of\ squares}$$

$$= 1 - \frac{\Sigma(y_i - \hat{y}_i)^2}{\Sigma(y_i - \bar{y})^2}$$

Fortunately, scikit-learn makes it easy to generate these statistics. In this section, we will build a linear regression model of land temperatures and use these statistics to evaluate it. We will work with data from the United States National Oceanic and Atmospheric Administration on average annual temperatures, elevation, and latitude at weather stations in 2019.

> **Note**
>
> The land temperature dataset contains the average temperature readings (in Celsius) in 2019 from over 12,000 stations across the world, though the majority of the stations are in the United States. The raw data was retrieved from the Global Historical Climatology Network integrated database. It has been made available for public use by the United States National Oceanic and Atmospheric Administration at `https://www.ncdc.noaa.gov/data-access/land-based-station-data/land-based-datasets/global-historical-climatology-network-monthly-version-4`.

Let's start building a linear regression model:

1.  We will start by loading the libraries we need and the land temperatures data. We will also create training and testing DataFrames:

    ```python
    import pandas as pd
    import numpy as np
    from sklearn.model_selection import train_test_split
    from sklearn.linear_model import LinearRegression
    import sklearn.metrics as skmet
    import matplotlib.pyplot as plt


    landtemps = pd.read_csv("data/landtemps2019avgs.csv")
    feature_cols = ['latabs','elevation']

    X_train, X_test, y_train, y_test =  \
       train_test_split(landtemps[feature_cols],\
       landtemps[['avgtemp']], test_size=0.3, random_state=0)
    ```

    > **Note**
    >
    > The `latabs` feature is the value of latitude without the North or South indicators; so, Cairo, Egypt, at approximately 30 degrees north, and Porto Alegre, Brazil, at about 30 degrees south, have the same value.

2.  Now, we scale our data:

    ```python
    scaler = StandardScaler()
    scaler.fit(X_train)
    X_train = \
    ```

```
    pd.DataFrame(scaler.transform(X_train),
      columns=feature_cols, index=X_train.index)
  X_test = \
    pd.DataFrame(scaler.transform(X_test),
      columns=feature_cols, index=X_test.index)

  scaler.fit(y_train)
  y_train, y_test = \
    pd.DataFrame(scaler.transform(y_train),
      columns=['avgtemp'], index=y_train.index),\
    pd.DataFrame(scaler.transform(y_test),
      columns=['avgtemp'], index=y_test.index)
```

3.  Next, we instantiate a scikit-learn `LinearRegression` object and fit a model on the training data. Our target is the annual average temperature (`avgtemp`), while the features are latitude (`latabs`) and `elevation`. The `coef_` attribute gives us the coefficient for each feature:

```
lr = LinearRegression()
lr.fit(X_train, y_train)
np.column_stack((lr.coef_.ravel(),
  X_test.columns.values))
array([[-0.8538957537748768, 'latabs'],
       [-0.3058979822791853, 'elevation']], dtype=object)
```

The interpretation of the `latabs` coefficient is that standardized average annual temperature will decline by 0.85 for every one standard deviation increase in latitude. (The `LinearRegression` module does not return p-values, a measure of the statistical significance of the coefficient estimate. You can use `statsmodels` instead to see a full summary of an ordinary least squares model.)

4.  Now, we can get predicted values. Let's also join the returned NumPy array with the features and the target from the testing data. Then, we can calculate the residuals by subtracting the predicted values from the actual values (`avgtemp`). The residuals do not look bad, though there is a little negative skew and excessive kurtosis:

```
pred = lr.predict(X_test)
preddf = pd.DataFrame(pred, columns=['prediction'],
  index=X_test.index).join(X_test).join(y_test)
preddf['resid'] = preddf.avgtemp-preddf.prediction
```

```
preddf.resid.agg(['mean','median','skew','kurtosis'])
mean            -0.021
median           0.032
skew            -0.641
kurtosis         6.816
Name: resid, dtype: float64
```

It is worth noting that we will be generating predictions and calculated residuals in this way most of the time we work with regression models in this book. If you feel a little unclear about what we just did in the preceding code block, it may be a good idea to go over it again.

5.  We should plot the residuals to get a better sense of how they are distributed.

```
Plt.hist(preddf.resid, color="blue")
plt.axvline(preddf.resid.mean(), color='red',
linestyle='dashed', linewidth=1)
plt.title("Histogram of Residuals for Temperature Model")
plt.xlabel("Residuals")
plt.ylabel("Frequency")
```

This produces the following plot:

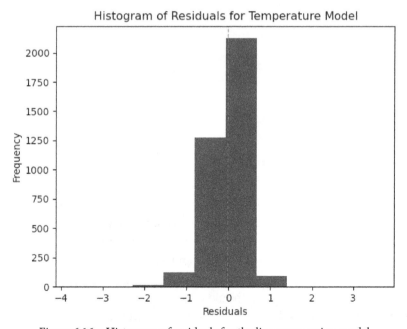

Figure 6.16 – Histogram of residuals for the linear regression model

This does not look too bad, but we have more positive residuals, where we have predicted a lower temperature in the testing data than the actual temperature, than negative residuals.

6.  Plotting our predictions by the residuals may give us a better sense of what is happening:

```
plt.scatter(preddf.prediction, preddf.resid,
color="blue")
plt.axhline(0, color='red', linestyle='dashed',
linewidth=1)
plt.title("Scatterplot of Predictions and Residuals")
plt.xlabel("Predicted Temperature")
plt.ylabel("Residuals")
```

This produces the following plot:

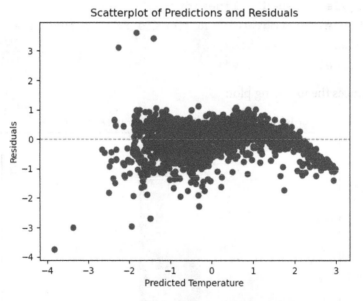

Figure 6.17 – Scatterplot of predictions by residuals for the linear regression model

This does not look horrible. The residuals hover somewhat randomly around 0. However, predictions between 1 and 2 standard deviations are much more likely to be too low (to have positive residuals) than too high. Above 2, the predictions are always too high (they have negative residuals). This model's assumption of linearity might not be sound. We should explore a couple of the transformations we discussed in *Chapter 4, Encoding, Transforming, and Scaling Features*, or try a non-parametric model such as KNN regression.

It is also likely that extreme values are tugging our coefficients around a fair bit. A good next move might be to remove outliers, as we discussed in the *Identifying extreme values and outliers* section of *Chapter 1, Examining the Distribution of Features and Targets*. We will not do that here, however.

7. Let's look at some evaluation measures. This can easily be done with scikit-learn's `metrics` library. We can call the same function to get RMSE as MSE. We just need to set the squared parameter to `False`:

```
mse = skmet.mean_squared_error(y_test, pred)
mse
0.18906346144036693


rmse = skmet.mean_squared_error(y_test, pred,
squared=False)
rmse
0.4348142838504353


mae = skmet.mean_absolute_error(y_test, pred)
mae
0.318307379728143


r2 = skmet.r2_score(y_test, pred)
r2
0.8162525715296725
```

An MSE of less than 0.2 of a standard deviation and an MAE of less than 0. 3 of a standard deviation look pretty decent, especially for such a sparse model. An R-squared above 80% is also fairly promising.

8. Let's see what we get if we use a KNN model instead:

```
knn = KNeighborsRegressor(n_neighbors=5)
knn.fit(X_train, y_train)
pred = knn.predict(X_test)


mae = skmet.mean_absolute_error(y_test, pred)
mae
0.2501829988751876
```

```
r2 = skmet.r2_score(y_test, pred)
r2
0.8631113217183314
```

This model is actually an improvement in both MAE and R-squared.

9.  We should also take a look at the residuals again:

```
preddf = pd.DataFrame(pred, columns=['prediction'],
    index=X_test.index).join(X_test).join(y_test)
preddf['resid'] = preddf.avgtemp-preddf.prediction

plt.scatter(preddf.prediction, preddf.resid,
color="blue")
plt.axhline(0, color='red', linestyle='dashed',
linewidth=1)
plt.title("Scatterplot of Predictions and Residuals with
KNN Model")
plt.xlabel("Predicted Temperature")
plt.ylabel("Residuals")
plt.show()
```

This produces the following plot:

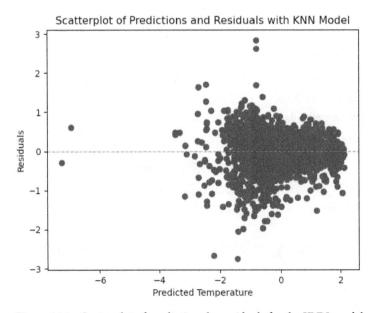

Figure 6.18 – Scatterplot of predictions by residuals for the KNN model

This plot of the residuals looks better as well. There are no parts of the target's distribution where we are much more likely to over-predict or under-predict.

This section has introduced key measures for evaluating regression models, and how to interpret them. It has also demonstrated how visualizations, particularly of model residuals, can improve that interpretation.

However, we have been limited so far, in both our use of regression and classification measures, by how we have constructed our training and testing DataFrames. What if, for some reason, the testing data is unusual in some way? More generally, what is our basis for concluding that our evaluation measures are accurate? We can be more confident in these measures if we use K-fold cross-validation, which we will cover in the next section.

# Using K-fold cross-validation

So far, we have held back 30% of our data for validation. This is not a bad strategy. It prevents us from peeking ahead to the testing data as we train our model. However, this approach does not take full advantage of all the available data, either for training or for testing. If we use K-fold cross-validation instead, we can use all of our data while also avoiding data leakage. Perhaps that seems too good to be true. But it's not because of a neat little trick.

**K-fold cross-validation** trains our model on all but one of the K folds, or parts, leaving one out for testing. This is repeated *k* times, each time excluding a different fold for testing. Performance metrics are then based on the average scores across the K folds.

Before we start, though, we need to think again about the possibility of data leakage. If we scale all of the data that we will use to train our model and then split it up into folds, we will be using information from all the folds in our training. To avoid this, we need to do the scaling, as well as any other Preprocessing, on just the training folds for each iteration. While we could do this manually, scikit-learn's `pipeline` library can do much of this work for us. We will go over how to use pipelines for cross-validation in this section.

Let's try evaluating the two models we specified in the previous section using K-fold cross-validation. While we are at it, let's also see how well a random forest regressor may work:

1.  In addition to the libraries we have worked with so far, we need scikit-learn's `make_pipeline`, `cross_validate`, and `Kfold` libraries:

    ```
    import pandas as pd
    from sklearn.model_selection import train_test_split
    from sklearn.preprocessing import StandardScaler
    from sklearn.linear_model import LinearRegression
    ```

```
from sklearn.neighbors import KNeighborsRegressor
from sklearn.ensemble import RandomForestRegressor
from sklearn.pipeline import make_pipeline
from sklearn.model_selection import cross_validate
from sklearn.model_selection import KFold
```

2.  We load the land temperatures data again and create training and testing DataFrames. We still want to leave some data out for final validation, but this time, we will only leave out 10%. We will do both training and testing with the remaining 90%:

```
landtemps = pd.read_csv("data/landtemps2019avgs.csv")
feature_cols = ['latabs','elevation']

X_train, X_test, y_train, y_test =  \
  train_test_split(landtemps[feature_cols],\
  landtemps[['avgtemp']],test_size=0.1,random_state=0)
```

3.  Now, we create a KFold object and indicate that we want five folds and for the data to be shuffled (shuffling the data is a good idea if it is not already sorted randomly):

```
kf = Kfold(n_splits=5, shuffle=True, random_state=0)
```

4.  Next, we define a function to create a pipeline. The function then runs cross_validate, which takes the pipeline and the KFold object we created earlier:

```
def getscores(model):
  pipeline = make_pipeline(StandardScaler(), model)
  scores = cross_validate(pipeline, X=X_train,
    y=y_train, cv=kf, scoring=['r2'], n_jobs=1)
  scorelist.append(dict(model=str(model),
    fit_time=scores['fit_time'].mean(),
    r2=scores['test_r2'].mean()))
```

5.  Now, we are ready to call the `getscores` function for the linear regression, random forest regression, and KNN regression models:

```
scorelist = []
getscores(LinearRegression())
getscores(RandomForestRegressor(max_depth=2))
getscores(KNeighborsRegressor(n_neighbors=5))
```

6.  We can print the `scorelist` list to see our results:

```
scorelist
[{'model': 'LinearRegression()',
  'fit_time': 0.004968833923339844,
  'r2': 0.8181125031214872},
 {'model': 'RandomForestRegressor(max_depth=2)',
  'fit_time': 0.28124608993530276,
  'r2': 0.7122492698889024},
 {'model': 'KNeighborsRegressor()',
  'fit_time': 0.006945991516113281,
  'r2': 0.8686733636724104}]
```

The KNN regressor model performs better than either the linear regression or random forest regression model, based on R-squared. The random forest regressor also has a significant disadvantage in that it has a much longer fit time.

# Preprocessing data with pipelines

We just scratched the surface of what we can do with scikit-learn pipelines in the previous section. We often need to fold all of our Preprocessing and feature engineering into a pipeline, including scaling, encoding, and handling outliers and missing values. This can be complicated as different features may need to be handled differently. We may need to impute the median for missing values with numeric features and the most frequent value for categorical features. We may also need to transform our target variable. We will explore how to do that in this section.

Follow these steps:

1. We will start by loading the libraries we have already worked with in this chapter. Then, we will add the `ColumnTransformer` and `TransformedTargetRe-gressor` classes. We will use those classes to transform our features and target, respectively:

```
import pandas as pd
import numpy as np
from sklearn.model_selection import train_test_split
from sklearn.preprocessing import StandardScaler
from sklearn.linear_model import LinearRegression
from sklearn.impute import SimpleImputer
from sklearn.pipeline import make_pipeline
from feature_engine.encoding import OneHotEncoder
from sklearn.impute import KNNImputer
from sklearn.model_selection import cross_validate, KFold
import sklearn.metrics as skmet
from sklearn.compose import ColumnTransformer
from sklearn.compose import TransformedTargetRegressor
```

2. The column transformer is quite flexible. We can even use it with the Preprocessing functions that we have defined ourselves. The following code block imports the `OutlierTrans` class from the `preprocfunc` module in the `helperfunctions` subfolder:

```
import os
import sys
sys.path.append(os.getcwd() + "/helperfunctions")
from preprocfunc import OutlierTrans
```

3. The `OutlierTrans` class identifies extreme values by distance from the interquartile range. This is a technique we demonstrated in *Chapter 3*, *Identifying and Fixing Missing Values*.

   To work in a scikit-learn pipeline, our class has to have fit and transform methods. We also need to inherit the `BaseEstimator` and `TransformerMixin` classes.

In this class, almost all of the action happens in the transform method. Any value that is more than 1.5 times the interquartile range above the third quartile or below the first quartile is assigned missing:

```python
class OutlierTrans(BaseEstimator,TransformerMixin):
  def __init__(self,threshold=1.5):
    self.threshold = threshold

  def fit(self,X,y=None):
    return self

  def transform(self,X,y=None):
    Xnew = X.copy()
    for col in Xnew.columns:
      thirdq, firstq = Xnew[col].quantile(0.75),\
        Xnew[col].quantile(0.25)
      inlierrange = self.threshold*(thirdq-firstq)
      outlierhigh, outlierlow = inlierrange+thirdq,\
        firstq-inlierrange
      Xnew.loc[(Xnew[col]>outlierhigh) | \
        (Xnew[col]<outlierlow),col] = np.nan
    return Xnew.values
```

Our OutlierTrans class can be used later in our pipeline in the same way we used StandardScaler in the previous section. We will do that later.

4.  Now, we are ready to load the data that needs to be processed. We will work with the NLS weekly wage data in this section. Weekly wages will be our target, and we will use high school GPA, mother's and father's highest grade completed, parent income, gender, and whether the individual completed a bachelor's degree as features.

We will create lists of features to handle in different ways here. This will be helpful later when we instruct our pipeline to carry out different operations on numerical, categorical, and binary features:

```python
nls97wages = pd.read_csv("data/nls97wagesb.csv")
nls97wages.set_index("personid", inplace=True)
nls97wages.dropna(subset=['wageincome'], inplace=True)

nls97wages.loc[nls97wages.motherhighgrade==95,
  'motherhighgrade'] = np.nan
```

```
nls97wages.loc[nls97wages.fatherhighgrade==95,
  'fatherhighgrade'] = np.nan

num_cols =
['gpascience','gpaenglish','gpamath','gpaoverall',
  'motherhighgrade','fatherhighgrade','parentincome']
cat_cols = ['gender']
bin_cols = ['completedba']

target = nls97wages[['wageincome']]
features = nls97wages[num_cols + cat_cols + bin_cols]
X_train, X_test, y_train, y_test =  \
  train_test_split(features,\
  target, test_size=0.2, random_state=0)
```

5.  Let's look at some descriptive statistics. Some variables have over a thousand missing values (gpascience, gpaenglish, gpamath, gpaoverall, and parentincome):

```
nls97wages[['wageincome'] + num_cols].
agg(['count','min','median','max']).T
```

|  | count | min | median | max |
|---|---|---|---|---|
| wageincome | 5,091 | 0 | 40,000 | 235,884 |
| gpascience | 3,521 | 0 | 284 | 424 |
| gpaenglish | 3,558 | 0 | 288 | 418 |
| gpamath | 3,549 | 0 | 280 | 419 |
| gpaoverall | 3,653 | 42 | 292 | 411 |
| motherhighgrade | 4,734 | 1 | 12 | 20 |
| fatherhighgrade | 4,173 | 1 | 12 | 29 |
| parentincome | 3,803 | -48,100 | 40,045 | 246,474 |

6.  Now, we can set up a column transformer. First, we will create pipelines for handling numerical data (standtrans), categorical data, and binary data.

For the numerical data, we want to assign outlier values as missing. Here, we will pass a value of 2 to the threshold parameter of OutlierTrans, indicating that we want values two times the interquartile range above or below that range to be set to missing. Recall that the default is 1.5, so we are being somewhat more conservative.

Then, we will create a `ColumnTransformer` object, passing to it the three pipelines we just created, and indicating which features to use with which pipeline:

```
standtrans = make_pipeline(OutlierTrans(2),
  StandardScaler())


cattrans = make_pipeline(SimpleImputer(strategy="most_
frequent"),
  OneHotEncoder(drop_last=True))


bintrans = make_pipeline(SimpleImputer(strategy="most_
frequent"))


coltrans = ColumnTransformer(
  transformers=[
    ("stand", standtrans, num_cols),
    ("cat", cattrans, ['gender']),
    ("bin", bintrans, ['completedba'])
  ]
)
```

7.  Now, we can add the column transformer to a pipeline that also includes the linear model that we would like to run. We will add KNN imputation to the pipeline to handle missing values.

    We also need to scale the target, which cannot be done in our pipeline. We will use scikit-learn's `TransformedTargetRegressor` for that. We will pass the pipeline we just created to the target regressor's `regressor` parameter:

```
lr = LinearRegression()


pipe1 = make_pipeline(coltrans,
  KNNImputer(n_neighbors=5), lr)


ttr=TransformedTargetRegressor(regressor=pipe1,
  transformer=StandardScaler())
```

8.  Let's do K-fold cross validation using this pipeline. We can pass our pipeline, via the target regressor, `ttr`, to the `cross_validate` function:

```
kf = KFold(n_splits=10, shuffle=True, random_state=0)

scores = cross_validate(ttr, X=X_train, y=y_train,
    cv=kf, scoring=('r2', 'neg_mean_absolute_error'),
    n_jobs=1)

print("Mean Absolute Error: %.2f, R-squared: %.2f" %
    (scores['test_neg_mean_absolute_error'].mean(),
    scores['test_r2'].mean()))
Mean Absolute Error: -23781.32, R-squared: 0.20
```

These scores are not very good, though that was not quite the point of this exercise. The key takeaway here is that we typically want to fold most of the Preprocessing we will do into a pipeline. This is the best way to avoid data leakage. The column transformer is an extremely flexible tool, allowing us to apply different transformations to different features.

# Summary

This chapter introduced key model evaluation measures and techniques so that they will be familiar when we make extensive use of them, and extend them, in the remaining chapters of this book. We examined the very different approaches to evaluation for classification and regression models. We also explored how to use visualizations to improve our analysis of our predictions. Finally, we used pipelines and cross-validation to get reliable estimates of model performance.

I hope this chapter also gave you a chance to get used to the general approach of this book going forward. Although a large number of algorithms will be discussed in the remaining chapters, we will continue to surface the Preprocessing issues we have discussed in the first few chapters. We will discuss the core concepts of each algorithm, of course. But, in a true *hands-on* fashion, we will also deal with the messiness of real-world data. Each chapter will go from relatively raw data to feature engineering to model specification and model evaluation, relying heavily on scikit-learn's pipelines to pull it all together.

We will discuss regression algorithms in the next few chapters – those algorithms that allow us to model a continuous target. We will explore some of the most popular regression algorithms – linear regression, support vector regression, K-nearest neighbors regression, and decision tree regression. We will also consider making modifications to regression models that address underfitting and overfitting, including nonlinear transformations and regularization.

# Section 3 – Modeling Continuous Targets with Supervised Learning

The final ten chapters of this book introduce a wide range of machine learning algorithms, for predicting both continuous or categorical targets, or when there is no target. We explore models for continuous targets in this chapter.

A persistent theme in these chapters is that finding the best possible model is partly about balancing variance and bias. When our models fit the training data too well, they may not be as generalizable as we need them to be. In cases like that, they may have low bias but high variance. For each algorithm we examine in these chapters, we discuss strategies for achieving this balance. These strategies range from regularization for linear regression and support vector regression models, to the value of k for k-nearest neighbors, to the maximum depth of decision trees.

We also get a chance to practice the preprocessing, feature selection, and model evaluation strategies we worked with in *Chapter 6, Preparing for Model Evaluation*. Each of the algorithms we discuss in this section requires different preprocessing for optimal results. For example, feature scaling is important for support vector regression, but not usually for a decision tree regression. We might use a polynomial transformation with a linear regression model, but that would also be unnecessary with a decision tree. We consider those choices in each chapter of this part.

This section comprises the following chapters:

- *Chapter 7, Linear Regression Models*
- *Chapter 8, Support Vector Regression*
- *Chapter 9, K-Nearest Neighbor, Decision Tree, Random Forest, and Gradient Boosted Regression*

# 7
# Linear Regression Models

Linear regression is perhaps the most well-known machine learning algorithm, having origins in statistical learning at least 200 years ago. If you took a statistics, econometrics, or psychometrics course in college, you were likely introduced to linear regression, even if you took that course long before machine learning was taught in undergraduate courses. As it turns out, many social and physical phenomena can be successfully modeled as a function of a linear combination of predictor variables. This is as useful for machine learning as it has been for statistical learning all these years, though, with machine learning, we care much less about the parameter values than we do about predictions.

Linear regression is a very good choice for modeling a continuous target, assuming that our features and target have certain qualities. In this chapter, we will go over the assumptions of linear regression models and construct a model using data that is largely consistent with these assumptions. However, we will also explore alternative approaches, such as nonlinear regression, which we use when these assumptions do not hold. We will conclude this chapter by looking at techniques that address the possibility of overfitting, such as lasso regression.

In this chapter, we will cover the following topics:

- Key concepts
- Linear regression and gradient descent
- Using classical linear regression
- Using lasso regression
- Using non-linear regression
- Regression with gradient descent

# Technical requirements

In this chapter, we will stick to the libraries that are available with most scientific distributions of Python – NumPy, pandas, and scikit-learn. The code for this chapter can be found in this book's GitHub repository at `https://github.com/PacktPublishing/Data-Cleaning-and-Exploration-with-Machine-Learning`.

# Key concepts

The typical analyst who has been doing predictive modeling for a while has constructed tens, perhaps hundreds, of linear regression models over the years. If you worked for a large accounting firm in the late 1980s, as I did, and you were doing forecasting, you may have spent your whole day, every day, specifying linear models. You would have run all conceivable permutations of independent variables and transformations of dependent variables, and diligently looked for evidence of heteroscedasticity (non-constant variance in residuals) or multicollinearity (highly correlated features). But most of all, you worked hard to identify key predictor variables and address any bias in your parameter estimates (your coefficients or weights).

## Key assumptions of linear regression models

Much of that effort still applies today, though there is now much more emphasis on the accuracy of predictions than on parameter estimates. We worry about overfitting now, in a way that we did not 30 years ago. We are also more likely to seek alternatives when the assumptions of linear regression models are violated. These assumptions are as follows:

- There there is a linear relationship between features (independent variables) and the target (dependent variable)

- That the residuals (the difference between actual and predicted values) are normally distributed

- That the residuals are independent across observations

- That the variance of residuals is constant

It is not unusual for one or more of these assumptions to be violated with real-world data. The relationship between a feature and target is often not linear. The influence of the feature may vary across the range of that feature. Anyone familiar with the expression "*too many cooks in the kitchen*" likely appreciates that the marginal increase in productivity with the fifth cook may not be as great as with the second or third.

Our residuals are sometimes not normally distributed. This can indicate that our model is less accurate along certain ranges of our target. For example, it is not unusual to have smaller residuals along the middle of the target's range, say the 25th to 75th percentile, and higher residuals at the extremes. This can happen when the relationship with the target is nonlinear.

There are several reasons why residuals may not be independent. This is often the case with time series data. For a model of daily stock price, the residuals may be correlated for adjacent days. This is referred to as autocorrelation. This can also be a problem with longitudinal or repeated measures data. For example, we may have test scores for 600 students in 20 different classrooms or annual wage income for 100 people. Our residuals would not be independent if our model failed to account for there being no variation in some features across a group – the classroom-determined and person-determined features in these examples.

Finally, it is not uncommon for our residuals to have greater variability along different ranges of a feature. If we are predicting temperatures at weather stations around the world, and latitude is one of the features we are using, there is a chance that there will be greater residuals at higher latitude values. This is known as heteroscedasticity. This may also be an indicator that our model has omitted important predictors.

Beyond these four key assumptions, another common challenge with linear regression is the high correlation among features. This is known as multicollinearity. As we discussed in *Chapter 5, Feature Selection*, we likely increase the risk of overfitting when our model struggles to isolate the independent effect of a particular feature because it moves so much with another feature. This will be familiar to any of you who have spent weeks building a model where the coefficients shift dramatically with each new specification.

When one or more of these assumptions is violated, we may still be able to use a traditional regression model. However, we may need to transform the data in some way. We will discuss techniques for identifying violations of these assumptions, the implications of those violations for model performance, and possible ways to address these issues throughout this chapter.

## Linear regression and ordinary least squares

The most common estimation technique for linear regression is **ordinary least squares** (**OLS**). OLS selects coefficients that minimize the sum of the squared distance between the actual target values and the predicted values. More precisely, OLS minimizes the following:

$$\sum_{i=1}^{N} (y_i - \hat{y}_i)^2$$

Here, $y_i$ is the actual value at the ith observation and $\hat{y}_i$ is the predicted value. As we have discussed, the difference between the actual target value and the predicted target value, $y_i - \hat{y}_i$, is known as the residual.

Graphically, OLS fits a line through our data that minimizes the vertical distance of data points from that line. The following plot illustrates a model with one feature, known as simple linear regression, with made-up data points. The vertical distance between each data point and the regression line is the residual, which can be positive or negative:

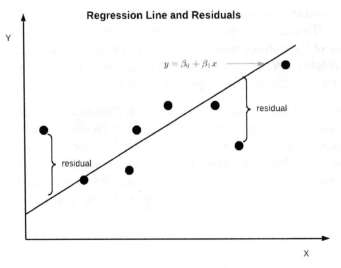

Figure 7.1 – Ordinary least squares regression line

The line, $y = \beta_0 + \beta_1 x$, gives us the predicted value of $y$ for each value of $x$. It is equal to the estimated intercept, $\beta_0$, plus the estimated coefficient for the feature times the feature value, $\beta_1 x$. This is the OLS line. Any other straight line through the data would result in a higher sum of squared residuals. This can be extended to multiple linear regression models – that is, those with more than one feature:

$$y = \beta_0 + \beta_1 x_1 + \beta_2 x_2 + \dots + \beta_n x_n + \varepsilon$$

Here, $y$ is the target, each $x$ is a feature, each $\beta$ is a coefficient (or the intercept), $n$ is the number of features, and $\varepsilon$ is an error term. Each coefficient is the estimated change in the target from a 1-unit change in the associated feature. This is a good place to notice that the coefficient is constant across the whole range of each feature; that is, an increase in the feature from 0 to 1 is assumed to have the same impact on the target as from 999 to 1000. However, this does not always make sense. Later in this chapter, we will discuss how to use transformations when the relationship between a feature and the target is not linear.

An important advantage of linear regression is that it is not as computationally expensive as other supervised regression algorithms. When linear regression performs well, based on metrics such as those we discussed in the previous chapter, it is a good choice. This is particularly true when you have large amounts of data to train or your business process does not permit large blocks of time for model training. The efficiency of the algorithm can also make it feasible to use more resource-intensive feature selection techniques, such as wrapper methods, which we discussed in *Chapter 5, Feature Selection*. As we saw there, you may not want to use exhaustive feature selection with a decision tree regressor. However, it may be perfectly fine with a linear regression model.

# Linear regression and gradient descent

We can use gradient descent, rather than ordinary least squares, to estimate our linear regression parameters. Gradient descent iterates over possible coefficient values to find those that minimize the residual sum of squares. It starts with random coefficient values and calculates the sum of squared errors for that iteration. Then, it generates new values for coefficients that yield smaller residuals than those from the previous step. We specify a learning rate when using gradient descent. The learning rate determines the amount of improvement in residuals at each step.

Gradient descent can often be a good choice when working with very large datasets. It may be the only choice if the full dataset does not fit into your machine's memory. We will use both OLS and gradient descent to estimate our parameters in the next section.

# Using classical linear regression

In this section, we will specify a fairly straightforward linear model. We will use it to predict the implied gasoline tax of a country based on several national economic and political measures. But before we specify our model, we need to do the pre-processing tasks we discussed in the first few chapters of this book.

## Pre-processing the data for our regression model

We will use pipelines to pre-process our data in this chapter, and throughout the rest of this book. We need to impute values where they are missing, identify and handle outliers, and encode and scale our data. We also need to do this in a way that avoids data leakage and cleans the training data without peeking ahead to the testing data. As we saw in *Chapter 6*, *Preparing for Model Evaluation*, scikit-learn's pipelines can help with these tasks.

The dataset we will use contains the implied gasoline tax for each country and some possible predictors, including national income per capita, government debt, fuel income dependency, extent of car use, and measures of democratic processes and government effectiveness.

> **Note**
>
> This dataset on implied gasoline tax by country is available for public use on the Harvard Dataverse at `https://dataverse.harvard.edu/dataset.xhtml?persistentId=doi:10.7910/DVN/RX4JGK`. It was compiled by *Paasha Mahdavi*, *Cesar B. Martinez-Alvarez*, and *Michael L. Ross*. The implied gasoline tax is calculated based on the difference between the world benchmark price and the local price for a liter of gas. A local price above the benchmark price represents a tax. When the benchmark price is higher, it can be considered a subsidy. We will use 2014 data for each country for this analysis.

Let's start by pre-processing the data:

1.  First, we load many of the libraries we worked with in the last few chapters. However, we also need two new libraries to build the pipeline for our data – `ColumnTransformer` and `TransformedTargetRegressor`. These libraries allow us to build a pipeline that does different pre-processing on numerical and categorical features, and that also transforms our target:

    ```
    import pandas as pd
    import numpy as np
    from sklearn.model_selection import train_test_split
    from sklearn.preprocessing import StandardScaler
    from sklearn.linear_model import LinearRegression
    ```

```
from sklearn.impute import SimpleImputer
from sklearn.pipeline import make_pipeline
from sklearn.compose import ColumnTransformer
from sklearn.compose import TransformedTargetRegressor
from sklearn.feature_selection import RFE
from sklearn.impute import KNNImputer
from sklearn.model_selection import cross_validate, KFold
import sklearn.metrics as skmet
import matplotlib.pyplot as plt
```

2.  We can extend the functionality of a scikit-learn pipeline by adding our own classes. Let's add a class to handle extreme values called OutlierTrans.

To include this class in a pipeline, it must inherit from the BaseEstimator class. We must also inherit from TransformerMixin, though there are other possibilities. Our class needs the fit and transform methods. We can put code for assigning extreme values as missing in the transform method.

But before we can use our class, we need to import it. To import it, we need to append the helperfunctions subfolder, since that is where we have placed the preprocfunc module that contains our class:

```
import os
import sys
sys.path.append(os.getcwd() + "/helperfunctions")
from preprocfunc import OutlierTrans
```

This imports the OutlierTrans class, which we can add to the pipelines we create:

```
class OutlierTrans(BaseEstimator,TransformerMixin):
  def __init__(self,threshold=1.5):
    self.threshold = threshold

  def fit(self,X,y=None):
    return self

  def transform(self,X,y=None):
    Xnew = X.copy()
    for col in Xnew.columns:
      thirdq, firstq = Xnew[col].quantile(0.75),\
        Xnew[col].quantile(0.25)
```

```
         interquartilerange = self.threshold*(thirdq-firstq)
         outlierhigh, outlierlow =
    interquartilerange+thirdq,\
         firstq-interquartilerange
      Xnew.loc[(Xnew[col]>outlierhigh) | \
         (Xnew[col]<outlierlow),col] = np.nan
   return Xnew.values
```

The OutlierTrans class uses a fairly standard univariate approach for identifying an outlier. It calculates the **interquartile range (IQR)** for each feature, then sets any value that is more than 1.5 times the IQR above the third quartile or below the first quartile to missing. We can change the threshold to something other than 1.5, such as 2.0, if we want to be more conservative. (We discussed this technique for identifying outliers in *Chapter 1, Examining the Distribution of Features and Targets*.)

3.  Next, we load the gasoline tax data for 2014. There are 154 rows – one for each country in the DataFrame. A few features have some missing values, but only one, motorization_rate, has double-digit missings. motorization_rate is the number of cars per person:

```
fftaxrate14 = pd.read_csv("data/fossilfueltaxrate14.csv")
fftaxrate14.set_index('countrycode', inplace=True)
fftaxrate14.info()
<class 'pandas.core.frame.DataFrame'>
Index: 154 entries, AFG to ZWE
Data columns (total 19 columns):
```

| # | Column | Non-Null Count | Dtype |
|---|--------|----------------|-------|
| 0 | country | 154 non-null | object |
| 1 | region | 154 non-null | object |
| 2 | region_wb | 154 non-null | object |
| 3 | year | 154 non-null | int64 |
| 4 | gas_tax_imp | 154 non-null | float64 |
| 5 | bmgap_diesel_spotprice_la | 146 non-null | float64 |
| 6 | fuel_income_dependence | 152 non-null | float64 |
| 7 | national_income_per_cap | 152 non-null | float64 |
| 8 | VAT_Rate | 151 non-null | float64 |
| 9 | gov_debt_per_gdp | 139 non-null | float64 |
| 10 | polity | 151 non-null | float64 |

```
11   democracy_polity        151 non-null    float64
12   autocracy_polity        151 non-null    float64
13   goveffect               154 non-null    float64
14   democracy_index         152 non-null    float64
15   democracy               154 non-null    int64
16   nat_oil_comp            152 non-null    float64
17   nat_oil_comp_state      152 non-null    float64
18   motorization_rate       127 non-null    float64
dtypes: float64(14), int64(2), object(3)
memory usage: 24.1+ KB
```

4.  Let's separate the features into numerical and binary features. We will put
    `motorization_rate` into a special category because we anticipate having to do a
    little more with it than with the other features:

    ```
    num_cols = ['fuel_income_dependence',
      'national_income_per_cap', 'VAT_Rate',
      'gov_debt_per_gdp', 'polity', 'goveffect',
      'democracy_index']
    dummy_cols = 'democracy_polity','autocracy_polity',
      'democracy','nat_oil_comp','nat_oil_comp_state']
    spec_cols = ['motorization_rate']
    ```

5.  We should look at some descriptives for the numeric features and the target. Our
    target, `gas_tax_imp`, has a median value of 0.52. Notice that some of the features
    have a very different range. More than half of the countries have a `polity` score
    of 7 or higher; 10 is the highest possible `polity` score, meaning most democratic.
    Most countries have a negative value for government effectiveness. `democracy_`
    `index` is a very similar measure to `polity`, though there is more variation:

    ```
    fftaxrate14[['gas_tax_imp'] + num_cols + spec_cols].\
      agg(['count','min','median','max']).T
    ```

    |                         | count | min     | median   | max        |
    | ----------------------- | ----- | ------- | -------- | ---------- |
    | gas_tax_imp             | 154   | -0.80   | 0.52     | 1.73       |
    | fuel_income_dependence  | 152   | 0.00    | 0.14     | 34.43      |
    | national_income_per_cap | 152   | 260.00  | 6,050.00 | 104,540.00 |
    | VAT_Rate                | 151   | 0.00    | 16.50    | 27.00      |
    | gov_debt_per_gdp        | 139   | 0.55    | 39.30    | 194.76     |
    | polity                  | 151   | -10.00  | 7.00     | 10.00      |

| goveffect | 154 | -2.04 | -0.15 | 2.18 |
| democracy_index | 152 | 0.03 | 0.57 | 0.93 |
| motorization_rate | 127 | 0.00 | 0.20 | 0.81 |

6.  Let's also look at the distribution of the binary features. We must set `normalize` to `True` to generate ratios rather than counts. The `democracy_polity` and `autocracy_polity` features are just binarized versions of the `polity` feature; very high `polity` scores get `democracy_polity` values of 1, while very low `polity` scores get `autocracy_polity` values of 1. Similarly, `democracy` is a dummy feature for those countries with high `democracy_index` values. Interestingly, nearly half of the countries (0.46) have a national oil company, and almost a quarter (0.23) have a state-owned national oil company:

```
fftaxrate14[dummy_cols].apply(pd.value_counts,
normalize=True).T
```

|  | 0 | 1 |
| --- | --- | --- |
| democracy_polity | 0.41 | 0.59 |
| autocracy_polity | 0.89 | 0.11 |
| democracy | 0.42 | 0.58 |
| nat_oil_comp | 0.54 | 0.46 |
| nat_oil_comp_state | 0.77 | 0.23 |

This all looks to be in fairly good shape. However, we will need to do some work on the missing values for several features. We also need to do some scaling, but there is no need to do any encoding because we can use the binary features as they are. Some features are correlated, so we need to do some feature elimination.

7.  We begin our pre-processing by creating training and testing DataFrames. We will only reserve 20% for testing:

```
target = fftaxrate14[['gas_tax_imp']]
features = fftaxrate14[num_cols + dummy_cols + spec_cols]

X_train, X_test, y_train, y_test =  \
  train_test_split(features,\
   target, test_size=0.2, random_state=0)
```

8.  We need to build a pipeline with column transformations so that we can do different pre-processing on numeric and categorical data. We will construct a pipeline, standtrans, for all of the numeric columns in num_cols. First, we want to set outliers to missing. We will define an outlier value as one that is more than two times the interquartile range above the third quartile, or below the first quartile, for that feature. We will use SimpleImputer to set missing values to the median.

We do not want to scale the binary features in dummy_cols, but we do want to use SimpleImputer to set missing values to the most frequent value for each categorical feature.

We won't use SimpleImputer for motorization_rate. Remember that motorization_rate is not in the num_cols list – it is in the spec_cols list. We set up a special pipeline, spectrans, for features in spec_cols. We will use **K-Nearest Neighbor (KNN)** imputation later to handle missing motorization_rate values:

```
standtrans = make_pipeline(OutlierTrans(2),
    SimpleImputer(strategy="median"), StandardScaler())
cattrans = make_pipeline(
    SimpleImputer(strategy="most_frequent"))
spectrans = make_pipeline(OutlierTrans(2),
    StandardScaler())
coltrans = ColumnTransformer(
    transformers=[
        ("stand", standtrans, num_cols),
        ("cat", cattrans, dummy_cols),
        ("spec", spectrans, spec_cols)
    ]
)
```

This sets up all of the pre-processing we want to do on the gasoline tax data. To do the transformations, all we need to do is call the fit method of the column transformer. However, we will not do that yet because we also want to add feature selection to the pipeline and get it to run a linear regression. We will do that in the next few steps.

# Running and evaluating our linear model

We will use **recursive feature elimination** (**RFE**) to select features for our model. RFE has the advantages of wrapper feature selection methods – it evaluates features based on a selected algorithm, and it considers multivariate relationships in that assessment. However, it can also be computationally expensive. Since we do not have many features or observations, that is not much of a problem in this case.

After selecting the features, we run a linear regression model and take a look at our predictions. Let's get started:

1. First, we create linear regression and recursive feature elimination instances and add them to the pipeline. We also create a `TransformedTargetRegressor` object since we still need to transform the target. We pass our pipeline to the regressor parameter of `TransformedTargetRegressor`.

   Now, we can call the target regressor's `fit` method. After that, the `support_` attribute of the pipeline's `rfe` step will give us the selected features. Similarly, we can get the coefficients by getting the `coef_` value of the `linearregression` step. The key here is that referencing `ttr.regressor` gets us to the pipeline:

   ```
   lr = LinearRegression()
   rfe = RFE(estimator=lr, n_features_to_select=7)

   pipe1 = make_pipeline(coltrans,
     KNNImputer(n_neighbors=5), rfe, lr)

   ttr=TransformedTargetRegressor(regressor=pipe1,
     transformer=StandardScaler())
   ttr.fit(X_train, y_train)

   selcols = X_train.columns[ttr.regressor_.named_
   steps['rfe'].support_]
   coefs = ttr.regressor_.named_steps['linearregression'].
   coef_

   np.column_stack((coefs.ravel(),selcols))

   array([[0.44753064726665703, 'VAT_Rate'],
           [0.12368913577287821, 'gov_debt_per_gdp'],
           [0.17926454403985687, 'goveffect'],
   ```

```
       [-0.22100930246392841, 'autocracy_polity'],
       [-0.15726572731003752, 'nat_oil_comp'],
       [-0.7013454686632653, 'nat_oil_comp_state'],
       [0.13855012574945422, 'motorization_rate']],
  dtype=object)
```

Our feature selection identified the VAT rate, government debt, a measure of government effectiveness (goveffect), whether the country is in the autocracy category, whether there is a national oil company and one that is state-owned, and the motorization rate as the top seven features. The number of features indicated is an example of a hyperparameter, and our choice of seven here is fairly arbitrary. We will discuss techniques for hyperparameter tuning in the next section.

Notice that of the several autocracy/democracy measures in the dataset, the one that seems to matter most is the autocracy dummy, which has a value of 1 for countries with very low polity scores. It is estimated to have a negative effect gasoline taxes; that is, to reduce them.

2.  Let's take a look at the predictions and the residuals. We can pass the features from the testing data to the transformer's/pipeline's predict method to generate the predictions. There is a little positive skew and some overall bias; the residuals are negative overall:

```
pred = ttr.predict(X_test)


preddf = pd.DataFrame(pred, columns=['prediction'],
   index=X_test.index).join(X_test).join(y_test)


preddf['resid'] = preddf.gas_tax_imp-preddf.prediction


preddf.resid.agg(['mean','median','skew','kurtosis'])


mean                    -0.09
median                  -0.13
skew                     0.61
kurtosis                 0.04
Name: resid, dtype: float64
```

3.  Let's also generate some overall model evaluation statistics. We get a mean absolute error of 0.23. That's not a great average error, given that the median value for the gas tax price is 0.52. The r-squared is decent, however:

```
print("Mean Absolute Error: %.2f, R-squared: %.2f" %
    (skmet.mean_absolute_error(y_test, pred),
    skmet.r2_score(y_test, pred)))
```

**Mean Absolute Error: 0.23, R-squared: 0.75**

4.  It is usually helpful to look at a plot of the residuals. Let's also draw a red dashed line at the average value of the residuals:

```
plt.hist(preddf.resid, color="blue", bins=np.arange(-
0.5,1.0,0.25))
plt.axvline(preddf.resid.mean(), color='red',
linestyle='dashed', linewidth=1)
plt.title("Histogram of Residuals for Gax Tax Model")
plt.xlabel("Residuals")
plt.ylabel("Frequency")
plt.xlim()
plt.show()
```

This produces the following plot:

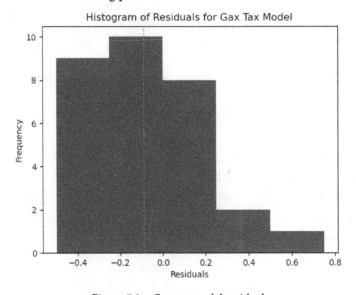

Figure 7.2 – Gas tax model residuals

This plot shows the positive skew. Moreover, our model is somewhat more likely to over-predict the gas tax than under-predict it. (The residual is negative when the prediction is greater than the actual target value.)

5.  Let's also look at a scatterplot of the predictions against the residuals. Let's draw a red dashed line at 0 on the *Y*-axis:

```
plt.scatter(preddf.prediction, preddf.resid,
color="blue")
plt.axhline(0, color='red', linestyle='dashed',
linewidth=1)
plt.title("Scatterplot of Predictions and Residuals")
plt.xlabel("Predicted Gax Tax")
plt.ylabel("Residuals")
plt.show()
```

This produces the following plot:

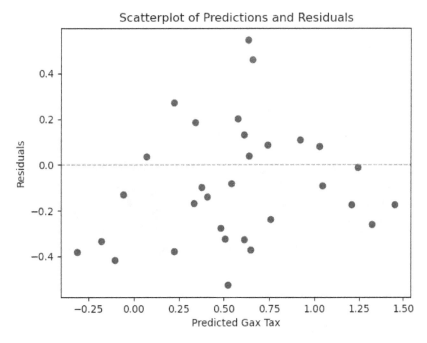

Figure 7.3 – Scatter plot of predictions against residuals

Here, overprediction occurs throughout the range of predicted values, but there are no underpredictions (positive residuals) with predictions below 0 or above 1. This should give us some doubts about our assumption of linearity.

# Improving our model evaluation

One problem with how we have evaluated our model so far is that we are not making great use of the data. We are only training on about 80% of the data. Our metrics are also quite dependent on the testing data being representative of the real world we want to predict. However, it might not be. We can improve our odds with k-fold cross-validation, as we discussed in the previous chapter.

Since we have been using pipelines for our analysis, we have already done much of the work we need for k-fold cross-validation. Recall from the previous chapter that the k-fold model evaluation divides our data into k equal parts. One of the folds is designated for testing and the rest for training. This is repeated k times, with a different fold being used for testing each time.

Let's try k-fold cross-validation with our linear regression model:

1.  We will start by creating new training and testing DataFrames, leaving just 10% for later validation. We do not need to retain as much data for validation, though it is a good idea to always hold a little back:

    ```
    X_train, X_test, y_train, y_test =  \
      train_test_split(features,\
      target, test_size=0.1, random_state=1)
    ```

2.  We also need to instantiate KFold and LinearRegression objects:

    ```
    kf = KFold(n_splits=3, shuffle=True, random_state=0)
    ```

3.  Now, we are ready to run our k-fold cross validation. We indicate that we want both r-squared and mean absolute error for each split. cross_validate automatically gives us fit and score times for each fold:

    ```
    scores = cross_validate(ttr, X=X_train, y=y_train,
      cv=kf, scoring=('r2', 'neg_mean_absolute_error'), n_
    jobs=1)

    print("Mean Absolute Error: %.2f, R-squared: %.2f" %
      (scores['test_neg_mean_absolute_error'].mean(),
      scores['test_r2'].mean()))

    Mean Absolute Error: -0.25, R-squared: 0.62
    ```

These scores are not very impressive. We do not end up explaining as much of the variance as we would like. R-squared scores average about 0.62 across the three folds. This is partly because the testing DataFrames of each fold are quite small, with about 40 observations in each. Nonetheless, we should explore modifications of the classical linear regression approach, such as regularization and non-linear regression.

One advantage of regularization is that we may be able to get similar results without going through a computationally expensive feature selection process. Regularization can also help us avoid overfitting. We will explore lasso regression with the same data in the next section. We will also look into non-linear regression strategies.

# Using lasso regression

A key characteristic of OLS is that it produces the parameter estimates with the least bias. However, OLS estimates may have a higher variance than we want. We need to be careful about overfitting when we use a classical linear regression model. One strategy to reduce the likelihood of overfitting is to use regularization. Regularization may also allow us to combine feature selection and model training. This may matter for datasets with a large number of features or observations.

Whereas OLS minimizes mean squared error, regularization techniques seek both minimal error and a reduced number of features. Lasso regression, which we explore in this section, uses L1 regularization, which penalizes the absolute value of the coefficients. Ridge regression is similar. It uses L2 regularization, which penalizes the squared values of the coefficients. Elastic net regression uses both L1 and L2 regularization.

Once again, we will work with the gasoline tax data from the previous section:

1.  We will start by importing the same libraries as in the previous section, except we will import the Lasso module rather than the linearregression module:

```
import pandas as pd
import numpy as np
from sklearn.model_selection import train_test_split
from sklearn.preprocessing import StandardScaler
from sklearn.linear_model import Lasso
from sklearn.impute import SimpleImputer
from sklearn.pipeline import make_pipeline
from sklearn.compose import ColumnTransformer
from sklearn.compose import TransformedTargetRegressor
from sklearn.model_selection import cross_validate, KFold
```

```
import sklearn.metrics as skmet
import matplotlib.pyplot as plt
```

2. We will also need the `OutlierTrans` class that we created:

```
import os
import sys
sys.path.append(os.getcwd() + "/helperfunctions")
from preprocfunc import OutlierTrans
```

3. Now, let's load the gasoline tax data and create testing and training DataFrames:

```
fftaxrate14 = pd.read_csv("data/fossilfueltaxrate14.csv")
fftaxrate14.set_index('countrycode', inplace=True)

num_cols = ['fuel_income_dependence','national_income_
per_cap',
  'VAT_Rate',  'gov_debt_per_gdp','polity','goveffect',
  'democracy_index']
dummy_cols = ['democracy_polity','autocracy_polity',
'democracy','nat_oil_comp','nat_oil_comp_state']
spec_cols = ['motorization_rate']

target = fftaxrate14[['gas_tax_imp']]
features = fftaxrate14[num_cols + dummy_cols + spec_cols]

X_train, X_test, y_train, y_test =  \
  train_test_split(features,\
  target, test_size=0.2, random_state=0)
```

4. We also need to set up the column transformations:

```
standtrans = make_pipeline(
  OutlierTrans(2), SimpleImputer(strategy="median"),
  StandardScaler())
cattrans = make_pipeline(SimpleImputer(strategy="most_
frequent"))
spectrans = make_pipeline(OutlierTrans(2),
StandardScaler())
coltrans = ColumnTransformer(
```

```
    transformers=[
      ("stand", standtrans, num_cols),
      ("cat", cattrans, dummy_cols),
      ("spec", spectrans, spec_cols)
    ]
  )
```

5.  Now, we are ready to fit our model. We will start with a fairly conservative alpha of 0.1. The higher the alpha, the greater the penalties for our coefficients. At 0, we get the same results as with linear regression. In addition to column transformation and lasso regression, our pipeline uses KNN imputation for missing values. We will also use the target transformer to scale the gasoline tax target. We will pass the pipeline we just created to the regressor parameter of the target transformer before we fit it:

```
lasso = Lasso(alpha=0.1,fit_intercept=False)

pipe1 = make_pipeline(coltrans, KNNImputer(n_neigh-
bors=5), lasso)
ttr=TransformedTargetRegressor(regressor=pipe1,trans-
former=StandardScaler())
ttr.fit(X_train, y_train)
```

6.  Let's take a look at the coefficients from lasso regression. If we compare them to the coefficients from linear regression in the previous section, we notice that we end up selecting the same features. Those features that were eliminated with recursive feature selection are largely the same ones that get near zero values with lasso regression:

```
coefs = ttr.regressor_['lasso'].coef_

np.column_stack((coefs.ravel(), num_cols + dummy_cols +
spec_cols))
array([['-0.0026505240129231175', 'fuel_income_
dependence'],
       ['0.0', 'national_income_per_cap'],
       ['0.43472262042825915', 'VAT_Rate'],
       ['0.10927136643326674', 'gov_debt_per_gdp'],
       ['0.006825858127837494', 'polity'],
       ['0.15823493727828816', 'goveffect'],
       ['0.09622123660935211', 'democracy_index'],
       ['0.0', 'democracy_polity'],
```

```
        ['-0.0', 'autocracy_polity'],
        ['0.0', 'democracy'],
        ['-0.0', 'nat_oil_comp'],
        ['-0.2199638245781246', 'nat_oil_comp_state'],
        ['0.016680304258453165', 'motorization_rate']],
   dtype='<U32')
```

7. Let's look at the predictions and residuals for this model. The residuals look decent, with little bias and not much skew:

```
pred = ttr.predict(X_test)

preddf = pd.DataFrame(pred, columns=['prediction'],
    index=X_test.index).join(X_test).join(y_test)

preddf['resid'] = preddf.gas_tax_imp-preddf.prediction

preddf.resid.agg(['mean','median','skew','kurtosis'])

mean                    -0.06
median                  -0.07
skew                     0.33
kurtosis                 0.10
Name: resid, dtype: float64
```

8. Let's also generate the mean absolute error and r-squared. These are not impressive scores. The r-squared is lower than with linear regression, but the mean absolute error is about the same:

```
print("Mean Absolute Error: %.2f, R-squared: %.2f" %
    (skmet.mean_absolute_error(y_test, pred),
    skmet.r2_score(y_test, pred)))

Mean Absolute Error: 0.24, R-squared: 0.68
```

9.  We should look at a histogram of the residuals. The distribution of the residuals is quite similar to the linear regression model:

```
plt.hist(preddf.resid, color="blue", bins=np.arange(-
0.5,1.0,0.25))
plt.axvline(preddf.resid.mean(), color='red',
linestyle='dashed', linewidth=1)
plt.title("Histogram of Residuals for Gax Tax Model")
plt.xlabel("Residuals")
plt.ylabel("Frequency")
plt.show()
```

This produces the following plot:

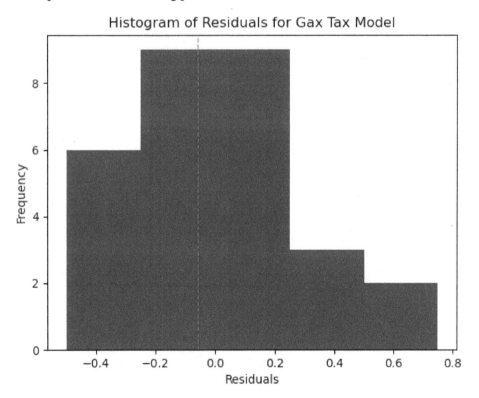

Figure 7.4 – Gas tax model residuals

10. Let's also look at a scatter plot of the predicted values on the residuals. Our model is likely to over-predict at the lower ranges and under-predict at the upper ranges. This is a change from the linear model, where we consistently over predicted at both extremes:

```
plt.scatter(preddf.prediction, preddf.resid,
color="blue")
plt.axhline(0, color='red', linestyle='dashed',
linewidth=1)
plt.title("Scatterplot of Predictions and Residuals")
plt.xlabel("Predicted Gax Tax")
plt.ylabel("Residuals")
plt.show()
```

This produces the following plot:

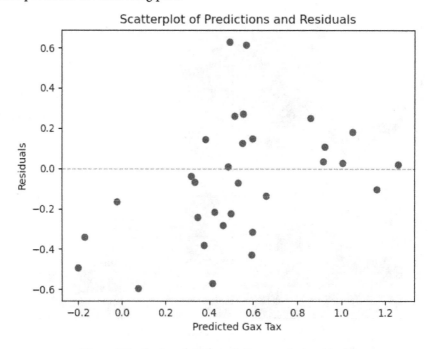

Figure 7.5 – Scatter plot of predictions against residuals

11. We will conclude by performing k-fold cross-validation on the model. The scores are lower than but close to those of the linear regression model:

```
X_train, X_test, y_train, y_test = \
    train_test_split(features,\
```

```
    target, test_size=0.1, random_state=22)

kf = KFold(n_splits=4, shuffle=True, random_state=0)

scores = cross_validate(ttr, X=X_train, y=y_train,
    cv=kf, scoring=('r2', 'neg_mean_absolute_error'), n_
jobs=1)

print("Mean Absolute Error: %.2f, R-squared: %.2f" %
    (scores['test_neg_mean_absolute_error'].mean(),
    scores['test_r2'].mean()))
```

**Mean Absolute Error: -0.27, R-squared: 0.57**

This gives us a model that is not any better than our original model, but it at least handles the feature selection process more efficiently. It is also possible that we could get better results if we tried different values for the alpha hyperparameter. Why not 0.05 or 1.0 instead? We will try to answer that in the next two steps.

## Tuning hyperparameters with grid searches

Figuring out the best value for a hyperparameter, such as the alpha value in the previous example, is known as hyperparameter tuning. One tool in scikit-learn for hyperparameter tuning is GridSearchCV. The CV suffix is for cross-validate.

Using GridSearchCV is very straightforward. If we already have a pipeline, as we do in this case, we pass it to a GridSearchCV object, along with a dictionary of parameters. GridSearchCV will try all combinations of parameters and return the best one. Let's try it on our lasso regression model:

1.  First, we will instantiate a lasso object and create a dictionary with the hyperparameters to be tuned. The dictionary, lasso_params, indicates that we want to try all the alpha values between 0.05 and 0.9 at 0.5 intervals. We cannot choose any name we want for the dictionary key. regressor__lasso__alpha is based on the names of the steps in the pipeline. Also, notice that we are using double underscores. Single underscores will return an error:

    ```
    lasso = Lasso()

    lasso_params = {'regressor__lasso__alpha':
    np.arange(0.05, 1, 0.05)}
    ```

2.  Now, we can run the grid search. We will pass the pipeline, which is a `TransformedTargetRegressor` in this case, and the dictionary to `GridSearchCV`. The `best_params_` attribute indicates that the best alpha is `0.05`. When we use that value, we get an r-squared of `0.60`:

```
gs = GridSearchCV(ttr,param_grid=lasso_params, cv=5)

gs.fit(X_train, y_train)

gs.best_params_
{'regressor__lasso__alpha': 0.05}
gs.best_score_
0.6028804486340877
```

The lasso regression model comes close to but does not do quite as well as the linear model in terms of mean absolute error and r-squared. One benefit of lasso regression is that we do not need to do a separate feature selection step before training our model. (Recall that for wrapper feature selection methods, the model needs to be trained during feature selection as well as after, as we discussed in *Chapter 5, Feature Selection.*)

# Using non-linear regression

Linear regression assumes that the relationship of a feature to the target is constant across the range of the feature. You may recall that the simple linear regression equation that we discussed at the beginning of this chapter had one slope estimate for each feature:

$$y = \beta_0 + \beta_1 x$$

Here, $y$ is the target, each $x$ is a feature, and each $\beta$ is a coefficient (or the intercept). If the true relationships between features and targets are nonlinear, our model will likely perform poorly.

Fortunately, we can still make good use of OLS when we cannot assume a linear relationship between the features and the target. We can use the same linear regression algorithm that we used in the previous section, but with a polynomial transformation of the features. This is referred to as polynomial regression.

We add a power to the feature to run a polynomial regression. This gives us the following equation:

$$y = \beta_0 + \beta_1 x + \beta_2 x^2 + \beta_3 x^3 + \dots + \beta_n x^n + \varepsilon$$

The following plot compares predicted values for linear versus polynomial regression. The polynomial curve seems to fit the fictional data points better than the linear regression line:

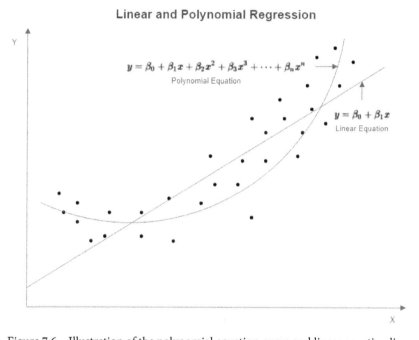

Figure 7.6 – Illustration of the polynomial equation curve and linear equation line

In this section, we will experiment with both a linear model and a non-linear model of average annual temperatures at weather stations across the world. We will use latitude and elevation as features. First, we will predict temperature using multiple linear regression, and then try a model with polynomial transformations. Follow these steps:

1.  We will start by importing the necessary libraries. These libraries will be familiar if you have been working through this chapter:

```
import pandas as pd
from sklearn.model_selection import train_test_split
from sklearn.preprocessing import StandardScaler,
PolynomialFeatures
from sklearn.linear_model import LinearRegression
from sklearn.pipeline import make_pipeline
from sklearn.model_selection import cross_validate
from sklearn.model_selection import KFold
from sklearn.impute import KNNImputer
import matplotlib.pyplot as plt
```

2.  We also need to import the module that contains our class for identifying outlier values:

```
import os
import sys
sys.path.append(os.getcwd() + "/helperfunctions")
from preprocfunc import OutlierTrans
```

3.  We load the land temperature data, identify the features we want, and generate some descriptive statistics. There are a few missing values for elevation and some extreme negative values for average annual temperature. The range of values for the target and features is very different, so we will probably want to scale our data:

```
landtemps = pd.read_csv("data/landtempsb2019avgs.csv")
landtemps.set_index('locationid', inplace=True)


feature_cols = ['latabs','elevation']


landtemps[['avgtemp'] + feature_cols].\
  agg(['count','min','median','max']).T
```

|           | count  | min     | median | max      |
|-----------|--------|---------|--------|----------|
| avgtemp   | 12,095 | -60.82  | 10.45  | 33.93    |
| latabs    | 12,095 | 0.02    | 40.67  | 90.00    |
| elevation | 12,088 | -350.00 | 271.30 | 4,701.00 |

4.  Next, we create the training and testing DataFrames:

```
X_train, X_test, y_train, y_test =  \
  train_test_split(landtemps[feature_cols],\
  landtemps[['avgtemp']], test_size=0.1, random_state=0)
```

5.  Now, we build a pipeline to handle our pre-processing – set outlier values to missing, do KNN imputation for all missing values, and scale the features – and then run a linear model. We do k-fold cross-validation with 10 folds and get an average r-squared of 0.79 and a mean absolute error of -2.8:

```
lr = LinearRegression()


knnimp = KNNImputer(n_neighbors=45)


pipe1 = make_pipeline(OutlierTrans(3),knnimp,
```

```
    StandardScaler(), lr)


  ttr=TransformedTargetRegressor(regressor=pipe1,
    transformer=StandardScaler())


  kf = KFold(n_splits=10, shuffle=True, random_state=0)


  scores = cross_validate(ttr, X=X_train, y=y_train,
    cv=kf, scoring=('r2', 'neg_mean_absolute_error'), n_
  jobs=1)


  scores['test_r2'].mean(), scores['test_neg_mean_absolute_
  error'].mean()
```

**(0.7933471824738406, -2.8047627785750913)**

Notice that we are quite conservative with our identification of outliers. We pass a threshold value of 3, meaning that a value needs to be more than three times the interquartile range above or below that range. Obviously, we would typically give much more thought to the identification of outliers. We demonstrate here only how to handle outliers in a pipeline once we have decided that that makes sense.

6.  Let's see the predictions and the residuals. There is almost no bias overall (the average of the residuals is 0), but there is some negative skew:

```
  ttr.fit(X_train, y_train)


  pred = ttr.predict(X_test)


  preddf = pd.DataFrame(pred, columns=['prediction'],
    index=X_test.index).join(X_test).join(y_test)


  preddf.resid.agg(['mean','median','skew','kurtosis'])
  mean                    0.00
  median                  0.50
  skew                   -1.13
  kurtosis                3.48
  Name: resid, dtype: float64
```

7.  It is easy to see this skew if we create a histogram of the residuals. There are some extreme negative residuals – that is, times when we over-predict the average temperature by a lot:

```
plt.hist(preddf.resid, color="blue")
plt.axvline(preddf.resid.mean(), color='red',
    linestyle='dashed', linewidth=1)
plt.title("Histogram of Residuals for Linear Model of
Temperature")
plt.xlabel("Residuals")
plt.ylabel("Frequency")
plt.show()
```

This produces the following plot:

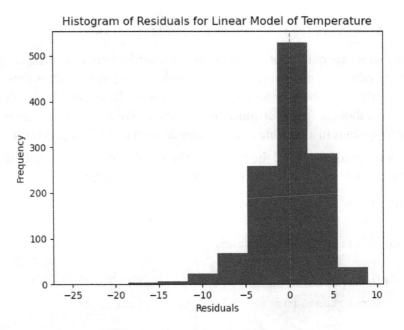

Figure 7.7 – Temperature model residuals

8.  It can also be helpful to plot the predicted values against the residuals:

```
plt.scatter(preddf.prediction, preddf.resid,
color="blue")
plt.axhline(0, color='red', linestyle='dashed',
linewidth=1)
plt.title("Scatterplot of Predictions and Residuals")
```

```
plt.xlabel("Predicted Temperature")
plt.ylabel("Residuals")
plt.xlim(-20,40)
plt.ylim(-27,10)
plt.show()
```

This produces the following plot:

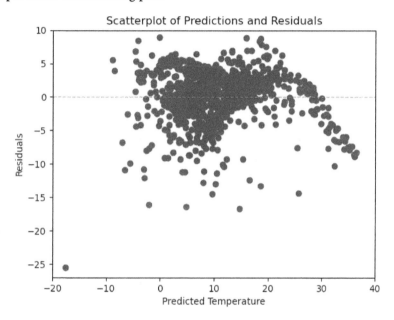

Figure 7.8 – Scatter plot of predictions against residuals

Our model over-predicts all predictions above approximately 28 degrees Celsius. It is also likely to underpredict when it predicts values between 18 and 28.

Let's see if we get any better results with polynomial regression:

9. First, we will create a `PolynomialFeatures` object with a `degree` of 4 and fit it. We can pass the original feature names to the `get_feature_names` method to get the names of the columns that will be returned after the transformation. Second, third, and fourth power values of each feature are created, as well as interaction effects between variables (such as `latabs * elevation`). We do not need to run the fit here since that will happen in the pipeline. We are just doing it here to get a feel for how it works:

```
polytrans = PolynomialFeatures(degree=4, include_
bias=False)
polytrans.fit(X_train.dropna())
```

```
featurenames = polytrans.get_feature_names(feature_cols)
featurenames
['latabs',
 'elevation',
 'latabs^2',
 'latabs elevation',
 'elevation^2',
 'latabs^3',
 'latabs^2 elevation',
 'latabs elevation^2',
 'elevation^3',
 'latabs^4',
 'latabs^3 elevation',
 'latabs^2 elevation^2',
 'latabs elevation^3',
 'elevation^4']
```

10. Next, we will create a pipeline for our polynomial regression. The pipeline is pretty much the same as with linear regression, except we insert the polynomial transformation step after the KNN imputation:

```
pipe2 = make_pipeline(OutlierTrans(3), knnimp,
    polytrans, StandardScaler(), lr)


ttr2 = TransformedTargetRegressor(regressor=pipe2,\
    transformer=StandardScaler())
```

11. Now, let's create predictions and residuals based on the polynomial model. There is a little less skew in the residuals than with the linear model:

```
ttr2.fit(X_train, y_train)


pred = ttr2.predict(X_test)


preddf = pd.DataFrame(pred, columns=['prediction'],
    index=X_test.index).join(X_test).join(y_test)


preddf['resid'] = preddf.avgtemp-preddf.prediction
```

```
preddf.resid.agg(['mean','median','skew','kurtosis'])
mean            0.01
median          0.20
skew           -0.98
kurtosis        3.34
Name: resid, dtype: float64
```

12. We should take a look at a histogram of the residuals:

```
plt.hist(preddf.resid, color="blue")
plt.axvline(preddf.resid.mean(), color='red',
linestyle='dashed', linewidth=1)
plt.title("Histogram of Residuals for Temperature Model")
plt.xlabel("Residuals")
plt.ylabel("Frequency")
plt.show()
```

This produces the following plot:

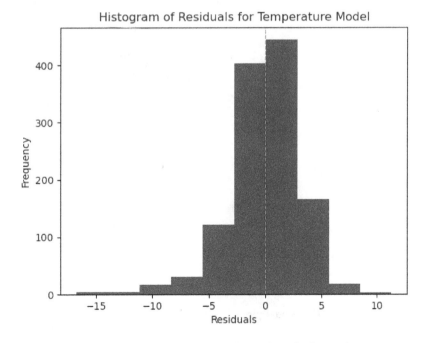

Figure 7.9 – Temperature model residuals

13. Let's also do another scatter plot of predicted values against residuals. These look a little bit better than the residuals with the linear model, particularly at the upper ranges of the predictions:

```
plt.scatter(preddf.prediction, preddf.resid,
color="blue")
plt.axhline(0, color='red', linestyle='dashed',
linewidth=1)
plt.title("Scatterplot of Predictions and Residuals")
plt.xlabel("Predicted Temperature")
plt.ylabel("Residuals")
plt.xlim(-20,40)
plt.ylim(-27,10)
plt.show()
```

This produces the following plot:

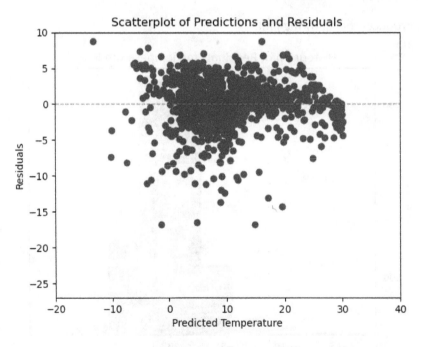

Figure 7.10 – Scatter plot of predictions against residuals

14. Let's do k-fold cross-validation once again and take the average r-squared value across the folds. There are improvements in both the r-squared and mean absolute error compared to the linear model:

```
scores = cross_validate(ttr2, X=X_train, y=y_train,
  cv=kf, scoring=('r2', 'neg_mean_absolute_error'),
  n_jobs=1)

scores['test_r2'].mean(), scores['test_neg_mean_absolute_
error'].mean()
```

```
(0.8323274036342788, -2.4035803290965507)
```

The polynomial transformation improved our overall results, particularly within certain ranges of our predictors. The residuals at high temperatures were noticeably lower. It is often a good idea to try a polynomial transformation when our residuals suggest that there might be a nonlinear relationship between our features and our target.

# Regression with gradient descent

Gradient descent can be a good alternative to ordinary least squares for optimizing the loss function of a linear model. This is particularly true when working with very large datasets. In this section, we will use gradient descent with the land temperatures dataset, mainly to demonstrate how to use it and to give us another opportunity to explore exhaustive grid searches. Let's get started:

1. First, we will load the same libraries we have been working with so far, plus the stochastic gradient descent regressor from scikit-learn:

```
import pandas as pd
import numpy as np
from sklearn.model_selection import train_test_split
from sklearn.preprocessing import StandardScaler
from sklearn.linear_model import SGDRegressor
from sklearn.compose import TransformedTargetRegressor
from sklearn.pipeline import make_pipeline
from sklearn.impute import KNNImputer
from sklearn.model_selection import GridSearchCV
import os
import sys
```

```
sys.path.append(os.getcwd() + "/helperfunctions")
from preprocfunc import OutlierTrans
```

2. Then, we will load the land temperatures dataset again and create training and testing DataFrames:

```
landtemps = pd.read_csv("data/landtempsb2019avgs.csv")
landtemps.set_index('locationid', inplace=True)


feature_cols = ['latabs','elevation']


X_train, X_test, y_train, y_test =  \
  train_test_split(landtemps[feature_cols],\
  landtemps[['avgtemp']], test_size=0.1, random_state=0)
```

3. Next, we will set up a pipeline to deal with outliers, impute values for missings, and scale the data before running the gradient descent:

```
knnimp = KNNImputer(n_neighbors=45)


sgdr = SGDRegressor()


pipe1 = make_pipeline(OutlierTrans(3),knnimp,Standard-
Scaler(), sgdr)


ttr=TransformedTargetRegressor(regressor=pipe1,trans-
former=StandardScaler())
```

4. Now, we need to create a dictionary to indicate the hyperparameters we want to tune and the values to try. We want to try values for alpha, the loss function, epsilon, and the penalty. We will give each key in the dictionary a prefix of `regressor__` `sgdregressor__` because that is where the stochastic gradient descent regressor can be found in the pipeline.

The `alpha` parameter determines the size of the penalty. The default is `0.0001`. We can choose L1, L2, or elastic net regularization. We will select `huber` and `epsilon_` `insensitive` as loss functions to include in the search. The default loss function is `squared_error`, but that would just give us ordinary least squares again.

The `huber` loss function is less sensitive to outliers than is OLS. How sensitive it is, is based on the value of epsilon we specify. With the `epsilon_insensitive` loss function, errors within a given range (epsilon) are not penalized (we will construct models with epsilon-insensitive tubes in the next chapter, where we'll examine support vector regression):

```
sgdr_params = {
 'regressor__sgdregressor__alpha': 10.0 ** -np.arange(1,
7),
 'regressor__sgdregressor__loss': ['huber','epsilon_
insensitive'],
 'regressor__sgdregressor__penalty': ['l2', 'l1',
'elasticnet'],
 'regressor__sgdregressor__epsilon': np.arange(0.1, 1.6,
0.1)
}
```

5.  Now, we are ready to run an exhaustive grid search. The best parameters are `0.001` for alpha, `1.3` for epsilon, `huber` for the loss function, and elastic net regularization:

```
gs = GridSearchCV(ttr,param_grid=sgdr_params, cv=5,
scoring="r2")


gs.fit(X_train, y_train)


gs.best_params_
{'regressor__sgdregressor__alpha': 0.001,
 'regressor__sgdregressor__epsilon': 1.3000000000000003,
 'regressor__sgdregressor__loss': 'huber',
 'regressor__sgdregressor__penalty': 'elasticnet'}


gs.best_score_
0.7941051735846133
```

6.  I usually find it helpful to look at the hyperparameters for some of the other grid search iterations. Huber loss models, with either elastic net or L2 regularization, perform best:

```
Results = \
  pd.DataFrame(gs.cv_results_['mean_test_score'], \
    columns=['meanscore']).\
```

```
      join(pd.DataFrame(gs.cv_results_['params'])).\
      sort_values(['meanscore'], ascending=False)


results.head(3).T
```

|                                      |      | 254      | 25       |
| ------------------------------------ | ---- | -------- | -------- |
|                                      | 2    | 534      |
| meanscore                            |      | 0.794105 | 0.794011 |
|                                      | 0.794009 |      |          |
| regressor__sgdregressor__alpha       |      | 0.001000 | 0.001000 |
|                                      | 0.000001 |      |          |
| regressor__sgdregressor__epsilon     |      | 1.300000 | 1.300000 |
|                                      | 1.500000 |      |          |
| regressor__sgdregressor__loss        |      | huber    | huber    |
|                                      | huber |      |          |
| regressor__sgdregressor__penalty     | elastic- |      |          |
|                                      | net  | 12       | 12       |

Stochastic gradient descent is a generalized approach to optimization and can be applied to many machine learning problems. It is often quite efficient, as we can see here. We were able to run an exhaustive grid search on penalty, penalty size, epsilon, and loss function fairly quickly.

# Summary

This chapter allowed us to explore a very well-known machine learning algorithm: linear regression. We examined the qualities of a feature space that makes it a good candidate for a linear model. We also explored how to improve a linear model, when necessary, with regularization and with transformations. Then, we looked at stochastic gradient descent as an alternative to OLS optimization. We also learned how to add our own classes to a pipeline and how to do hyperparameter tuning.

In the next chapter, we will explore support vector regression.

# 8
# Support Vector Regression

**Support vector regression** (**SVR**) can be an excellent option when the assumptions of linear regression models do not hold, such as when the relationship between our features and our target is too complicated to be described by a linear combination of weights. Even better, SVR allows us to model that complexity without having to expand the feature space.

Support vector machines identify the hyperplane that maximizes the margin between two classes. The support vectors are the data points closest to the margin that *support* it, if you will. This turns out to be as useful for regression modeling as it is for classification. SVR finds the hyperplane containing the greatest number of data points. We will discuss how that works in the first section of this chapter.

Rather than minimizing the sum of the squared residuals, as ordinary least squares regression does, SVR minimizes the coefficients within an acceptable error range. Like ridge and lasso regression, this can reduce model variance and the risk of overfitting. SVR works best when we are working with a small- to medium-sized dataset.

The algorithm is also quite flexible, allowing us to specify the acceptable error range, use kernels to model nonlinear relationships, and adjust hyperparameters to get the best bias-variance tradeoff possible. We will demonstrate that in this chapter.

In this chapter, we will cover the following topics:

- Key concepts of SVR
- SVR with a linear model
- Using kernels for nonlinear SVR

## Technical requirements

In this chapter, we will be working with the scikit-learn and `matplotlib` libraries. You can use `pip` to install these packages.

## Key concepts of SVR

We will start this section by discussing how support vector machines are used for classification. We will not go into much detail here, leaving a detailed discussion of support vector classification to *Chapter 13, Support Vector Machine Classification*. But starting with support vector machines for classification will lead nicely to an explanation of SVR.

As I discussed at the beginning of this chapter, support vector machines find the hyperplane that maximizes the margin between classes. When there are only two features present, that hyperplane is just a line. Consider the following example plot:

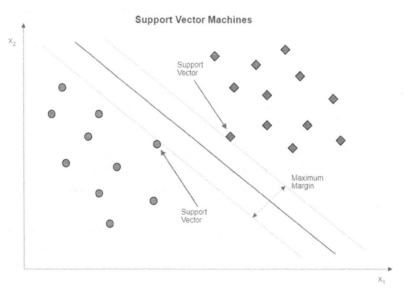

Figure 8.1 – Support vector machine classification based on two features

The two classes in this diagram, represented by red circles and blue squares, are **linearly separable** using the two features, x1 and x2. The bold line is the decision boundary. It is the line that is furthest away from border data points for each class, or the maximum margin. These points are known as the support vectors.

Since the data in the preceding plot is linearly separable, we can use what is known as **hard margin classification** without problems; that is, we can be strict about all the observations for each class being on the correct side of the decision boundary. But what if our data points look like what's shown in the following plot?

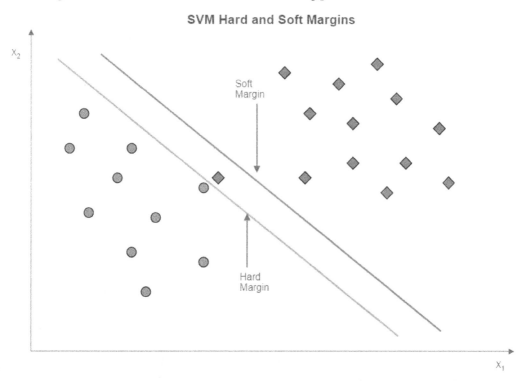

Figure 8.2 – Support vector machine classification with soft margins

These data points are not linearly separable. In this case, we can choose **soft margin classification** and ignore the outlier red circles.

We will discuss support vector classification in much greater detail in *Chapter 13, Support Vector Machine Classification*, but this illustrates some of the key support vector machine concepts. These concepts can be applied well to models involving a continuous target. This is called **support vector regression** or **SVR**.

When building an SVR model, we decide on the acceptable amount of prediction error, $\varepsilon$. Errors within $\varepsilon$ of our prediction, $\beta_0 + \beta_1 x$, in a one-feature model are not penalized. This is sometimes referred to as the epsilon-insensitive tube. SVR minimizes the coefficients consistent with all data points falling within that range. This is illustrated in the following plot:

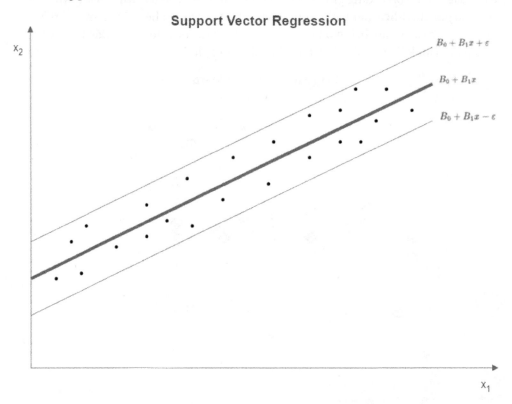

Figure 8.3 – SVR with an acceptable error range

Stated more precisely, SVR minimizes the square of the coefficients, subject to the constraint that the error, $\varepsilon$, does not exceed a given amount.

It minimizes $\frac{1}{2}\|\omega\|^2$ with the constraint that $|y_i - \hat{y}_i| \le \varepsilon$, where $\omega$ is a vector of weights (or coefficients), $y_i - \hat{y}_i$ is the actual target value minus the predicted value, and $\varepsilon$ is the acceptable amount of error.

Of course, it is not reasonable to expect all the data points to fall within the desired range. But we can still seek to minimize that deviation. Let's denote the distance of the wayward points from the margin as $\xi$. This gives us a new objective function.

We minimize $\frac{1}{2}\|\omega\|^2 + C\sum_{i=1}^{N}|\xi_i|$ with the constraint that $|y_i - \hat{y}_i| \leq \varepsilon + |\xi_i|$, where $C$ is a hyperparameter indicating how tolerant the model should be of errors outside the margin. A value of 0 for $C$ means that it is not at all tolerant of those large errors. This is equivalent to the original objective function:

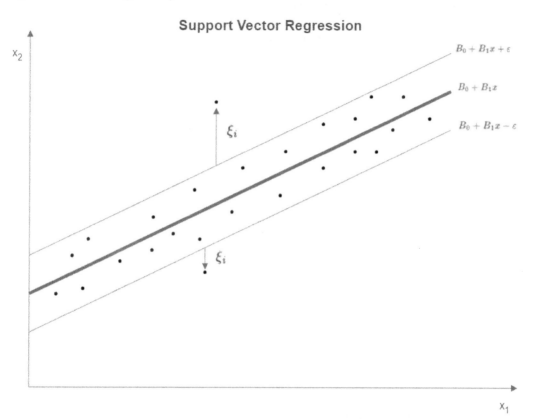

Figure 8.4 – SVR with data points outside the acceptable range

Here, we can see several advantages of SVR. It is sometimes more important that our errors will not exceed a certain amount, than picking a model with the lowest absolute error. It may matter more if we are often off by a little but rarely by a lot than if we are often spot on but occasionally way off. Since this approach also minimizes our weights, it has the same advantages as regularization, and we reduce the likelihood of overfitting.

# Nonlinear SVR and the kernel trick

We have not yet fully addressed the issue of linear separability with SVR. For simplicity, we will return to a classification problem involving two features. Let's look at a plot of two features against a categorical target. The target has two possible values, represented by the dots and squares. x1 and x2 are numeric and have negative values:

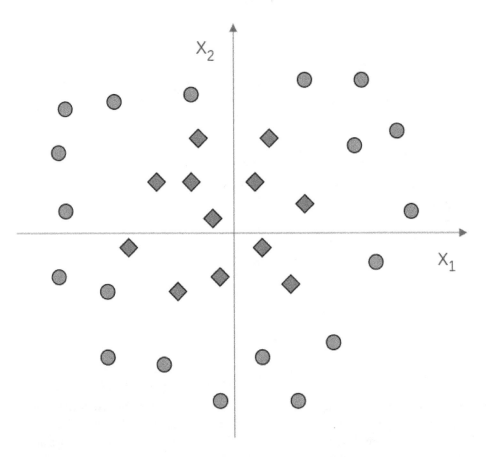

Figure 8.5 – Class labels not linearly separable with two features

What can we do in a case like this to identify a margin between the classes? It is often the case that a margin can be identified at a higher dimension. In this example, we can use a polynomial transformation, as illustrated in the following plot:

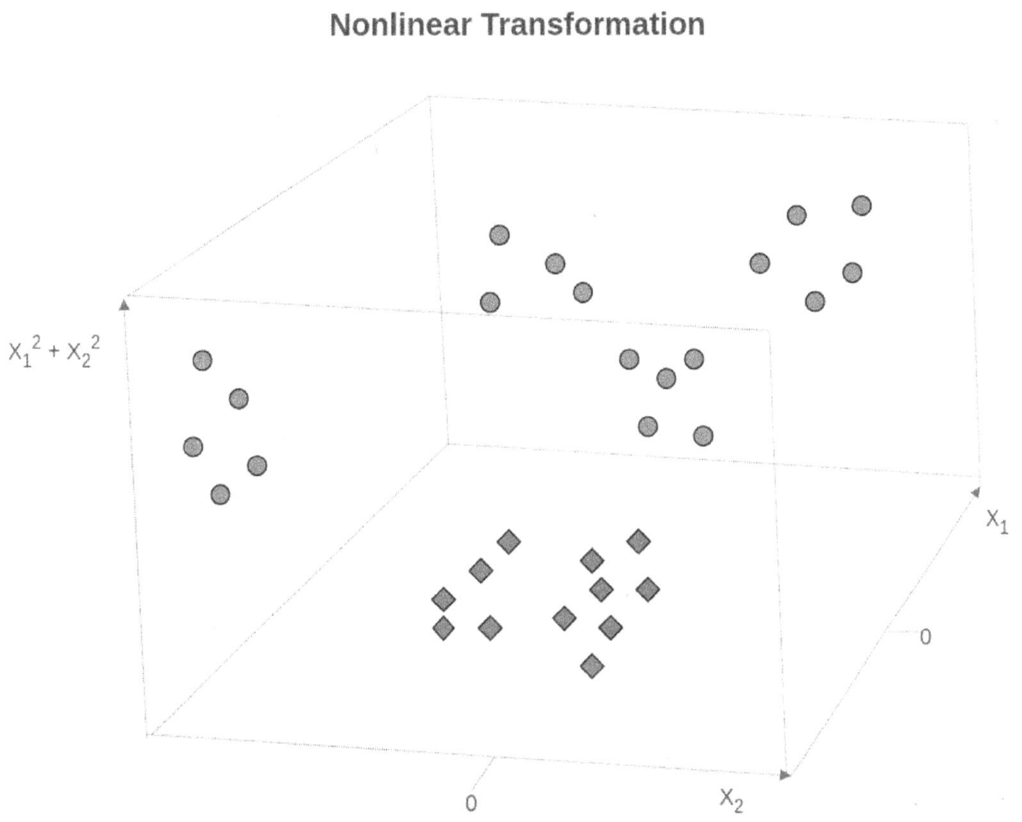

## Nonlinear Transformation

Figure 8.6 – Using polynomial transformation to establish the margin

There is now a third dimension, which is the sum of the squares of x1 and x2. The dots are all higher than the squares. This is similar to how we used polynomial transformation in the previous chapter to specify a nonlinear regression model.

One drawback of this approach is that we can quickly end up with too many features for our model to perform well. This is where the **kernel trick** comes in very handy. SVR can use a kernel function to expand the feature space implicitly without actually creating more features. This is done by creating a vector of values that can be used to fit a nonlinear margin.

While this allows us to fit a polynomial transformation such as the hypothetical one illustrated in the preceding plot, the most frequently used kernel function with SVR is the **radial basis function** (**RBF**). RBF is popular because it is faster than the other common kernel functions and because its gamma parameter makes it very flexible. We will explore how to use it in the last section of this chapter.

But for now, let's start with a relatively straightforward linear model to see SVR in action.

# SVR with a linear model

We often have enough domain knowledge to take an approach that is more nuanced than simply minimizing prediction errors in our training data. Using this knowledge may allow us to accept more bias in our model, when small amounts of bias do not matter much substantively, to reduce variance. With SVR, we can adjust hyperparameters such as epsilon (the acceptable error range) and *C* (which adjusts the tolerance for errors outside of that range) to improve our model's performance.

If a linear model can perform well on your data, linear SVR might be a good choice. We can build a linear SVR model with scikit-learn's `LinearSVR` class. Let's try creating a linear SVR model with the gasoline tax data that we used in the previous chapter:

1.  We need many of the same libraries that we used in the previous chapter to create the training and testing DataFrames and to preprocess the data. We also need to import the `LinearSVR` and `uniform` modules from scikit-learn and scipy, respectively:

    ```
    import pandas as pd
    import numpy as np
    from sklearn.model_selection import train_test_split
    from sklearn.preprocessing import StandardScaler
    from sklearn.svm import LinearSVR
    from scipy.stats import uniform
    from sklearn.impute import SimpleImputer
    from sklearn.pipeline import make_pipeline
    from sklearn.compose import ColumnTransformer
    from sklearn.compose import TransformedTargetRegressor
    from sklearn.impute import KNNImputer
    from sklearn.model_selection import cross_validate, \
        KFold, GridSearchCV, RandomizedSearchCV
    import sklearn.metrics as skmet
    import matplotlib.pyplot as plt
    ```

2.  We also need to import the `OutlierTrans` class, which we first discussed in *Chapter 7*, *Linear Regression Models*, to handle outliers:

    ```
    import os
    import sys
    sys.path.append(os.getcwd() + "/helperfunctions")
    from preprocfunc import OutlierTrans
    ```

3.  Next, we load the gasoline tax data and create training and testing DataFrames. We create lists for numerical and binary features, as well as a separate list for `motorization_rate`. As we saw when we looked at the data in the previous chapter, we need to do a little more preprocessing with `motorization_rate`.

    This dataset contains gasoline tax data for each country in 2014, as well as fuel income dependence and measures of the strength of the democratic institutions: `polity`, `democracy_polity`, and `autocracy_polity`. `democracy_polity` is a binarized `polity` variable, taking on a value of 1 for countries with high `polity` scores. `autocracy_polity` has a value of 1 for countries with low `polity` scores. The `polity` feature is a measure of how democratic a country is:

    ```
    fftaxrate14 = pd.read_csv("data/fossilfueltaxrate14.csv")
    fftaxrate14.set_index('countrycode', inplace=True)

    num_cols = ['fuel_income_dependence',
      'national_income_per_cap', 'VAT_Rate',
      'gov_debt_per_gdp', 'polity','goveffect',
      'democracy_index']
    dummy_cols = 'democracy_polity','autocracy_polity',
      'democracy', 'nat_oil_comp','nat_oil_comp_state']
    spec_cols = ['motorization_rate']

    target = fftaxrate14[['gas_tax_imp']]
    features = fftaxrate14[num_cols + dummy_cols + spec_cols]

    X_train, X_test, y_train, y_test =  \
      train_test_split(features,\
        target, test_size=0.2, random_state=0)
    ```

4.  Let's look at summary statistics for the training data. We will need to standardize the data since there are dramatically different ranges and SVR performs much better on standardized data. Also, notice that `motorization_rate` has a lot of missing values. We may want to do better than simple imputation with that feature. We have decent non-missing counts for the dummy columns:

    ```
    X_train.shape
    (123, 13)

    X_train[num_cols + spec_cols].\
    ```

```
agg(['count','min','median','max']).T
```

|                         | count | min    | median   | max        |
|-------------------------|-------|--------|----------|------------|
| fuel_income_dependence  | 121   | 0.00   | 0.10     | 34.23      |
| national_income_per_cap | 121   | 260.00 | 6,110.00 | 104,540.00 |
| VAT_Rate                | 121   | 0.00   | 16.00    | 27.00      |
| gov_debt_per_gdp        | 112   | 1.56   | 38.45    | 194.76     |
| polity                  | 121   | -10.00 | 6.00     | 10.00      |
| goveffect               | 123   | -2.04  | -0.10    | 2.18       |
| democracy_index         | 121   | 0.03   | 0.54     | 0.93       |
| motorization_rate       | 100   | 0.00   | 0.20     | 0.81       |

```
X_train[dummy_cols].apply(pd.value_counts,
normalize=True).T
```

|                   | 0.00 | 1.00 |
|-------------------|------|------|
| democracy_polity  | 0.42 | 0.58 |
| autocracy_polity  | 0.88 | 0.12 |
| democracy         | 0.41 | 0.59 |
| nat_oil_comp      | 0.54 | 0.46 |
| nat_oil_comp_state| 0.76 | 0.24 |

```
X_train[dummy_cols].count()
```

| democracy_polity   | 121 |
|--------------------|-----|
| autocracy_polity   | 121 |
| democracy          | 123 |
| nat_oil_comp       | 121 |
| nat_oil_comp_state | 121 |

5.  We need to build a column transformer to handle different data types. We can use `SimpleImputer` for the categorical features and numerical features, except for `motorization_rate`. We will use KNN imputation for the `motorization_rate` feature later:

```
standtrans = make_pipeline(OutlierTrans(2),
  SimpleImputer(strategy="median"), StandardScaler())
cattrans = make_pipeline(SimpleImputer(strategy="most_
frequent"))
spectrans = make_pipeline(OutlierTrans(2),
StandardScaler())
coltrans = ColumnTransformer(
  transformers=[
    ("stand", standtrans, num_cols),
    ("cat", cattrans, dummy_cols),
    ("spec", spectrans, spec_cols)
  ]
)
```

6.  Now, we are ready to fit our linear SVR model. We will choose a value of `0.2` for `epsilon`. This means that we are fine with any error within 0.2 standard deviations of the actual value (we use `TransformedTargetRegressor` to standardize the target). We will leave $C$ – the hyperparameter determining our model's tolerance for values outside of epsilon – at its default value of 1.0.

Before we fit our model, we still need to handle missing values for `motorization_rate`. We will add the KNN imputer to a pipeline after the column transformations. Since `motorization_rate` will be the only feature with missing values after the column transformations, the KNN imputer only changes values for that feature.

We need to use the target transformer because the column transformer will only change the features, not the target. We will pass the pipeline we just created to the target transformer's `regressor` parameter to do the feature transformations, and indicate that we just want to do standard scaling for the target.

Note that the default loss function for linear SVR is L1, but we could have chosen L2 instead:

```
svr = LinearSVR(epsilon=0.2, max_iter=10000,
  random_state=0)
```

```
pipe1 = make_pipeline(coltrans,
  KNNImputer(n_neighbors=5), svr)

ttr=TransformedTargetRegressor(regressor=pipe1,
  transformer=StandardScaler())
ttr.fit(X_train, y_train)
```

7. We can use `ttr.regressor_` to access all the elements of the pipeline, including the `linearsvr` object. This is how we get to the `coef_` attribute. The coefficients that are substantially different from 0 are `VAT_Rate` and the autocracy and national oil company dummies. Our model estimates a positive relationship between value-added tax rates and gasoline taxes, all else being equal. It estimates a negative relationship between having an autocracy or having a national oil company, and gasoline taxes:

```
coefs = ttr.regressor_['linearsvr'].coef_
np.column_stack((coefs.ravel(), num_cols + dummy_cols +
spec_cols))
array([['-0.03040694175014407', 'fuel_income_
dependence'],
       ['0.10549935644031803', 'national_income_per_
cap'],
       ['0.49519936241642026', 'VAT_Rate'],
       ['0.0857845735264331', 'gov_debt_per_gdp'],
       ['0.018198547504343885', 'polity'],
       ['0.12656984468734492', 'goveffect'],
       ['-0.09889163752261303', 'democracy_index'],
       ['-0.036584519840546594', 'democracy_polity'],
       ['-0.5446613604546718', 'autocracy_polity'],
       ['0.033234557366924815', 'democracy'],
       ['-0.2048732386478349', 'nat_oil_comp'],
       ['-0.6142887840649164', 'nat_oil_comp_state'],
       ['0.14488410358761755', 'motorization_rate']],
dtype='<U32')
```

Notice that we have not done any feature selection here. Instead, we are relying on the L1 regularization to push feature coefficients to near 0. If we had many more features, or we were more concerned about computation time, it would be important to think about our feature selection strategy more carefully.

8.  Let's do some cross-validation on this model. The mean absolute error and r-squared are not great, though that is certainly impacted by the small sample size:

```
kf = KFold(n_splits=3, shuffle=True, random_state=0)


ttr.fit(X_train, y_train)


scores = cross_validate(ttr, X=X_train, y=y_train,
   cv=kf, scoring=('r2', 'neg_mean_absolute_error'),
     n_jobs=1)


print("Mean Absolute Error: %.2f, R-squared: %.2f" %
   (scores['test_neg_mean_absolute_error'].mean(),
     scores['test_r2'].mean()))


Mean Absolute Error: -0.26, R-squared: 0.57
```

We have not done any hyperparameter tuning yet. We do not know if our values for epsilon and *C* are the best ones for our model. Therefore, we need to do a grid search to experiment with different hyperparameter values. We will start with an exhaustive grid search, which often is not practical ( I recommend not running the next few steps on your machine unless you have a fairly high-performing one). After the exhaustive grid search, we will do a randomized grid search, which is usually substantially easier on system resources.

9.  We will start by creating a LinearSVR object without the epsilon hyperparameter specified, and we will recreate the pipeline. Then, we will create a dictionary, svr_params, with values to check for epsilon and *C*, called regressor_linearsvr_epsilon and regressor_linearsvr_C, respectively.

Remember from our grid search from the previous chapter that the names of the keys must correspond with our pipeline steps. Our pipeline, which in this case can be accessed as the transformed target's regressor object, has a linearsvr object with attributes for epsilon and *C*.

We will pass the svr_params dictionary to a GridSearchCV object and indicate that we want the scoring to be based on r-squared (if we wanted to base scoring on the mean absolute error, we could have also done that).

Then, we will run the `fit` method of the grid search object. I should repeat the warning I mentioned previously that you may not want to run an exhaustive grid search unless you are using a high-performing machine, or you do not mind letting it run while you go get a cup of coffee. Note that it takes about 26 seconds to run each time on my machine:

```
svr = LinearSVR(max_iter=100000, random_state=0)

pipe1 = make_pipeline(coltrans,
   KNNImputer(n_neighbors=5), svr)

ttr=TransformedTargetRegressor(regressor=pipe1,
   transformer=StandardScaler())

svr_params = {
   'regressor__linearsvr__epsilon': np.arange(0.1, 1.6,
0.1),
   'regressor__linearsvr__C': np.arange(0.1, 1.6, 0.1)
}

gs = GridSearchCV(ttr,param_grid=svr_params, cv=3,
   scoring='r2')

%timeit gs.fit(X_train, y_train)
26.2 s ± 50.7 ms per loop (mean ± std. dev. of 7 runs, 1
loop each)
```

10. Now, we can use the `best_params_` attribute of the grid search to get the hyperparameters associated with the highest score. We can see the score with those parameters with the `best_scores_` attribute. This tells us that we get the highest r-squared, which is 0.6, with a *C* of 0.1 and an `epsilon` value of 0.2:

```
gs.best_params_
{'regressor__linearsvr__C': 0.1, 'regressor__linearsvr__
epsilon': 0.2}
gs.best_score_
0.599751107082899
```

It is good to know which values to choose for our hyperparameters. However, the exhaustive grid search was quite expensive computationally. Let's try a randomized search instead.

11. We will indicate that the random values for both `epsilon` and *C* should come from a uniform distribution with values between 0 and 1.5. Then, we will pass that dictionary to a `RandomizedSearchCV` object. This runs substantially faster than the exhaustive grid search – a little over 1 second per iteration. This gives us higher `epsilon` and *C* values than the exhaustive grid search – that is, 0.23 and 0.7, respectively. The r-squared value is a little lower, however:

```
svr_params = {
  'regressor__linearsvr__epsilon': uniform(loc=0,
scale=1.5),
  'regressor__linearsvr__C': uniform(loc=0, scale=1.5)
}


rs = RandomizedSearchCV(ttr, svr_params, cv=3,
scoring='r2')


%timeit rs.fit(X_train, y_train)
1.21 s ± 24.5 ms per loop (mean ± std. dev. of 7 runs, 1
loop each)


rs.best_params_
{'regressor__linearsvr__C': 0.23062453444814285,
  'regressor__linearsvr__epsilon': 0.6976844872643301}


rs.best_score_
0.5785452537781279
```

12. Let's look at the predictions based on the best model from the randomized grid search. The randomized grid search object's `predict` method can generate those predictions for us:

```
pred = rs.predict(X_test)


preddf = pd.DataFrame(pred, columns=['prediction'],
    index=X_test.index).join(X_test).join(y_test)


preddf['resid'] = preddf.gas_tax_imp-preddf.prediction
```

13. Now, let's look at the distribution of our residuals:

```
plt.hist(preddf.resid, color="blue", bins=np.arange(-
0.5,1.0,0.25))
plt.axvline(preddf.resid.mean(), color='red',
linestyle='dashed', linewidth=1)
plt.title("Histogram of Residuals for Gas Tax Model")
plt.xlabel("Residuals")
plt.ylabel("Frequency")
plt.xlim()
plt.show()
```

This produces the following plot:

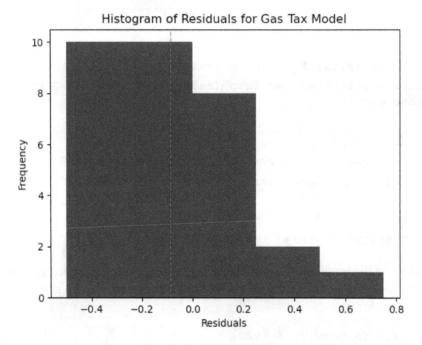

Figure 8.7 – Residual distribution for the gasoline tax linear SVR model

Here, there is a little bit of bias (some overpredicting overall) and some positive skew.

14. Let's also view a scatterplot of the predicted values against the residuals:

```
plt.scatter(preddf.prediction, preddf.resid,
color="blue")
plt.axhline(0, color='red', linestyle='dashed',
linewidth=1)
```

```
plt.title("Scatterplot of Predictions and Residuals")
plt.xlabel("Predicted Gas Tax")
plt.ylabel("Residuals")
plt.show()
```

This produces the following plot:

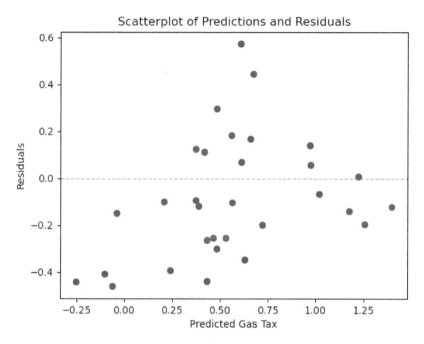

Figure 8.8 – Scatterplot of predictions and residuals for the gasoline tax linear SVR model

These residuals are problematic. We are always overpredicting (predicted values are higher than actual values) at the lower and upper range of the predicted values. This is not what we want and is perhaps warning us of an unaccounted-for nonlinear relationship.

When our data is linearly separable, linear SVR can be an efficient choice. It can be used in many of the same situations where we would have used linear regression or linear regression with regularization. Its relative efficiency means we are not as concerned about using it with datasets that contain more than 10,000 observations as we are with nonlinear SVR. However, when linear separability is not possible, we should explore nonlinear models.

# Using kernels for nonlinear SVR

Recall from our discussion at the beginning of this chapter that we can use a kernel function to fit a nonlinear epsilon-insensitive tube. In this section, we will run a nonlinear SVR with the land temperatures data that we worked with in the previous chapter. But first, we will construct a linear SVR with the same data for comparison.

We will model the average temperature for weather stations as a function of latitude and elevation. Follow these steps:

1.  We will begin by loading the familiar libraries. The only new class is SVR from scikit-learn:

    ```
    import pandas as pd
    import numpy as np
    from sklearn.model_selection import train_test_split
    from sklearn.preprocessing import StandardScaler
    from sklearn.svm import LinearSVR, SVR
    from scipy.stats import uniform
    from sklearn.impute import SimpleImputer
    from sklearn.pipeline import make_pipeline
    from sklearn.compose import TransformedTargetRegressor
    from sklearn.impute import KNNImputer
    from sklearn.model_selection import RandomizedSearchCV
    import sklearn.metrics as skmet
    import matplotlib.pyplot as plt
    import os
    import sys
    sys.path.append(os.getcwd() + "/helperfunctions")
    from preprocfunc import OutlierTrans
    ```

2.  Next, we will load the land temperatures data and create training and testing DataFrames. We will also take a look at some descriptive statistics. There are several missing values for elevation and the ranges of the two features are very different. There are also some exceedingly low average temperatures:

    ```
    landtemps = pd.read_csv("data/landtempsb2019avgs.csv")
    landtemps.set_index('locationid', inplace=True)


    feature_cols = ['latabs','elevation']
    ```

```
landtemps[['avgtemp'] + feature_cols].\
  agg(['count','min','median','max']).T
              count       min      median      max
avgtemp       12,095      -61      10          . 34
latabs        12,095      0        41          90
elevation     12,088      -350     271         4,701


X_train, X_test, y_train, y_test =  \
  train_test_split(landtemps[feature_cols],\
  landtemps[['avgtemp']], test_size=0.1, random_state=0)
```

3.  Let's start with a linear SVR model of average temperatures. We can be fairly conservative with how we handle the outliers, only setting them to missing when the interquartile range is more than three times above or below the interquartile range. (We created the OutlierTrans class in *Chapter 7, Linear Regression Models*.) We will use KNN imputation for the missing elevation values and scale the data. Remember that we need to use the target transformer to scale the target variable.

Just as we did in the previous section, we will use a dictionary, svr_params, to indicate that we want to sample values from a uniform distribution for our hyperparameters – that is, epsilon and *C*. We will pass this dictionary to the RandomizedSearchCV object.

After running fit, we can get the best parameters for epsilon and *C*, and the mean absolute error for the best model. The mean absolute error is fairly decent at about 2.8 degrees:

```
svr = LinearSVR(epsilon=1.0, max_iter=100000)


knnimp = KNNImputer(n_neighbors=45)


pipe1 = make_pipeline(OutlierTrans(3), knnimp,
StandardScaler(), svr)


ttr=TransformedTargetRegressor(regressor=pipe1,
  transformer=StandardScaler())


svr_params = {
  'regressor__linearsvr__epsilon': uniform(loc=0,
```

```
   scale=1.5),
    'regressor__linearsvr__C': uniform(loc=0, scale=20)
}


rs = RandomizedSearchCV(ttr, svr_params, cv=10,
scoring='neg_mean_absolute_error')
rs.fit(X_train, y_train)


rs.best_params_
{'regressor__linearsvr__C': 15.07662849482442,
 'regressor__linearsvr__epsilon': 0.06750238486004034}


rs.best_score_
-2.769283402595076
```

4.  Let's look at the predictions:

```
pred = rs.predict(X_test)


preddf = pd.DataFrame(pred, columns=['prediction'],
   index=X_test.index).join(X_test).join(y_test)


preddf['resid'] = preddf.avgtemp-preddf.prediction


plt.scatter(preddf.prediction, preddf.resid,
color="blue")
plt.axhline(0, color='red', linestyle='dashed',
linewidth=1)
plt.title("Scatterplot of Predictions and Residuals")
plt.xlabel("Predicted Gas Tax")
plt.ylabel("Residuals")
plt.show()
```

This produces the following plot:

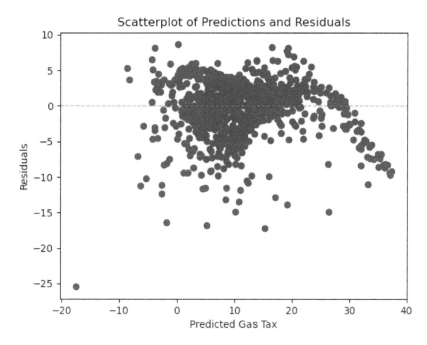

Figure 8.9 – Scatterplot of predictions and residuals for the land temperatures linear SVR model

There is a good amount of overpredicting at the upper range of the predicted values. We typically underpredict values just below that, between predicted gas tax values from 15 to 25 degrees. Perhaps we can improve the fit with a nonlinear model.

5.  We do not have to change much to run a nonlinear SVR. We just need to create an SVR object and choose a kernel function. `rbf` is typically selected. (You may not want to fit this model on your machine unless you are using good hardware, or do not mind doing something else for a while and coming back for your results.) Take a look at the following code:

```
svr = SVR(kernel='rbf')

pipe1 = make_pipeline(OutlierTrans(3), knnimp,
StandardScaler(), svr)

ttr=TransformedTargetRegressor(regressor=pipe1,
  transformer=StandardScaler())

svr_params = {
```

```
'regressor__svr__epsilon': uniform(loc=0, scale=5),
'regressor__svr__C': uniform(loc=0, scale=20),
'regressor__svr__gamma': uniform(loc=0, scale=100)
}


rs = RandomizedSearchCV(ttr, svr_params, cv=10,
scoring='neg_mean_absolute_error')
rs.fit(X_train, y_train)


rs.best_params_
{'regressor__svr__C': 5.3715128489311255,
 'regressor__svr__epsilon': 0.03997496426101643,
 'regressor__svr__gamma': 53.867632383007994}


rs.best_score_
-2.1319240416548775
```

There is a noticeable improvement in terms of the mean absolute error. Here, we can see that the gamma and C hyperparameters are doing a fair bit of work for us. If we are okay being about 2 degrees off on average, this model gets us there.

We go into much more detail regarding the gamma and C hyperparameters in *Chapter 13, Support Vector Machine Classification*. We also explore other kernels besides the rbf kernel.

6.  Let's look again at the residuals to see if there is something problematic in how our errors are distributed, as was the case with our linear model:

```
pred = rs.predict(X_test)


preddf = pd.DataFrame(pred, columns=['prediction'],
    index=X_test.index).join(X_test).join(y_test)


preddf['resid'] = preddf.avgtemp-preddf.prediction


plt.scatter(preddf.prediction, preddf.resid,
color="blue")
plt.axhline(0, color='red', linestyle='dashed',
linewidth=1)
plt.title("Scatterplot of Predictions and Residuals")
```

```
plt.xlabel("Predicted Gas Tax")
plt.ylabel("Residuals")
plt.show()
```

This produces the following plot:

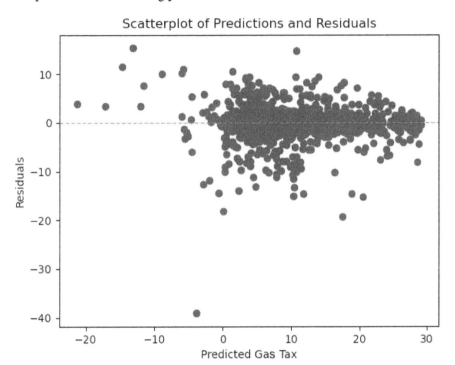

Figure 8.10 – Scatterplot of predictions and residuals for the land temperatures nonlinear SVR model

These residuals look substantially better than those for the linear model.

This illustrates how using a kernel function can increase the complexity of our model without us having to increase the feature space. By using the rbf kernel and adjusting the C and gamma hyperparameters, we address some of the underfitting we saw with the linear model. This is one of the great advantages of nonlinear SVR. The disadvantage, as we also saw, was that it was quite taxing on system resources. A dataset that contains 12,000 observations is at the upper limit of what can be handled easily with nonlinear SVR, particularly with a grid search for the best hyperparameters.

# Summary

The examples in this chapter illustrated some of the advantages of SVR. The algorithm allows us to adjust hyperparameters to address underfitting or overfitting. This can be done without increasing the number of features. SVR is also less sensitive to outliers than methods such as linear regression.

When we can build a good model with linear SVR, it is a perfectly reasonable choice. It can be trained much faster than a nonlinear model. However, we can often improve performance with a nonlinear SVR, as we saw in the last section of this chapter.

This discussion leads us to what we will explore in the next chapter, where we will look at two popular non-parametric regression algorithms: k-nearest neighbors and decision tree regression. These two algorithms make almost no assumptions about the distribution of our features and targets. Similar to SVR, they can capture complicated relationships in the data without increasing the feature space.

# 9

# K-Nearest Neighbors, Decision Tree, Random Forest, and Gradient Boosted Regression

As is true for support vector machines, K-nearest neighbors and decision tree models are best known as classification models. However, they can also be used for regression and present some advantages over classical linear regression. K-nearest neighbors and decision trees can handle nonlinearity well and no assumptions regarding the Gaussian distribution of features need to be made. Moreover, by adjusting our value of $k$ for **K-nearest neighbors** (**KNN**) or maximal depth for decision trees, we can avoid fitting the training data too precisely.

This brings us back to a theme from the previous two chapters – how to increase model complexity, including accounting for nonlinearity, without overfitting. We have seen how allowing some bias can reduce variance and give us more reliable estimates of model performance. We will continue to explore that balance in this chapter.

Specifically, we will cover the following main topics:

- Key concepts for K-nearest neighbors regression
- K-nearest neighbors regression
- Key concepts for decision tree and random forest regression
- Decision tree and random forest regression
- Using gradient boosted regression

# Technical requirements

In this chapter, we will work with the scikit-learn and `matplotlib` libraries. We will also work with XGBoost. You can use `pip` to install these packages.

# Key concepts for K-nearest neighbors regression

Part of the appeal of the KNN algorithm is that it is quite straightforward and easy to interpret. For each observation where we need to predict the target, KNN finds the $k$ training observations whose features are most similar to those of that observation. When the target is categorical, KNN selects the most frequent value of the target for the $k$ training observations. (We often select an odd value for $k$ for classification problems to avoid ties.)

When the target is numeric, KNN gives us the average value of the target for the $k$ training observations. By *training* observation, I mean those observations that have known target values. No real training is done with KNN, as it is what is called a lazy learner. I will discuss that in more detail later in this section.

*Figure 9.1* illustrates using K-nearest neighbors for classification with values of 1 and 3 for $k$. When $k$ is 1, our new observation will be assigned the red label. When $k$ is 3, it will be assigned blue:

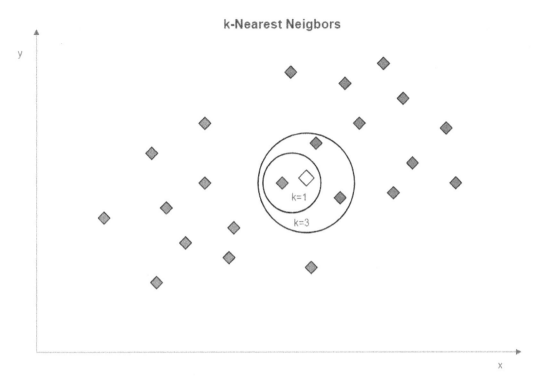

Figure 9.1 – K-nearest neighbors with a k of 1 and 3

But what do we mean by similar, or nearest, observations? There are several ways to measure similarity, but a common measure is the Euclidean distance. The Euclidean distance is the sum of the squared difference between two points. This may remind you of the Pythagorean theorem. The Euclidean distance from point a to point b is as follows:

$$D(a, b) = \sqrt{\sum_{i=1}^{n}(a_i - b_i)^2}$$

A reasonable alternative to Euclidean distance is Manhattan distance. The Manhattan distance from point a to point b is as follows:

$$D(a, b) = \sum_{i=1}^{n}|a_i - b_i|$$

Manhattan distance is sometimes called taxicab distance. This is because it reflects the distance between two points along a path on a grid. *Figure 9.2* illustrates the Manhattan distance and compares it to the Euclidean distance:

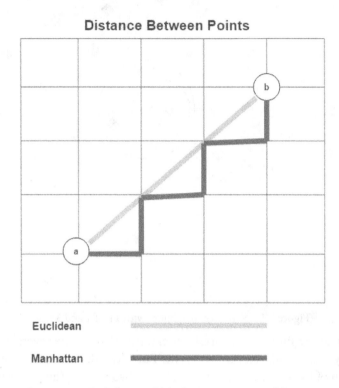

Figure 9.2 – Euclidean and Manhattan measures of distance

Using Manhattan distance can yield better results when features are very different in terms of type or scale. However, we can treat the choice of distance measure as an empirical question; that is, we can try both (or other distance measures) and see which gives us the best-performing model. We will demonstrate this with a grid search in the next section.

As you likely suspect, KNN models are sensitive to the choice of $k$. Lower values of $k$ will result in a model that attempts to identify subtle distinctions between observations. Of course, there is a substantial risk of overfitting at very low values of $k$. But at higher values of $k$, our model may not be flexible enough. We are once again confronted with the variance-bias trade-off. Lower $k$ values result in less bias and more variance, while high values result in the opposite.

There is no definitive answer to the choice of $k$. A good rule of thumb is to start with the square root of the number of observations. However, just as we would do for the distance measure, we should test a model's performance at different values of $k$.

K-nearest neighbors is a lazy learner algorithm, as I have already mentioned. No calculations are performed at training time. The learning happens mainly during testing. This has some disadvantages. KNN may not be a good choice when there are many instances or dimensions in the data, and the speed of predictions matters. It also tends not to perform well when we have sparse data – that is, datasets with many 0 values.

K-nearest neighbors is a non-parametric algorithm. No assumptions are made about the attributes of the underlying data, such as linearity or normally distributed features. It can often give us decent results when a linear model would not. We will build a KNN regression model in the next section.

# K-nearest neighbors regression

As mentioned previously, K-nearest neighbors can be a good alternative to linear regression when the assumptions of ordinary least squares do not hold, and the number of observations and dimensions is small. It is also very easy to specify, so even if we do not use it for our final model, it can be valuable for diagnostic purposes.

In this section, we will use KNN to build a model of the ratio of female to male incomes at the level of country. We will base this on labor force participation rates, educational attainment, teenage birth frequency, and female participation in politics at the highest level. This is a good dataset to experiment with because the small sample size and feature space mean that it is not likely to tax your system's resources. The small number of features also makes it easier to interpret. The drawback is that it might be hard to find significant results. That being said, let's see what we find.

> **Note**
>
> We will be working with the income gap dataset throughout this chapter. The dataset has been made available for public use by the United Nations Development Program at `https://www.kaggle.com/datasets/undp/human-development`. There is one record per country with aggregate employment, income, and education data by gender for 2015.

Let's start building our model:

1.  First, we must import some of the same `sklearn` libraries we used in the previous two chapters. We must also import `KNeighborsRegressor` and an old friend from *Chapter 5, Feature Selection* – that is, `SelectFromModel`. We will use `SelectFromModel` to add feature selection to the pipeline we will construct:

    ```
    import pandas as pd
    import numpy as np
    ```

```
from sklearn.model_selection import train_test_split
from sklearn.preprocessing import StandardScaler
from sklearn.impute import SimpleImputer
from sklearn.pipeline import make_pipeline
from sklearn.model_selection import RandomizedSearchCV
from sklearn.neighbors import KNeighborsRegressor
from sklearn.linear_model import LinearRegression
from sklearn.feature_selection import SelectFromModel
import seaborn as sns
import matplotlib.pyplot as plt
```

2. We also need the OutlierTrans class that we created in *Chapter 7, Linear Regression Models*. We will use it to identify outliers based on the interquartile range, as we first discussed in *Chapter 3, Identifying and Fixing Missing Values*:

```
import os
import sys
sys.path.append(os.getcwd() + "/helperfunctions")
from preprocfunc import OutlierTrans
```

3. Next, we must load the income data. We also need to construct a series for the ratio of female to male incomes, the years of education ratio, the labor force participation ratio, and the human development index ratio. Lower values for any of these measures suggest a possible advantage for males, assuming a positive relationship between these features and the ratio of female to male incomes. For example, we would expect the income ratio to improve – that is, to get closer to 1.0 – as the labor force participation ratio gets closer to 1.0– that is, when the labor force participation of women equals that of men.

4. We must drop rows where our target, incomeratio, is missing:

```
un_income_gap = pd.read_csv("data/un_income_gap.csv")
un_income_gap.set_index('country', inplace=True)
un_income_gap['incomeratio'] = \
  un_income_gap.femaleincomepercapita / \
    un_income_gap.maleincomepercapita
un_income_gap['educratio'] = \
  un_income_gap.femaleyearseducation / \
    un_income_gap.maleyearseducation
un_income_gap['laborforcepartratio'] = \
```

```
    un_income_gap.femalelaborforceparticipation / \
        un_income_gap.malelaborforceparticipation
un_income_gap['humandevratio'] = \
    un_income_gap.femalehumandevelopment / \
        un_income_gap.malehumandevelopment
un_income_gap.dropna(subset=['incomeratio'],
inplace=True)
```

5. Let's look at a few rows of data:

```
num_cols = ['educratio','laborforcepartratio',
'humandevratio','genderinequality','maternalmortality',
    'adolescentbirthrate','femaleperparliament',
'incomepercapita']

gap_sub = un_income_gap[['incomeratio'] + num_cols]

gap_sub.head()
incomeratio  educratio  laborforcepartratio  humandevra-
tio\
country
Norway           0.78       1.02       0.89       1.00
Australia        0.66       1.02       0.82       0.98
Switzerland      0.64       0.88       0.83       0.95
Denmark          0.70       1.01       0.88       0.98
Netherlands      0.48       0.95       0.83       0.95


genderinequality  maternalmortality  adolescentbirthrate\
country
Norway           0.07       4.00       7.80
Australia        0.11       6.00       12.10
Switzerland      0.03       6.00       1.90
Denmark          0.05       5.00       5.10
Netherlands      0.06       6.00       6.20


                 femaleperparliament  incomepercapita
country
Norway           39.60       64992
```

```
Australia      30.50    42261
Switzerland    28.50    56431
Denmark        38.00    44025
Netherlands    36.90    45435
```

6.  Let's also look at some descriptive statistics:

```
gap_sub.\
  agg(['count','min','median','max']).T
                      count   min    median     max
incomeratio           177.00  0.16   0.60       0.93
educratio             169.00  0.24   0.93       1.35
laborforcepartratio   177.00  0.19   0.75       1.04
humandevratio         161.00  0.60   0.95       1.03
genderinequality      155.00  0.02   0.39       0.74
maternalmortality     174.00  1.00   60.00      1,100.00
adolescentbirthrate   177.00  0.60   40.90      204.80
femaleperparliament   174.00  0.00   19.35      57.50
incomepercapita       177.00  581.00 10,512.00  123,124.00
```

We have 177 observations with our target variable, incomeratio. A couple
of features, humandevratio and genderinequality, have more than 15
missing values. We will need to impute some reasonable values there. We will also
need to do some scaling as some features have very different ranges than others,
from incomeratio and incomepercapita on one end to educratio and
humandevratio on the other.

> **Note**
>
> The dataset has separate human development indices for women and men. The
> index is a measure of health, access to knowledge, and standard of living. The
> humandevratio feature, which we calculated earlier, divides the score for
> women by the score for men. The genderinequality feature is an index
> of health and labor market policies in countries that have a disproportionate
> impact on women. femaleperparliament is the percentage of the
> highest national legislative body that is female.

7.  We should also look at a heatmap of the correlations of features and the features with the target. It is a good idea to keep the higher correlations (either negative or positive) in mind when we are doing our modeling. The higher positive correlations are represented with the warmer colors. `laborforcepartratio`, `humandevratio`, and `maternalmortality` are all positively correlated with our target, the latter somewhat surprisingly. `humandevratio` and `laborforcepartratio` are also correlated, so our model may have some trouble disentangling the influence of each. Some feature selection should help us figure out which feature is more important. (We will need to use a wrapper or embedded feature selection method to tease that out well. We discuss those methods in detail in *Chapter 5*, *Feature Selection*.) Look at the following code:

```
corrmatrix = gap_sub.corr(method="pearson")
corrmatrix
```

```
sns.heatmap(corrmatrix, xticklabels=corrmatrix.columns,
yticklabels=corrmatrix.columns, cmap="coolwarm")
plt.title('Heat Map of Correlation Matrix')
plt.tight_layout()
plt.show()
```

This produces the following plot:

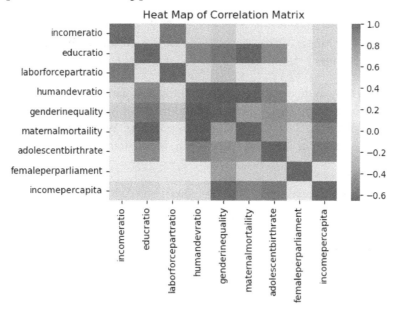

Figure 9.3 – Correlation matrix

8. Next, we must set up the training and testing DataFrames:

```
X_train, X_test, y_train, y_test =  \
  train_test_split(gap_sub[num_cols],\
  gap_sub[['incomeratio']], test_size=0.2, random_
state=0)
```

We are now ready to set up the KNN regression model. We will also build a pipeline to handle outliers, do an imputation based on the median value of each feature, scale features, and do some feature selection with scikit-learn's `SelectFromModel`.

9. We will use linear regression for our feature selection, but we can choose any algorithm that will return feature importance values. We will set the feature importance threshold to 80% of the mean feature importance. The mean is the default. Our choice here is fairly arbitrary, but I like the idea of keeping features that are just below the average feature importance level, in addition to those with higher importance of course:

```
knnreg = KNeighborsRegressor()


feature_sel = SelectFromModel(LinearRegression(),
threshold="0.8*mean")


pipe1 = make_pipeline(OutlierTrans(3), \
  SimpleImputer(strategy="median"), StandardScaler(), \
  feature_sel, knnreg)
```

10. We are now ready to do a grid search to find the best parameters. First, we will create a dictionary, `knnreg_params`, to indicate that we want the KNN model to select values of $k$ from 3 to 19, skipping even numbers. We also want the grid search to find the best distance measure – Euclidean, Manhattan, or Minkowski:

```
knnreg_params = {
  'kneighborsregressor__n_neighbors': \
      np.arange(3, 21, 2),
  'kneighborsregressor__metric': \
      ['euclidean','manhattan','minkowski']
}
```

11. We will pass those parameters to the RandomizedSearchCV object and then fit
    the model. We can use the best_params_ attribute of RandomizedSearchCV
    to see the selected hyperparameters for our feature selection and KNN regression.
    These results suggest that the best hyperparameter values are 11 for *k* for KNN and
    Manhattan for the distance metric:

    The best model has a negative mean squared error of -0.05. This is fairly
    decent, given the small sample size. It is less than 10% of the median value of
    incomeratio, which is 0.6:

    ```
    rs = RandomizedSearchCV(pipe1, knnreg_params, cv=4, n_
    iter=20, \
        scoring='neg_mean_absolute_error', random_state=1)
    rs.fit(X_train, y_train)


    rs.best_params_
    {'kneighborsregressor__n_neighbors': 11,
     'kneighborsregressor__metric': 'manhattan'}


    rs.best_score_
    -0.05419731104389228
    ```

12. Let's take a look at the features that were selected during the feature selection step
    of the pipeline. Only two features were selected – laborforcepartratio and
    humandevratio. Note that this step is not necessary to run our model. It just
    helps us interpret it:

    ```
    selected = rs.best_estimator_['selectfrommodel'].get_
    support()
    np.array(num_cols)[selected]
    array(['laborforcepartratio', 'humandevratio'],
    dtype='<U19')
    ```

13. This is a tad easier if you are using *scikit-learn* 1.0 or later. You can use the get_
    feature_names_out method in that case:

    ```
    rs.best_estimator_['selectfrommodel'].\
        get_feature_names_out(np.array(num_cols))
    array(['laborforcepartratio', 'humandevratio'],
    dtype=object)
    ```

14. We should also take a peek at some of the other top results. There is a model that uses euclidean distance that performs nearly as well as the best model:

```
results = \
  pd.DataFrame(rs.cv_results_['mean_test_score'], \
    columns=['meanscore']).\
  join(pd.DataFrame(rs.cv_results_['params'])).\
  sort_values(['meanscore'], ascending=False)

results.head(3).T
                    13          1        3
Meanscore      -0.05      -0.05    -0.05
regressor__kneighborsregressor__n_neighbors  11  13  9
regressor__kneighborsregressor__
metric   manhattan   manhattan   euclidean
```

15. Let's look at the residuals for this model. We can use the predict method of the RandomizedSearchCV object to generate predictions on the testing data. The residuals are nicely balanced around 0. There is a little bit of negative skew but that's not bad either. There is low kurtosis, but we are good with there not being much in the way of tails in this case. It likely reflects not very much in the way of outlier residuals:

```
pred = rs.predict(X_test)


preddf = pd.DataFrame(pred, columns=['prediction'],
    index=X_test.index).join(X_test).join(y_test)


preddf['resid'] = preddf.incomeratio-preddf.prediction


preddf.resid.agg(['mean','median','skew','kurtosis'])
mean              -0.01
median            -0.01
skew              -0.61
kurtosis           0.23
Name: resid, dtype: float64
```

16. Let's plot the residuals:

```
plt.hist(preddf.resid, color="blue")
plt.axvline(preddf.resid.mean(), color='red',
linestyle='dashed', linewidth=1)
plt.title("Histogram of Residuals for Gax Tax Model")
plt.xlabel("Residuals")
plt.ylabel("Frequency")
plt.xlim()
plt.show()
```

This produces the following plot:

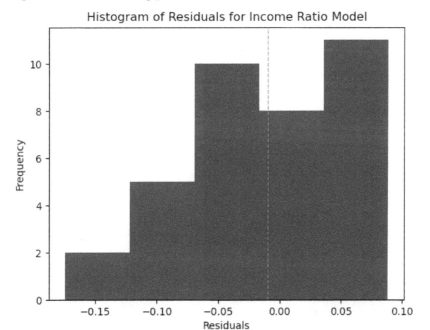

Figure 9.4 – Residuals for the income ratio model with KNN regression

The residuals also look pretty decent when we plot them. There are a couple of countries, however, where we are more than 0.1 off in our prediction. We over-predict in both of those cases. (The dashed red line is the average residual amount.)

17. Let's also look at a scatterplot. Here, we can see that the two large over-predictions are at different ends of the predicted range. In general, the residuals are fairly constant across the predicted income ratio range. We just may want to do something with the two outliers:

```
plt.scatter(preddf.prediction, preddf.resid,
color="blue")
plt.axhline(0, color='red', linestyle='dashed',
linewidth=1)
plt.title("Scatterplot of Predictions and Residuals")
plt.xlabel("Predicted Income Gap")
plt.ylabel("Residuals")
plt.show()
```

This produces the following plot:

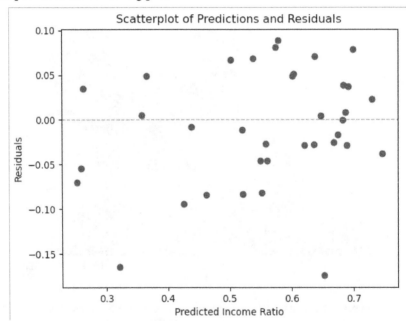

Figure 9.5 – Scatterplot of predictions and residuals for the income ratio model with KNN regression

We should take a closer look at the countries where there were high residuals. Our model does not do a good job of predicting income ratios for either Afghanistan or the Netherlands, over-predicting a fair bit in both cases. Recall that our feature selection step gave us a model with just two predictors: laborforcepartratio and humandevratio.

For Afghanistan, the labor force participation ratio (the participation of females relative to that of males) is very near the minimum of 0.19 and the human development ratio is at the minimum. This still does not get us close to predicting the very low income ratio (the income of women relative to that of men), which is also at the minimum.

For the Netherlands, the labor force participation ratio of 0.83 is a fair bit above the median of 0.75, but the human development ratio is right at the median. This is why our model predicts an income ratio a little above the median of 0.6. The actual income ratio for the Netherlands is, then, surprisingly low:

```
preddf.loc[np.abs(preddf.resid)>=0.1,
    ['incomeratio', 'prediction', 'resid',
    'laborforcepartratio', 'humandevratio']].T
```

| country | Afghanistan | Netherlands |
| --- | --- | --- |
| incomeratio | 0.16 | 0.48 |
| prediction | 0.32 | 0.65 |
| resid | -0.16 | -0.17 |
| laborforcepartratio | 0.20 | 0.83 |
| humandevratio | 0.60 | 0.95 |

Here, we can see some of the advantages of KNN regression. We can get okay predictions on difficult-to-model data without spending a lot of time specifying a model. Other than some imputation and scaling, we did not do any transformations or create interaction effects. We did not need to worry about nonlinearity either. KNN regression can handle that fine.

But this approach would probably not scale very well. A lazy learner was fine in this example. For more industrial-level work, however, we often need to turn to an algorithm with many of the advantages of KNN, but without some of the disadvantages. We will explore decision trees and random forest regression in the remainder of this chapter.

# Key concepts for decision tree and random forest regression

Decision trees are an exceptionally useful machine learning tool. They have some of the same advantages as KNN – they are non-parametric, easy to interpret, and can work with a wide range of data – but without some of the limitations.

Decision trees group the observations in a dataset based on the values of their features. This is done with a series of binary decisions, starting from an initial split at the root node, and ending with a leaf for each grouping. All observations with the same values, or the same range of values, along the branches from the root node to that leaf, get the same predicted value for the target. When the target is numeric, that is the average value for the target for the training observations at that leaf. *Figure 9.6* illustrates this:

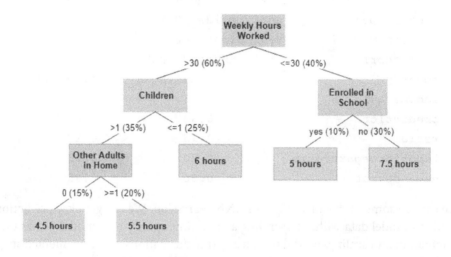

Figure 9.6 – Decision tree model of nightly hours of sleep

This is a model of nightly hours of sleep for individuals based on weekly hours worked, number of children, number of other adults in the home, and whether the person is enrolled in school. (These results are based on hypothetical data.) The root node is based on weekly hours worked and splits the data into observations with hours worked greater than 30 and 30 or less. The numbers in parentheses are the percentage of the training data that reaches that node. 60% of the observations have hours worked greater than 30. On the left-hand side of the tree, our model further splits the data by the number of children and then by the number of other adults in the home. On the other side of the tree, which represents observations with hours worked less than or equal to 30, the only additional split is by enrollment in school.

I realize now that all readers will not see this in color. We can navigate up the tree from each leaf to describe how the tree has segmented the data. 15% of observations have 0 other adults in the home, more than 1 child, and weekly hours worked greater than 30. These observations have an average nightly hours slept value of 4.5 hours. This will be the predicted value for new observations with the same characteristics.

You might be wondering how the decision tree algorithm selected the threshold amounts for the numeric features. Why greater than 30 for weekly hours worked or greater than 1 for the number of children, for example? The algorithm selects the split at each level, starting with the root, which minimizes the sum of squared errors. More precisely, splits are chosen that minimize:

$$RSS = \frac{1}{n}\sum_{i=1}^{n}(y_i - \hat{y}_i)^2$$

You may have noticed the similarity with optimization for linear regression. But there are several advantages of decision tree regression over linear regression. Decision trees can be used to model both linear and nonlinear relationships without us having to modify features. We also can avoid feature scaling with decision trees, as the algorithm can deal with very different ranges in our features.

The main disadvantage of decision trees is their high variance. Depending on the characteristics of our data, we can get a very different model each time we fit a decision tree. We can use ensemble methods, such as bagging or random forest, to address this issue.

## Using random forest regression

Random forests, perhaps not surprisingly, are collections of decision trees. But this would not distinguish a random forest from bootstrap aggregating, commonly referred to as bagging. Bagging is often used to reduce the variance of machine learning algorithms that have high variances, such as decision trees. With bagging, we generate random samples from our dataset. Then, we run our model, such as a decision tree regression, on each of those samples, averaging the predictions.

However, the samples generated with bagging can be correlated, and the resulting decision trees may have many similarities. This is more likely to be the case when there are just a few features that explain much of the variation. Random forests address this issue by limiting the number of features that can be selected for each tree. A good rule of thumb is to divide the number of features available by 3 to determine the number of features to use for each split for each decision tree. For example, if there are 21 features, we would use seven for each split. We will build both decision tree and random forest regression models in the next section.

# Decision tree and random forest regression

We will use a decision tree and a random forest in this section to build a regression model with the same income gap data we worked with earlier in this chapter. We will also use tuning to identify the hyperparameters that give us the best-performing model, just as we did with KNN regression. Let's get started:

1.  We must load many of the same libraries as we did with KNN regression, plus `DecisionTreeRegressor` and `RandomForestRegressor` from scikit-learn:

```
import pandas as pd
import numpy as np
from sklearn.model_selection import train_test_split
from sklearn.impute import SimpleImputer
from sklearn.pipeline import make_pipeline
from sklearn.model_selection import RandomizedSearchCV
from sklearn.tree import DecisionTreeRegressor, plot_tree
from sklearn.ensemble import RandomForestRegressor
from sklearn.linear_model import LinearRegression
from sklearn.feature_selection import SelectFromModel
```

2.  We must also import our class for handling outliers:

```
import os
import sys
sys.path.append(os.getcwd() + "/helperfunctions")
from preprocfunc import OutlierTrans
```

3.  We must load the same income gap data that we worked with previously and create testing and training DataFrames:

```
un_income_gap = pd.read_csv("data/un_income_gap.csv")
un_income_gap.set_index('country', inplace=True)
un_income_gap['incomeratio'] = \
  un_income_gap.femaleincomepercapita / \
    un_income_gap.maleincomepercapita
un_income_gap['educratio'] = \
  un_income_gap.femaleyearseducation / \
    un_income_gap.maleyearseducation
un_income_gap['laborforcepartratio'] = \
```

```
    un_income_gap.femalelaborforceparticipation / \
        un_income_gap.malelaborforceparticipation
un_income_gap['humandevratio'] = \
    un_income_gap.femalehumandevelopment / \
        un_income_gap.malehumandevelopment
un_income_gap.dropna(subset=['incomeratio'],
    inplace=True)

num_cols = ['educratio','laborforcepartratio',
    'humandevratio', 'genderinequality',
    'maternalmortality', 'adolescentbirthrate',
    'femaleperparliament', 'incomepercapita']

gap_sub = un_income_gap[['incomeratio'] + num_cols]

X_train, X_test, y_train, y_test =  \
    train_test_split(gap_sub[num_cols],\
    gap_sub[['incomeratio']], test_size=0.2,
        random_state=0)
```

Let's start with a relatively simple decision tree– one without too many levels. A simple tree can easily be shown on one page.

## A decision tree example with interpretation

Before we build our decision tree regressor, let's just look at a quick example with maximum depth set to a low value. Decision trees are more difficult to explain and plot as the depth increases. Let's get started:

1.  We start by instantiating a decision tree regressor, limiting the depth to three, and requiring that each leaf has at least five observations. We create a pipeline that only preprocesses the data and passes the resulting NumPy array, X_train_imp, to the fit method of the decision tree regressor:

    ```
    dtreg_example = DecisionTreeRegressor(
        min_samples_leaf=5,
        max_depth=3)

    pipe0 = make_pipeline(OutlierTrans(3),
    ```

```
  SimpleImputer(strategy="median"))

X_train_imp = pipe0.fit_transform(X_train)

dtreg_example.fit(X_train_imp, y_train)

plot_tree(dtreg_example,
    feature_names=X_train.columns,
    label="root", fontsize=10)
```

This generates the following plot:

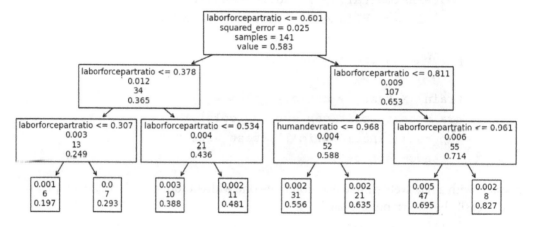

Figure 9.7 – Decision tree example with a maximum depth of 3

We will not go over all nodes on this tree. We can get the general idea of how to interpret a decision tree regression plot by describing the path down to a couple of leaf nodes:

- **Interpreting the leaf node with labor force participation ratio <= 0.307:**

    The root node split is based on labor force participation ratios less than or equal to 0.601. (Recall that the labor force participation ratio is the ratio of female participation rates to male participation rates.) 34 countries fall into that category. (True values for the split test are to the left. False values are to the right.) There is another split after that that is also based on the labor force participation ratio, this time with the split at 0.378. There are 13 countries with values less than or equal to that. Finally, we reach the leaf node furthest to the left for countries with a labor force participation ratio less than or equal to 0.307. Six countries have labor force participation ratios that low. Those six countries have an average income ratio of 0.197. Our decision tree regressor would then predict 0.197 for the income ratio for testing instances with labor force participation ratios less than or equal to 0.307.

- **Interpreting the leaf node with labor force participation ratio between 0.601 and 0.811, and humandevratio <= 0.968:**

  There are 107 countries with labor force participation ratios greater than 0.601. This is shown on the right-hand side of the tree. There is another binary split when the labor force participation ratio is less than or equal to 0.811, which is split further based on the human development ratio being less than or equal to 0.968. This takes us to a leaf node that has 31 countries, those with human development ratio less than or equal to 0.968, and a labor force participation ratio less than or equal to 0.811, but greater than 0.601. The decision tree regressor would predict the average value for income ratio for those 31 countries, 0.556, for all testing instances with values for human development ratio and labor force participation ratio in those ranges.

Interestingly, we have not done any feature selection yet, but this first effort to build a decision tree model already suggests that income ratio can be predicted with just two features: `laborforcepartratio` and `humandevratio`.

Although the simplicity of this model makes it very easy to interpret, we have not done the work we need to do to find the best hyperparameters yet. Let's do that next.

## Building and interpreting our actual model

Follow these steps:

1. First, we instantiate a new decision tree regressor and create a pipeline that uses it. We also create a dictionary for some of the hyperparameters– that is, for the maximum tree depth and the minimum number of samples (observations) for each leaf. Notice that we do not need to scale either our features or the target, as that is not necessary with a decision tree:

```
dtreg = DecisionTreeRegressor()

feature_sel = SelectFromModel(LinearRegression(),
  threshold="0.8*mean")

pipe1 = make_pipeline(OutlierTrans(3),
  SimpleImputer(strategy="median"),
  feature_sel, dtreg)

dtreg_params={
  "decisiontreeregressor__max_depth": np.arange(2, 20),
```

```
"decisiontreeregressor__min_samples_leaf": np.arange(5,
11)
}
```

2. Next, we must set up a randomized search based on the dictionary from the previous step. The best parameters for our decision tree are minimum samples of 5 and a maximum depth of 9:

```
rs = RandomizedSearchCV(pipe1, dtreg_params, cv=4, n_
iter=20,
  scoring='neg_mean_absolute_error', random_state=1)


rs.fit(X_train, y_train.values.ravel())

rs.best_params_
{'decisiontreeregressor__min_samples_leaf': 5,
 'decisiontreeregressor__max_depth': 9}


rs.best_score_
-0.05268976358459662
```

As we discussed in the previous section, decision trees have many of the advantages of KNN for regression. They are easy to interpret and do not make many assumptions about the underlying data. However, decision trees can still work reasonably well with large datasets. A less important, but still helpful, advantage of decision trees is that they do not require feature scaling.

But decision trees do have high variance. It is often worth sacrificing the interpretability of decision trees for a related method, such as random forest, which can substantially reduce that variance. We discussed the random forest algorithm conceptually in the previous section. We'll try it out with the income gap data in the next section.

# Random forest regression

Recall that random forests can be thought of as decision trees with bagging; they improve bagging by reducing the correlation between samples. This sounds complicated but it is as easy to implement as decision trees are. Let's take a look:

1.  We will start by instantiating a random forest regressor and creating a dictionary for the hyperparameters. We will also create a pipeline for the pre-processing and the regressor:

    ```
    rfreg = RandomForestRegressor()

    rfreg_params = {
      'randomforestregressor__max_depth': np.arange(2, 20),
      'randomforestregressor__max_features': ['auto', 'sqrt'],
      'randomforestregressor__min_samples_leaf':  np.arange(5,
    11)
    }

    pipe2 = make_pipeline(OutlierTrans(3),
        SimpleImputer(strategy="median"),
        feature_sel, rfreg)
    ```

2.  We will pass the pipeline and the hyperparameter dictionary to the RandomizedSearchCV object to run the grid search. There is a minor improvement in terms of the score:

    ```
    rs = RandomizedSearchCV(pipe2, rfreg_params, cv=4, n_
    iter=20,
        scoring='neg_mean_absolute_error', random_state=1)

    rs.fit(X_train, y_train.values.ravel())

    rs.best_params_
    {'randomforestregressor__min_samples_leaf': 5,
      'randomforestregressor__max_features': 'auto',
      'randomforestregressor__max_depth': 9}

    rs.best_score_
    -0.04930503752638253
    ```

3. Let's take a look at the residuals:

```
pred = rs.predict(X_test)

preddf = pd.DataFrame(pred, columns=['prediction'],
  index=X_test.index).join(X_test).join(y_test)

preddf['resid'] = preddf.incomegap-preddf.prediction

plt.hist(preddf.resid, color="blue", bins=5)
plt.axvline(preddf.resid.mean(), color='red',
linestyle='dashed', linewidth=1)
plt.title("Histogram of Residuals for Income Gap")
plt.xlabel("Residuals")
plt.ylabel("Frequency")
plt.xlim()
plt.show()
```

This produces the following plot:

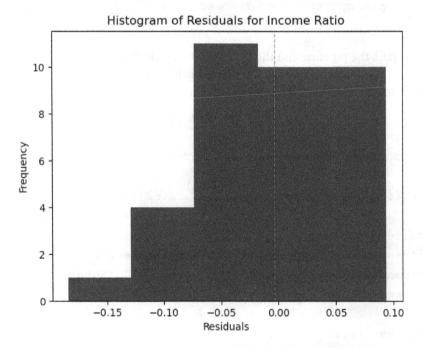

Figure 9.8 – Histogram of residuals for the random forest model on income ratio

4. Let's also take a look at a scatterplot of residuals by predictions:

```
plt.scatter(preddf.prediction, preddf.resid,
color="blue")
plt.axhline(0, color='red', linestyle='dashed',
linewidth=1)
plt.title("Scatterplot of Predictions and Residuals")
plt.xlabel("Predicted Income Gap")
plt.ylabel("Residuals")
plt.show()
```

This produces the following plot:

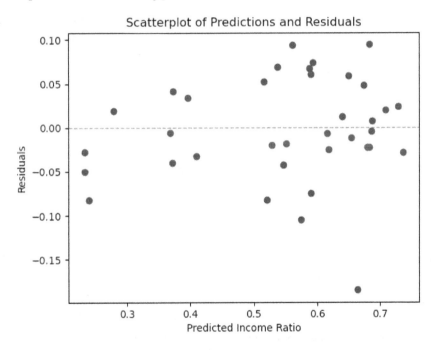

Figure 9.9 – Scatterplot of predictions by residuals for the random forest model on income ratio

5. Let's take a look a closer look at the one significant outlier where we are seriously over-predicting:

```
preddf.loc[np.abs(preddf.resid)>=0.12,
    ['incomeratio','prediction','resid',
    'laborforcepartratio', 'humandevratio']].T
country                 Netherlands
incomeratio                    0.48
```

| | |
|---|---|
| prediction | 0.66 |
| resid | -0.18 |
| laborforcepartratio | 0.83 |
| humandevratio | 0.95 |

We still have trouble with the Netherlands, but the fairly even distribution of residuals suggests that this is anomalous. It is actually good news, in terms of our ability to predict an income ratio for new instances, showing that our model is not working too hard to fit this unusual case.

# Using gradient boosted regression

We can sometimes improve upon random forest models by using gradient boosting instead. Similar to random forests, gradient boosting is an ensemble method that combines learners, typically trees. But unlike random forests, each tree is built to learn from the errors of previous trees. This can significantly improve our ability to model complexity.

Although gradient boosting is not particularly prone to overfitting, we have to be even more careful with our hyperparameter tuning than we have to be with random forest models. We can slow the learning rate, also known as shrinkage. We can also adjust the number of estimators (trees). The choice of learning rate influences the number of estimators needed. Typically, if we slow the learning rate, our model will require more estimators.

There are several tools for implementing gradient boosting. We will work with two of them: gradient boosted regression from scikit-learn and XGBoost.

We will work with data on housing prices in this section. We will try to predict housing prices in Kings County in Washington State in the United States, based on the characteristics of the home and of nearby homes.

> **Note**
>
> This dataset on housing prices in Kings County can be downloaded by the public at `https://www.kaggle.com/datasets/harlfoxem/housesalesprediction`. It has several bedrooms, bathrooms, and floors, the square feet of the home and the lot, the condition of the home, the square feet of the 15 nearest homes, and more as features.

Let's start working on the model:

1.  We will start by importing the modules we will need. The two new ones are
    GradientBoostingRegressor and XGBRegressor from XGBoost:

```
import pandas as pd
import numpy as np
from sklearn.model_selection import train_test_split
from sklearn.impute import SimpleImputer
from sklearn.pipeline import make_pipeline
from sklearn.preprocessing import OneHotEncoder
from sklearn.preprocessing import MinMaxScaler
from sklearn.compose import ColumnTransformer
from sklearn.model_selection import RandomizedSearchCV
from sklearn.ensemble import GradientBoostingRegressor
from xgboost import XGBRegressor
from sklearn.linear_model import LinearRegression
from sklearn.feature_selection import SelectFromModel
import matplotlib.pyplot as plt
from scipy.stats import randint
from scipy.stats import uniform
import os
import sys
sys.path.append(os.getcwd() + "/helperfunctions")
from preprocfunc import OutlierTrans
```

2.  Let's load the housing data and look at a few instances:

```
housing = pd.read_csv("data/kc_house_data.csv")
housing.set_index('id', inplace=True)

num_cols = ['bedrooms', 'bathrooms', 'sqft_living',
    'sqft_lot', 'floors', 'view', 'condition',
    'sqft_above', 'sqft_basement', 'yr_built',
    'yr_renovated', 'sqft_living15', 'sqft_lot15']
cat_cols = ['waterfront']

housing[['price'] + num_cols + cat_cols].\
```

```
head(3).T
```

| id | 7129300520 | 6414100192 | 5631500400 |
|---|---|---|---|
| price | 221,900 | 538,000 | 180,000 |
| bedrooms | 3 | 3 | 2 |
| bathrooms | 1 | 2 | 1 |
| sqft_living | 1,180 | 2,570 | 770 |
| sqft_lot | 5,650 | 7,242 | 10,000 |
| floors | 1 | 2 | 1 |
| view | 0 | 0 | 0 |
| condition | 3 | 3 | 3 |
| sqft_above | 1,180 | 2,170 | 770 |
| sqft_basement | 0 | 400 | 0 |
| yr_built | 1,955 | 1,951 | 1,933 |
| yr_renovated | 0 | 1,991 | 0 |
| sqft_living15 | 1,340 | 1,690 | 2,720 |
| sqft_lot15 | 5,650 | 7,639 | 8,062 |
| waterfront | 0 | 0 | 0 |

3.  We should also look at some descriptive statistics. We do not have any missing values. Our target variable, price, has some extreme values, not surprisingly. This will probably present a problem for modeling. We also need to handle some extreme values for our features:

```
housing[['price'] + num_cols].\
  agg(['count','min','median','max']).T
```

|  | count | min | median | max |
|---|---|---|---|---|
| price | 21,613 | 75,000 | 450,000 | 7,700,000 |
| bedrooms | 21,613 | 0 | 3 | 33 |
| bathrooms | 21,613 | 0 | 2 | 8 |
| sqft_living | 21,613 | 290 | 1,910 | 13,540 |
| sqft_lot | 21,613 | 520 | 7,618 | 1,651,359 |
| floors | 21,613 | 1 | 2 | 4 |
| view | 21,613 | 0 | 0 | 4 |
| condition | 21,613 | 1 | 3 | 5 |
| sqft_above | 21,613 | 290 | 1,560 | 9,410 |
| sqft_basement | 21,613 | 0 | 0 | 4,820 |
| yr_built | 21,613 | 1,900 | 1,975 | 2,015 |

```
yr_renovated    21,613    0         0         2,015
sqft_living15   21,613    399       1,840     6,210
sqft_lot15      21,613    651       7,620     871,200
```

4.  Let's create a histogram of housing prices:

```
plt.hist(housing.price/1000)
plt.title("Housing Price (in thousands)")
plt.xlabel('Price')
plt.ylabel("Frequency")
plt.show()
```

This generates the following plot:

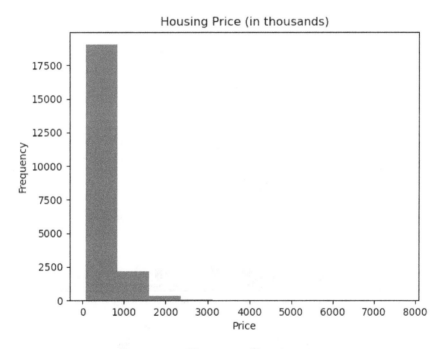

Figure 9.10 – Histogram of housing prices

5.    We may have better luck if we use a log transformation of our target for our modeling, as we tried in *Chapter 4, Encoding, Transforming, and Scaling Features* with the COVID total cases data.

```
housing['price_log'] = np.log(housing['price'])

plt.hist(housing.price_log)
plt.title("Housing Price Log")
plt.xlabel('Price Log')
plt.ylabel("Frequency")
plt.show()
```

This produces the following plot:

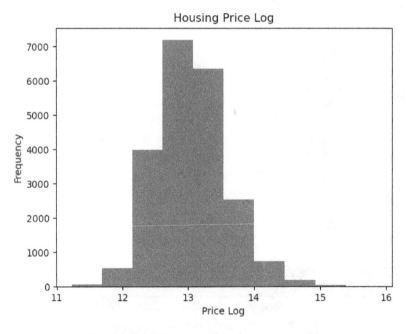

Figure 9.11 – Histogram of the housing price log

6.    This looks better. Let's take a look at the skew and kurtosis for both the price and price log. The log looks like a big improvement:

```
housing[['price','price_log']].agg(['kurtosis','skew'])
                     price        price_log
kurtosis            34.59             0.69
skew                 4.02             0.43
```

7.  We should also look at some correlations. The square feet of the living area, the square feet of the area above the ground level, the square feet of the living area of the nearest 15 homes, and the number of bathrooms are the features that are most correlated with price. The square feet of the living area and the square feet of the living area above ground level are very highly correlated. We will likely need to decide between one or the other in our model:

```
corrmatrix = housing[['price_log'] + num_cols].\
    corr(method="pearson")


sns.heatmap(corrmatrix,
  xticklabels=corrmatrix.columns,
  yticklabels=corrmatrix.columns, cmap="coolwarm")
plt.title('Heat Map of Correlation Matrix')
plt.tight_layout()
plt.show()
```

This produces the following plot:

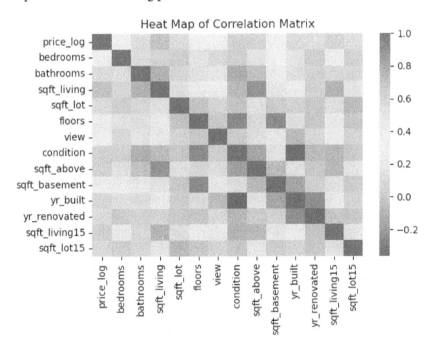

Figure 9.12 – Correlation matrix of the housing features

8. Next, we create training and testing DataFrames:

```
target = housing[['price_log']]
features = housing[num_cols + cat_cols]

X_train, X_test, y_train, y_test =  \
  train_test_split(features,\
  target, test_size=0.2, random_state=0)
```

9. We also need to set up our column transformations. For all of the numeric features, which is every feature except for waterfront, we will check for extreme values and then scale the data:

```
ohe = OneHotEncoder(drop='first', sparse=False)

standtrans = make_pipeline(OutlierTrans(2),
  SimpleImputer(strategy="median"),
  MinMaxScaler())
cattrans = make_pipeline(ohe)
coltrans = ColumnTransformer(
  transformers=[
    ("stand", standtrans, num_cols),
    ("cat", cattrans, cat_cols)
  ]
)
```

10. Now, we are ready to set up a pipeline for our pre-processing and our model. We will instantiate a GradientBoostingRegressor object and set up feature selection. We will also create a dictionary of hyperparameters to use in the randomized grid search we will do in the next step:

```
gbr = GradientBoostingRegressor(random_state=0)

feature_sel = SelectFromModel(LinearRegression(),
  threshold="0.6*mean")

gbr_params = {
  'gradientboostingregressor__learning_rate':
uniform(loc=0.01, scale=0.5),
```

```
    'gradientboostingregressor__n_estimators': randint(500,
    2000),
    'gradientboostingregressor__max_depth': randint(2, 20),
    'gradientboostingregressor__min_samples_leaf':
    randint(5, 11)
    }

    pipe1 = make_pipeline(coltrans, feature_sel, gbr)
```

11. Now, we are ready to run a randomized grid search. We get a pretty decent mean squared error score, given that the average for `price_log` is about 13:

```
    rs1 = RandomizedSearchCV(pipe1, gbr_params, cv=5, n_
    iter=20,
      scoring='neg_mean_squared_error', random_state=0)

    rs1.fit(X_train, y_train.values.ravel())
    rs1.best_params_
    {'gradientboostingregressor__learning_rate':
    0.118275177212,
     'gradientboostingregressor__max_depth': 2,
     'gradientboostingregressor__min_samples_leaf': 5,
     'gradientboostingregressor__n_estimators': 1577}

    rs1.best_score_
    -0.10695077555421204

    y_test.mean()
    price_log    13.03
    dtype: float64
```

12. Unfortunately, the mean fit time was quite long:

```
    print("fit time: %.3f, score time: %.3f"  %
      (np.mean(rs1.cv_results_['mean_fit_time']),\
      np.mean(rs1.cv_results_['mean_score_time'])))
    fit time: 35.695, score time: 0.152
```

13. Let's try XGBoost instead:

```
xgb = XGBRegressor()


xgb_params = {
 'xgbregressor__learning_rate': uniform(loc=0.01,
scale=0.5),
 'xgbregressor__n_estimators': randint(500, 2000),
 'xgbregressor__max_depth': randint(2, 20)
}


pipe2 = make_pipeline(coltrans, feature_sel, xgb)
```

14. We do not get a better score, but the mean fit time and score time have improved dramatically:

```
rs2 = RandomizedSearchCV(pipe2, xgb_params, cv=5, n_
iter=20,
   scoring='neg_mean_squared_error', random_state=0)


rs2.fit(X_train, y_train.values.ravel())
rs2.best_params_
{'xgbregressor__learning_rate': 0.019394900218177573,
 'xgbregressor__max_depth': 7,
 'xgbregressor__n_estimators': 1256}
rs2.best_score_
-0.10574300757906044
print("fit time: %.3f, score time: %.3f" %
   (np.mean(rs2.cv_results_['mean_fit_time']),\
   np.mean(rs2.cv_results_['mean_score_time'])))
fit time: 3.931, score time: 0.046
```

XGBoost has become a very popular gradient boosting tool for many reasons, some of which you have seen in this example. It can produce very good results, very quickly, with little model specification. We do need to carefully tune our hyperparameters to get the preferred variance-bias trade-off, but this is also true with other gradient boosting tools, as we have seen.

# Summary

In this chapter, we explored some of the most popular non-parametric regression algorithms: K-nearest neighbors, decision trees, and random forests. Models built with these algorithms can perform well, with a few limitations. We discussed some of the advantages and limitations of each of these techniques, including dimension and observation limits, as well as concerns about the time required for training, for KNN models. We discussed the key challenge with decision trees, which is high variance, but also how that can be addressed by a random forest model. We explored gradient boosted regression trees as well and discussed some of their advantages. We continued to improve our skills regarding hyperparameter tuning since each algorithm required a somewhat different strategy.

We discuss supervised learning algorithms where the target is categorical over the next few chapters, starting with perhaps the most familiar classification algorithm, logistic regression.

# Section 4 – Modeling Dichotomous and Multiclass Targets with Supervised Learning

There are a good number of high performing algorithms for predicting categorical targets. We will examine the most popular classification algorithms in this part. We will also consider why we might choose one algorithm over any of the others given the attributes our data and our domain knowledge.

We are as concerned with underfitting and overfitting with classification models as we were with regression models in the previous part. When the relationship between features and the target is complicated, we need to use an algorithm that can capture that complexity. But there is often a non-trivial risk of overfitting. We will discuss strategies for modeling complexity without overfitting in the chapters in this part. This usually involves some form of regularization for logistic regression models, limits on tree depth for decision trees, and adjusting the tolerance for margin violations with support vector classification.

If we are trying to model complexity without overfitting we have to be prepared to spend a good chunk of time doing hyperparameter tuning. We will definitely spend a fair bit of time on that in these chapters. Related to that, we also get really good at cross validation and generating and interpreting evaluation metrics. We will discuss accuracy, precision, sensitivity, and specificity in each of the next five chapters. We will also get very used to staring at a confusion matrix.

We will also examine how these algorithms can be extended to multiclass targets. This is straightforward with k-nearest neighbors and decision tress, but requires extension to the algorithm for logistic regression and support vector regression. We go over that in these chapters.

This section comprises the following chapters:

- *Chapter 10, Logistic Regression*
- *Chapter 11, Decision Trees and Random Forest Classification*
- *Chapter 12, K-Nearest Neighbors for Classification*
- *Chapter 13, Support Vector Machine Classification*
- *Chapter 14, Naive Bayes Classification*

# 10

# Logistic Regression

In this and the next few chapters, we will explore models for classification. These involve targets with two or several class values, such as whether a student will pass a class or not or whether a customer will choose chicken, beef, or tofu at a restaurant with only these three choices. There are several machine learning algorithms for these kinds of classification problems. We will take a look at some of the most popular ones in this chapter.

Logistic regression has been used to build models with binary targets for decades. Traditionally, it has been used to generate estimates of the impact of an independent variable or variables on the odds of a dichotomous outcome. Since our focus is on prediction, rather than the effect of each feature, we will also explore regularization techniques, such as lasso regression. These techniques can improve the accuracy of our classification predictions. We will also examine strategies for predicting a multiclass target (when there are more than two possible target values).

In this chapter, we will cover the following topics:

- Key concepts of logistic regression
- Binary classification with logistic regression
- Regularization with logistic regression
- Multinomial logistic regression

# Technical requirements

In this chapter, we will stick to the libraries that are available in most scientific distributions of Python: pandas, NumPy, and scikit-learn. All the code in this chapter will run fine with scikit-learn versions 0.24.2 and 1.0.2.

# Key concepts of logistic regression

If you are familiar with linear regression, or read *Chapter 7, Linear Regression Models*, of this book, you have probably anticipated some of the issues we will discuss in this chapter – regularization, linearity among regressors, and normally distributed residuals. If you have built supervised machine learning models in the past or worked through the last few chapters of this book, then you have also likely anticipated that we will spend some time discussing the bias-variance tradeoff and how that influences our choice of model.

I remember being introduced to logistic regression 35 years ago in a college course. It is often presented in undergraduate texts almost as a special case of linear regression; that is, linear regression with a binary dependent variable coupled with some transformation to keep predictions between 0 and 1.

It does share many similarities with linear regression of a numeric target variable. Logistic regression is relatively easy to train and interpret. Optimization techniques for both linear and logistic regression are efficient and can generate low bias predictors.

Also like linear regression, logistic regression predicts a target based on weights assigned to each feature. But to constrain the predicted probability to between 0 and 1, we use the sigmoid function. This function takes any value and maps it to a value between 0 and 1:

$$f(x) = \frac{1}{1 + e^{-x}}$$

As $x$ approaches infinity, $f(x)$ gets closer to 1. As $x$ approaches negative infinity, $f(x)$ gets closer to 0.

The following plot illustrates a sigmoid function:

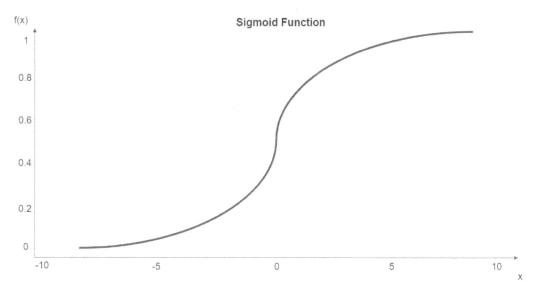

Figure 10.1 – Sigmoid function

We can plug the familiar equation for linear regression, $\beta_0 + \beta_1x_1 + \cdots +\beta_nx_n$, into the sigmoid function to predict the probability of class membership:

$$P(y = 1) = \frac{1}{1 + e^{-(\beta_0+\beta_1x_1+ \dots +\beta_nx_n)}}$$

Here $P(y = 1)$ is the predicted probability of class membership in the binary case. The coefficients (the betas) can be converted into odds ratios for interpretation, as follows:

$$r = e^{\beta}$$

Here, $r$ is the odds ratio and $\beta$ is the coefficient. A 1-unit increase in the value of a feature multiplies the odds of class membership by $\epsilon^{\beta}$. Similarly, for a binary feature, a true value has $\epsilon^{\beta}$ times the odds of class membership as does a false value for that feature, all else being equal.

Logistic regression has several advantages as an algorithm for classification problems. Features can be dichotomous, categorical, or numeric, and do not need to be normally distributed. The target variable can have more than two possible values, as we will discuss later, and it can be nominal or ordinal. Another key advantage is that the relationship between features and the target is not assumed to be linear.

The nomenclature here is a tad confusing. Why are we using a regression algorithm for a classification problem? Well, logistic regression predicts the probability of class membership. We apply a decision rule to those probabilities to predict membership. The default threshold is often 0.5 with binary targets. Instances with predicted probabilities greater than or equal to 0.5 get a positive class or 1 or True; those less than 0.5 are assigned 0 or False.

# Logistic regression extensions

We will consider two key extensions of logistic regression in this chapter. We will explore multiclass models – that is, those where the target has more than two values. We will also examine the regularization of logistic models to improve (lessen) variance.

A popular choice when constructing multiclass models is **multinomial logistic regression** (**MLR**). With MLR, the prediction probability distribution is a multinomial probability distribution. We can replace the equation we used for the binary classifier with a softmax function:

$$P\left(y = j \middle| z^{(i)}\right) = \frac{e^{z^{(i)}}}{\sum_{j=0}^{k} e^{z_k^{(i)}}}$$

Here, $z = \beta_0 + \beta_1 x_1 + \cdots + \beta_n x_n$. This calculates a probability for each class label, $j$, where $k$ is the number of classes.

An alternative to multinomial logistic regression when we have more than two classes is **one-versus-rest** (**OVR**) logistic regression. This extension to logistic regression turns the multiclass problem into a binary problem, estimating the probability of class membership versus membership in all of the other classes. The key assumption here is that membership in each class is independent. We will use MLR in an example in this chapter. One advantage it has over OVR is that the predicted probabilities are more reliable.

As mentioned previously, logistic regression has some of the same challenges as linear regression, including that the low bias of our predictions comes with high variance. This is more likely to be a problem when several features are highly correlated. Fortunately, we can deal with this with regularization, just as we saw in *Chapter 7, Linear Regression Models*.

Regularization adds a penalty to the loss function. We still seek to minimize the error, but also constrain the size of our parameters. **L1** regularization, also referred to as lasso regression, penalizes the absolute value of the weights (or coefficients):

$$\lambda \sum_{j=1}^{p} |\beta_j|$$

Here, $p$ is the number of features and $\lambda$ determines the strength of the regularization. **L2** regularization, also referred to as ridge regression, penalizes the squared values of the weights:

$$\lambda \sum_{j=1}^{p} \beta_j^2$$

Both L1 and L2 regularization push the weights toward 0, though L1 regularization is more likely to lead to sparse models. In scikit-learn, we use the $C$ parameter to adjust the value of $\lambda$, where $C$ is just the inverse of $\lambda$:

$$C = \frac{1}{\lambda}$$

We can get a balance between L1 and L2 with elastic net regression. With elastic net regression, we adjust the L1 ratio. A value of 0.5 uses L1 and L2 equally. We can use hyperparameter tuning to choose the best value for the L1 ratio.

Regularization can result in a model with lower variance, which is a good tradeoff when we are less concerned about our coefficients than we are with our predictions.

Before building a model with regularization, we will construct a fairly straightforward logistic model with a binary target. We will also spend a good amount of time evaluating that model. This will be the first classification model we will build in this book and model evaluation looks very different for those models than it does for regression models.

# Binary classification with logistic regression

Logistic regression is often used to model health outcomes when the target is binary, such as whether the person gets a disease or not. We will go through an example of that in this section. We will build a model to predict if an individual will have heart disease based on personal characteristics such as smoking and alcohol drinking habits; health features, including BMI, asthma, diabetes, and skin cancer; and age.

> **Note**
>
> In this chapter, we will work exclusively with data on heart disease that's available for public download at `https://www.kaggle.com/datasets/kamilpytlak/personal-key-indicators-of-heart-disease`. This dataset is derived from the United States Center for Disease Control data on more than 400,000 individuals from 2020. Data columns include whether respondents ever had heart disease, body mass index, ever smoked, heavy alcohol drinking, age, diabetes, and kidney disease. We will work with a 30,000 individual sample in this section to speed up the processing, but the full dataset is available in the same folder in this book's GitHub repository.

We will also do a little more preprocessing in this chapter than we have in previous chapters. We will integrate much of this work with our pipeline. This will make it easier to reuse this code in the future and lessens the likelihood of data leakage. Follow these steps:

1.  We will start by importing the same libraries we have worked with in the last few chapters. We will also import the `LogisticRegression` and `metrics` modules. We will use the `metrics` module from scikit-learn to evaluate each of our classification models in this part of this book. In addition to `matplotlib` for visualizations, we will also use `seaborn`:

```python
import pandas as pd
import numpy as np
from sklearn.model_selection import train_test_split
from sklearn.preprocessing import StandardScaler
from sklearn.preprocessing import OneHotEncoder
from sklearn.pipeline import make_pipeline
from sklearn.impute import SimpleImputer
from sklearn.compose import ColumnTransformer
from sklearn.model_selection import StratifiedKFold
from sklearn.feature_selection import RFECV
from sklearn.linear_model import LogisticRegression
import sklearn.metrics as skmet
import matplotlib.pyplot as plt
import seaborn as sns
```

2.  We are also going to need several custom classes to handle the preprocessing. We
    have already seen the `OutlierTrans` class. Here, we have added a couple of new
    classes – `MakeOrdinal` and `ReplaceVals`:

```
import os
import sys
sys.path.append(os.getcwd() + "/helperfunctions")
from preprocfunc import OutlierTrans,\
  MakeOrdinal, ReplaceVals
```

The `MakeOrdinal` class takes a character feature and assigns numeric values based
on an alphanumeric sort. For example, a feature that has three possible values – not
well, okay, and well – would be transformed into an ordinal feature with values of 0,
1, and 2, respectively.

Recall that scikit-learn pipeline transformers must have `fit` and `transform`
methods, and must inherit from `BaseEstimator`. They often also inherit from
`TransformerMixin`, though there are other options.

All the action in the `MakeOrdinal` class happens in the `transform` method. We
loop over all of the columns that are passed to it by the column transformer. For
each column, we find all the unique values and sort them alphanumerically, storing
the unique values in a NumPy array that we name `cats`. Then, we use a lambda
function and NumPy's `where` method to find the index of `cats` associated with
each feature value:

```
class MakeOrdinal(BaseEstimator,TransformerMixin):
  def fit(self,X,y=None):
    return self

  def transform(self,X,y=None):
    Xnew = X.copy()
    for col in Xnew.columns:
      cats = np.sort(Xnew[col].unique())
      Xnew[col] = Xnew.\
        apply(lambda x: int(np.where(cats==\
        x[col])[0]), axis=1)
    return Xnew.values
```

MakeOrdinal will work fine when the alphanumeric order matches a meaningful order, as with the previous example. When that is not true, we can use ReplaceVals to assign appropriate ordinal values. This class replaces values in any feature with alternative values based on a dictionary passed to it.

We could have just used the pandas replace method without putting it in a pipeline, but this way, it is easier to integrate our recoding with other pipeline steps, such as feature scaling:

```
class ReplaceVals(BaseEstimator,TransformerMixin):
  def __init__(self,repdict):
    self.repdict = repdict

  def fit(self,X,y=None):
    return self

  def transform(self,X,y=None):
    Xnew = X.copy().replace(self.repdict)
  return Xnew.values
```

Do not worry if you do not fully understand how we will use these classes yet. It will be clearer when we add them to our column transformations.

3.  Next, we will load the heart disease data and take a look at a few rows. Several string features are conceptually binary, such as alcoholdrinkingheavy, which is Yes when the person is a heavy drinker and No otherwise. We will need to encode these features before running a model.

    The agecategory feature is character data that represents the age interval. We will need to convert that feature into numeric:

```
healthinfo = pd.read_csv("data/healthinfo.csv")
healthinfo.set_index("personid", inplace=True)
healthinfo.head(2).T
```

| personid             | 299391 | 252786 |
|----------------------|--------|--------|
| heartdisease         | Yes    | No     |
| bmi                  | 28.48  | 25.24  |
| smoking              | Yes    | Yes    |
| alcoholdrinkingheavy | No     | No     |
| stroke               | No     | No     |
| physicalhealthbaddays| 7      | 0      |

| | | |
|---|---|---|
| mentalhealthbaddays | 0 | 2 |
| walkingdifficult | No | No |
| gender | Male | Female |
| agecategory | 70-74 | 65-69 |
| ethnicity | White | White |
| diabetic     No, borderline diabetes | | No |
| physicalactivity | Yes | Yes |
| genhealth | Good | Very good |
| sleeptimenightly | 8 | 8 |
| asthma | No | No |
| kidneydisease | No | No |
| skincancer | No | Yes |

4. Let's look at the size of the DataFrame and how many missing values we have. There are 30,000 instances, but there are no missings for any of the 18 data columns. That's great. We won't have to worry about that when we construct our pipeline:

```
healthinfo.shape
(30000, 18)
healthinfo.isnull().sum()
```

| | |
|---|---|
| heartdisease | 0 |
| bmi | 0 |
| smoking | 0 |
| alcoholdrinkingheavy | 0 |
| stroke | 0 |
| physicalhealthbaddays | 0 |
| mentalhealthbaddays | 0 |
| walkingdifficult | 0 |
| gender | 0 |
| agecategory | 0 |
| ethnicity | 0 |
| diabetic | 0 |
| physicalactivity | 0 |
| genhealth | 0 |
| sleeptimenightly | 0 |
| asthma | 0 |
| kidneydisease | 0 |

```
skincancer                    0
dtype: int64
```

5. Let's change the `heartdisease` variable, which will be our target, into a 0 and 1 variable. This will give us one less thing to worry about later. One thing to notice right away is that the target's values are quite imbalanced. Less than 10% of our observations have heart disease. That, of course, is good news, but it presents some challenges for modeling that we will need to handle:

```
healthinfo.heartdisease.value_counts()
No          27467
Yes         2533
Name: heartdisease, dtype: int64

healthinfo['heartdisease'] = \
  np.where(healthinfo.heartdisease=='No',0,1).\
    astype('int')

healthinfo.heartdisease.value_counts()
0           27467
1           2533
Name: heartdisease, dtype: int64
```

6. We should organize our features by the preprocessing we will be doing with them. We will be scaling the numeric features and doing one-hot encoding with the categorical features. We want to make the `agecategory` and `genhealth` features, which are currently strings, into ordinal features.

   We need to do a specific cleanup of the `diabetic` feature. Some individuals indicate no, but that they were borderline. For our purposes, we will consider them a *no*. Some individuals had diabetes during their pregnancies only. We will consider them a *yes*. For both `genhealth` and `diabetic`, we will set up a dictionary that will indicate how feature values should be replaced. We will use that dictionary in the `ReplaceVals` transformer of our pipeline:

```
num_cols = ['bmi','physicalhealthbaddays',
    'mentalhealthbaddays','sleeptimenightly']
binary_cols = ['smoking','alcoholdrinkingheavy',
    'stroke','walkingdifficult','physicalactivity',
    'asthma','kidneydisease','skincancer']
```

```
cat_cols = ['gender','ethnicity']
spec_cols1 = ['agecategory']
spec_cols2 = ['genhealth']
spec_cols3 = ['diabetic']

rep_dict = {
  'genhealth': {'Poor':0,'Fair':1,'Good':2,
    'Very good':3,'Excellent':4},
  'diabetic': {'No':0,
    'No, borderline diabetes':0,'Yes':1,
    'Yes (during pregnancy)':1}
}
```

7. We should take a look at some frequencies for the binary features, as well as other categorical features. A large percentage of the individuals (42%) report that they have been smokers. 14% report that they have difficulty walking:

```
healthinfo[binary_cols].\
  apply(pd.value_counts, normalize=True).T
```

|                     | No   | Yes  |
|---------------------|------|------|
| smoking             | 0.58 | 0.42 |
| alcoholdrinkingheavy | 0.93 | 0.07 |
| stroke              | 0.96 | 0.04 |
| walkingdifficult    | 0.86 | 0.14 |
| physicalactivity    | 0.23 | 0.77 |
| asthma              | 0.87 | 0.13 |
| kidneydisease       | 0.96 | 0.04 |
| skincancer          | 0.91 | 0.09 |

8. Let's also look at frequencies for the other categorical features. There are nearly equal numbers of men and women. Most people report excellent or very good health:

```
for col in healthinfo[cat_cols +
['genhealth','diabetic']].columns:
  print(col, "----------------------",
  healthinfo[col].value_counts(normalize=True).\
    sort_index(), sep="\n", end="\n\n")
```

This produces the following output:

```
gender
---------------------
Female   0.52
Male     0.48
Name: gender, dtype: float64


ethnicity
---------------------
American Indian/Alaskan Native   0.02
Asian                            0.03
Black                            0.07
Hispanic                         0.09
Other                            0.03
White                            0.77
Name: ethnicity, dtype: float64


genhealth
---------------------
Excellent    0.21
Fair         0.11
Good         0.29
Poor         0.04
Very good    0.36
Name: genhealth, dtype: float64


diabetic
---------------------
No                       0.84
No, borderline diabetes  0.02
Yes                      0.13
Yes (during pregnancy)   0.01
Name: diabetic, dtype: float64
```

9.  We should also look at some descriptive statistics for the numerical features. The median value for both bad physical health and mental health days is 0; that is, at least half of the observations report no bad physical health days, and at least half report no bad mental health days over the previous month:

```
healthinfo[num_cols].\
  agg(['count','min','median','max']).T
                          count    min    median  max
bmi                       30,000   12     27      92
physicalhealthbaddays     30,000   0      0       30
mentalhealthbaddays       30,000   0      0       30
sleeptimenightly          30,000   1      7       24
```

We will need to do some scaling. We will also need to do some encoding of the categorical features. There are also some extreme values for the numerical features. A sleeptimenightly value of 24 seems unlikely! It is probably a good idea to deal with them.

10. Now, we are ready to build our pipeline. Let's create the training and testing DataFrames:

```
X_train, X_test, y_train, y_test =  \
  train_test_split(healthinfo[num_cols +
    binary_cols + cat_cols + spec_cols1 +
    spec_cols2 + spec_cols3],\
  healthinfo[['heartdisease']], test_size=0.2,
    random_state=0)
```

11. Next, we will set up the column transformations. We will create a one-hot encoder instance that we will use for all of the categorical features. For the numeric columns, we will remove extreme values using the OutlierTrans object and then impute the median.

We will convert the agecategory feature into an ordinal one using the MakeOrdinal transformer and code the genhealth and diabetic features using the ReplaceVals transformer.

We will add the column transformation to our pipeline in the next step:

```
ohe = OneHotEncoder(drop='first', sparse=False)

standtrans = make_pipeline(OutlierTrans(3),
  SimpleImputer(strategy="median"),
```

```
    StandardScaler())
spectrans1 = make_pipeline(MakeOrdinal(),
    StandardScaler())
spectrans2 = make_pipeline(ReplaceVals(rep_dict),
    StandardScaler())
spectrans3 = make_pipeline(ReplaceVals(rep_dict))
bintrans = make_pipeline(ohe)
cattrans = make_pipeline(ohe)
coltrans = ColumnTransformer(
    transformers=[
        ("stand", standtrans, num_cols),
        ("spec1", spectrans1, spec_cols1),
        ("spec2", spectrans2, spec_cols2),
        ("spec3", spectrans3, spec_cols3),
        ("bin", bintrans, binary_cols),
        ("cat", cattrans, cat_cols),
    ]
)
```

12. Now, we are ready to set up and fit our pipeline. First, we will instantiate logistic regression and stratified k-fold objects, which we will use with recursive feature elimination. Recall that recursive feature elimination needs an estimator. We use stratified k-fold to get approximately the same target value distribution in each fold.

Now, we must create another logistic regression instance for our model. We will set the class_weight parameter to balanced. This should improve the model's ability to deal with the class imbalance. Then, we will add the column transformation, recursive feature elimination, and logistic regression instance to our pipeline, and then fit it:

```
lrsel = LogisticRegression(random_state=1,
    max_iter=1000)


kf = StratifiedKFold(n_splits=5, shuffle=True)


rfecv = RFECV(estimator=lrsel, cv=kf)


lr = LogisticRegression(random_state=1,
    class_weight='balanced', max_iter=1000)
```

```
pipe1 = make_pipeline(coltrans, rfecv, lr)

pipe1.fit(X_train, y_train.values.ravel())
```

13. We need to do a little work to recover the column names from the pipeline after the fit. We can use the `get_feature_names` method of the one-hot encoder for the `bin` transformer and the `cat` transformer for this. This gives us the column names for the binary and categorical features after the encoding. The names of the numerical features remain unchanged. We will use the feature names later:

```
new_binary_cols = \
  pipe1.named_steps['columntransformer'].\
  named_transformers_['bin'].\
  named_steps['onehotencoder'].\
  get_feature_names(binary_cols)
new_cat_cols = \
  pipe1.named_steps['columntransformer'].\
  named_transformers_['cat'].\
  named_steps['onehotencoder'].\
  get_feature_names(cat_cols)

new_cols = np.concatenate((np.array(num_cols +
  spec_cols1 + spec_cols2 + spec_cols3),
  new_binary_cols, new_cat_cols))

new_cols
array(['bmi', 'physicalhealthbaddays',
       'mentalhealthbaddays', 'sleeptimenightly',
       'agecategory', 'genhealth', 'diabetic',
       'smoking_Yes', 'alcoholdrinkingheavy_Yes',
       'stroke_Yes', 'walkingdifficult_Yes',
       'physicalactivity_Yes', 'asthma_Yes',
       'kidneydisease_Yes', 'skincancer_Yes',
       'gender_Male', 'ethnicity_Asian',
       'ethnicity_Black', 'ethnicity_Hispanic',
       'ethnicity_Other', 'ethnicity_White'],
      dtype=object)
```

14. Now, let's look at the results from the recursive feature elimination. We can use the `ranking_` attribute of the `rfecv` object to get the ranking of each feature. Those with a *1* for ranking will be selected for our model.

If we use the `get_support` method or the `support_` attribute of the `rfecv` object instead of the `ranking_` attribute, we get just those features that will be used in our model – that is, those with a ranking of 1. We will do that in the next step:

```
rankinglabs = \
 np.column_stack((pipe1.named_steps['rfecv'].ranking_,
 new_cols))
pd.DataFrame(rankinglabs,
 columns=['rank','feature']).\
 sort_values(['rank','feature']).\
 set_index("rank")
```

| rank | feature |
| --- | --- |
| 1 | agecategory |
| 1 | alcoholdrinkingheavy_Yes |
| 1 | asthma_Yes |
| 1 | diabetic |
| 1 | ethnicity_Asian |
| 1 | ethnicity_Other |
| 1 | ethnicity_White |
| 1 | gender_Male |
| 1 | genhealth |
| 1 | kidneydisease_Yes |
| 1 | smoking_Yes |
| 1 | stroke_Yes |
| 1 | walkingdifficult_Yes |
| 2 | ethnicity_Hispanic |
| 3 | skincancer_Yes |
| 4 | bmi |
| 5 | physicalhealthbaddays |
| 6 | sleeptimenightly |
| 7 | mentalhealthbaddays |
| 8 | physicalactivity_Yes |
| 9 | ethnicity_Black |

15. We can get the odds ratios from the coefficients from the logistic regression. Recall that the odds ratio is the exponentiated coefficient. There are 13 coefficients, which makes sense because we learned in the previous step that 13 features got a ranking of 1.

We will use the `get_support` method of the `rfecv` step to get the names of the selected features and create a NumPy array with those names and the odds ratios, `oddswithlabs`. We then create a pandas DataFrame and sort by the odds ratio in descending order.

Not surprisingly, those who had a stroke and older individuals are substantially more likely to have heart disease. If the individual had a stroke, they had three times the odds of having heart disease, controlling for everything else. The odds of having heart disease increase by 2.88 times for each increase in age category. On the other hand, the odds of having heart disease decline by about half (57%) for every increase in general health; from, say, fair to good. Surprisingly, heavy alcohol drinking is associated with lower odds of heart disease, controlling for everything else:

```
oddsratios = np.exp(pipe1.\
  named_steps['logisticregression'].coef_)

oddsratios.shape
(1, 13)

selcols = new_cols[pipe1.\
  named_steps['rfecv'].get_support()]

oddswithlabs = np.column_stack((oddsratios.\
  ravel(), selcols))

pd.DataFrame(oddswithlabs,
  columns=['odds','feature']).\
  sort_values(['odds'], ascending=False).\
  set_index('odds')
```

```
                        feature
odds
3.01                  stroke_Yes
2.88                  agecategory
```

```
2.12              gender_Male
1.97         kidneydisease_Yes
1.75                  diabetic
1.55               smoking_Yes
1.52                asthma_Yes
1.30        walkingdifficult_Yes
1.27            ethnicity_Other
1.22            ethnicity_White
0.72            ethnicity_Asian
0.61    alcoholdrinkingheavy_Yes
0.57                 genhealth
```

Now that we have fit our logistic regression model, we are ready to evaluate it. In the next section, we will spend some time looking into various performance measures, including accuracy and sensitivity. We will use many of the concepts that we introduced in *Chapter 6, Preparing for Model Evaluation.*

## Evaluating a logistic regression model

The most intuitive measure of a classification model's performance is its accuracy – that is, how often our predictions are correct. In some cases, however, we might be at least as concerned about sensitivity – the percent of positive cases that we predict correctly – as accuracy; we may even be willing to lose a little accuracy to improve sensitivity. Predictive models of diseases often fall into that category. But whenever there is a class imbalance, measures such as accuracy and sensitivity can give us very different estimates of the performance of our model.

In addition to being concerned about accuracy or sensitivity, we might be worried about our model's **specificity** or **precision**. We may want a model that can identify negative cases with high reliability, even if that means it does not do as good a job of identifying positives. Specificity is a measure of the percentage of all negatives identified by the model.

Precision, which is the percentage of predicted positives that are positives, is another important measure. For some applications, it is important to limit false positives, even if we have to tolerate lower sensitivity. An apple grower, using image recognition to identify bad apples, may prefer a high-precision model to a more sensitive one, not wanting to discard apples unnecessarily.

This can be made more clear by looking at a confusion matrix:

| Confusion Matrix | | | |
|---|---|---|---|
| | | **Predicted Value** | |
| | | **Negative** | **Positive** |
| **Actual Value** | **Negative** | True Negative (TN) | False Positive (FP) |
| | **Positive** | False Negative (FN) | True Positive (TP) |

Figure 10.2 – Confusion matrix of actual by predicted values for a binary target

The confusion matrix helps us conceptualize accuracy, sensitivity, specificity, and precision. Accuracy is the percentage of observations for which our prediction was correct. This can be stated more precisely as follows:

$$Accuracy = \frac{\sum TP + \sum TN}{Number\ of\ Observations}$$

Sensitivity is the number of times we predicted positives correctly divided by the number of positives. It might be helpful to glance at the confusion matrix again and confirm that actual positive values can either be predicted positives (TP) or predicted negatives (FN). Sensitivity is also referred to as **recall** or the **true positive rate**:

$$Sensitivity = \frac{\sum TP}{\sum TP + \sum FN}$$

Specificity is the number of times we correctly predicted a negative value (TN) divided by the number of actual negative values (TN + FP). Specificity is also known as the **true negative rate**:

$$Specificity = \frac{\sum TN}{\sum TN + \sum FP}$$

Precision is the number of times we correctly predicted a positive value (TP) divided by the number of positive values predicted:

$$Precision = \frac{\sum TP}{\sum TP + \sum FP}$$

We went over these concepts in more detail in *Chapter 6, Preparing for Model Evaluation*. In this section, we will examine the accuracy, sensitivity, specificity, and precision of our logistic regression model of heart disease:

1.  We can use the `predict` method of the pipeline we fitted in the previous section to generate predictions from our logistic regression. Then, we can generate a confusion matrix:

```
pred = pipe1.predict(X_test)

cm = skmet.confusion_matrix(y_test, pred)
cmplot = skmet.ConfusionMatrixDisplay(confusion_
matrix=cm, display_labels=['Negative', 'Positive'])
cmplot.plot()
cmplot.ax_.set(title='Heart Disease Prediction Confusion
Matrix',
    xlabel='Predicted Value', ylabel='Actual Value')
```

    This produces the following plot:

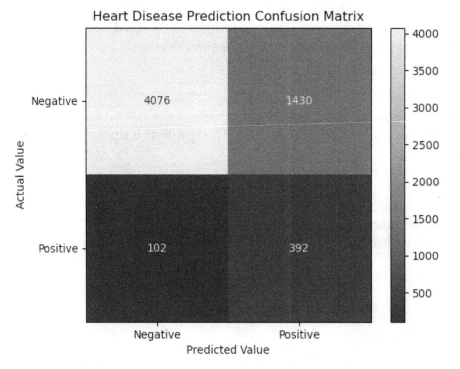

Figure 10.3 – A confusion matrix for heart disease prediction

The first thing to notice here is that most of the action is in the top-left quadrant, where we correctly predict actual negative values in the testing data. That is going to help our accuracy a fair bit. Nonetheless, we have a fair number of false positives. We predict heart disease 1,430 times (out of 5,506 negative instances) when there is no heart disease. We do seem to do an okay job of identifying positive heart disease instances, correctly classifying 392 instances (out of 494) that were positive.

2. Let's calculate the accuracy, sensitivity, specificity, and precision. The overall accuracy is not great, at 74%. Sensitivity is pretty decent though, at 79%. (Of course, how *decent* the sensitivity is depends on the domain and judgment. For something such as heart disease, we likely want it to be higher.) This can be seen in the following code:

```
tn, fp, fn, tp = skmet.confusion_matrix(y_test.values.
ravel(), pred).ravel()


tn, fp, fn, tp
(4076, 1430, 102, 392)


accuracy = (tp + tn) / pred.shape[0]
accuracy
0.7446666666666667


sensitivity = tp / (tp + fn)
sensitivity
0.7935222672064778


specificity = tn / (tn+fp)
specificity
0.7402833272793317


precision = tp / (tp + fp)
precision
0.21514818880351264
```

3.  We can do these calculations in a more straightforward way using the `metrics` module (I chose a more roundabout approach in the previous step to illustrate how the calculations are done):

```
print("accuracy: %.2f, sensitivity: %.2f, specificity:
%.2f, precision: %.2f"  %
  (skmet.accuracy_score(y_test.values.ravel(), pred),
   skmet.recall_score(y_test.values.ravel(), pred),
   skmet.recall_score(y_test.values.ravel(), pred,
   pos_label=0),
   skmet.precision_score(y_test.values.ravel(), pred)))
accuracy: 0.74, sensitivity: 0.79, specificity: 0.74,
precision: 0.22
```

The biggest problem with our model is the very low level of precision – that is, 22%. This is due to the large number of false positives. The majority of the time that our model predicts positive, it is wrong.

In addition to the four measures that we have already calculated, it can also be helpful to get the false positive rate. The false positive rate is the propensity of our model to predict positive when the actual value is negative:

$$False\ positive\ rate = \frac{\sum FP}{\sum TN + \sum FP}$$

4.  Let's calculate the false positive rate:

```
falsepositiverate = fp / (tn + fp)
falsepositiverate
0.25971667272066834
```

So, 26% of the time that a person does not have heart disease, we predicted that they do. While we certainly want to limit the number of false positives, this often means sacrificing some sensitivity. We will demonstrate why this is true later in this section.

5.  We should take a closer look at the prediction probabilities generated by our model. Here, the threshold for a positive class prediction is 0.5, which is often the default with logistic regression. (Recall that logistic regression predicts a probability of class membership. We need an accompanying decision rule, such as the 0.5 threshold, to predict the class.) This can be seen in the following code:

```
pred_probs = pipe1.predict_proba(X_test)[:, 1]

probdf = \
```

```
pd.DataFrame(zip(pred_probs, pred,
    y_test.values.ravel()),
    columns=(['prob','pred','actual'])))

probdf.groupby(['pred'])['prob'].\
    agg(['min','max','count'])
        min         max         count
pred
0       0.01        0.50        4178
1       0.50        0.99        1822
```

6.  We can use a **kernel density estimate** (**KDE**) plot to visualize these probabilities. We can also see how a different decision rule may impact our predictions. For example, we could move the threshold from 0.5 to 0.25. At a glance, that has some advantages. The area between the two possible thresholds has somewhat more heart disease cases than no heart disease cases. We would be getting the brown area between the dashed lines right, predicting heart disease correctly where we would not have with the 0.5 threshold. That is a larger area than the green area between the lines, where we turn some of the true negative predictions at the 0.5 threshold into false positives at the 0.25 threshold:

```
sns.kdeplot(probdf.loc[probdf.actual==1].prob,
    shade=True, color='red',label="Heart Disease")
sns.kdeplot(probdf.loc[probdf.actual==0].prob,
    shade=True,color='green',label="No Heart Disease")
plt.axvline(0.25, color='black', linestyle='dashed',
    linewidth=1)
plt.axvline(0.5, color='black', linestyle='dashed',
    linewidth=1)
plt.title("Predicted Probability Distribution")
plt.legend(loc="upper left")
```

This generates the following plot:

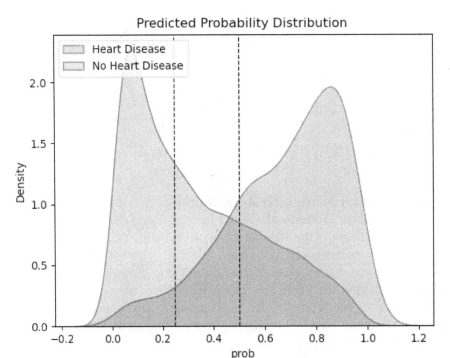

Figure 10.4 – Heart disease predicted probability distribution

Let's consider the tradeoff between precision and sensitivity a little more carefully than we have so far. Remember that precision is the rate at which we are right when we predict a positive class value. Sensitivity, also referred to as recall or the true positive rate, is the rate at which we identify an actual positive as positive.

7.    We can plot precision and sensitivity curves as follows:

```
prec, sens, ths = skmet.precision_recall_curve(y_test,
pred_probs)
sens = sens[1:-20]
prec = prec[1:-20]
ths  = ths[:-20]

fig, ax = plt.subplots()
ax.plot(ths, prec, label='Precision')
ax.plot(ths, sens, label='Sensitivity')
ax.set_title('Precision and Sensitivity by Threshold')
```

```
ax.set_xlabel('Threshold')
ax.set_ylabel('Precision and Sensitivity')
ax.legend()
```

This generates the following plot:

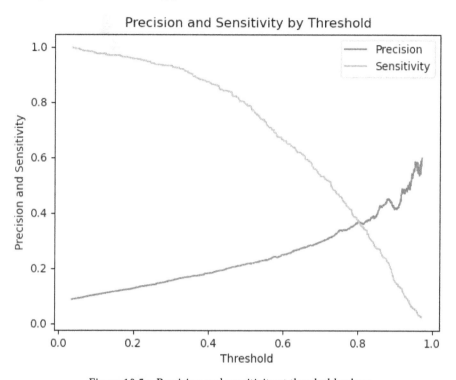

Figure 10.5 – Precision and sensitivity at threshold values

As the threshold increases beyond 0.2, there is a sharper decrease in sensitivity than there is an increase in precision.

8.  It is often also helpful to look at the false positive rate with the sensitivity rate. The false positive rate is the propensity of our model to predict positive when the actual value is negative. One way to see that relationship is with a ROC curve:

```
fpr, tpr, ths = skmet.roc_curve(y_test, pred_probs)
ths = ths[1:]
fpr = fpr[1:]
tpr = tpr[1:]

fig, ax = plt.subplots()
ax.plot(fpr, tpr, linewidth=4, color="black")
```

```
ax.set_title('ROC curve')
ax.set_xlabel('False Positive Rate')
ax.set_ylabel('Sensitivity')
```

This produces the following plot:

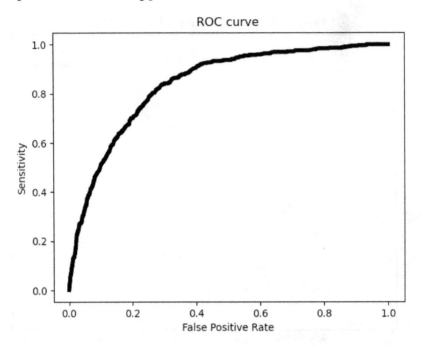

Figure 10.6 – ROC curve

Here, we can see that increasing the false positive rate buys us less increase in sensitivity the higher the false positive rate is. Beyond a false positive rate of 0.5, there is not much payoff at all.

9.  It may also be helpful to just plot the false positive rate and sensitivity by threshold:

```
fig, ax = plt.subplots()
ax.plot(ths, fpr, label="False Positive Rate")
ax.plot(ths, tpr, label="Sensitivity")
ax.set_title('False Positive Rate and Sensitivity by
Threshold')
ax.set_xlabel('Threshold')
ax.set_ylabel('False Positive Rate and Sensitivity')
ax.legend()
```

This produces the following plot:

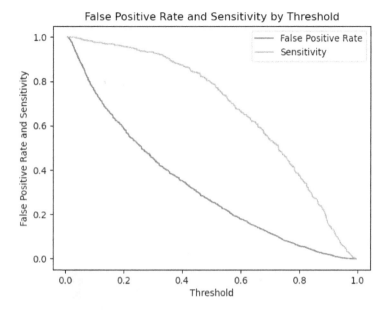

Figure 10.7 – Sensitivity and false positive rate

Here, we can see that as we lower the threshold below 0.25, the false positive rate increases more rapidly than sensitivity.

These last two visualizations hint at the possibility of finding an optimal threshold value – that is, one with the best tradeoff between sensitivity and the false positive rate; at least mathematically, ignoring domain knowledge.

10.  We will calculate the **Youden J** statistic to find this threshold value. We get this by passing a vector, which is the difference between the true positive and false positive rates at each threshold, to NumPy's argmax function. We want the value of the threshold at that index. The optimal threshold according to this calculation is 0.46, which isn't very different from the default:

```
jthresh = ths[np.argmax(tpr - fpr)]
jthresh
0.45946882675453804
```

11.  We can redo the confusion matrix based on this alternative threshold:

```
pred2 = np.where(pred_probs>=jthresh,1,0)
cm = skmet.confusion_matrix(y_test, pred2)
cmplot = skmet.ConfusionMatrixDisplay(
```

```
    confusion_matrix=cm,
    display_labels=['Negative', 'Positive'])
  cmplot.plot()
  cmplot.ax_.set(
    title='Heart Disease Prediction Confusion Matrix',
    xlabel='Predicted Value', ylabel='Actual Value')
```

This produces the following plot:

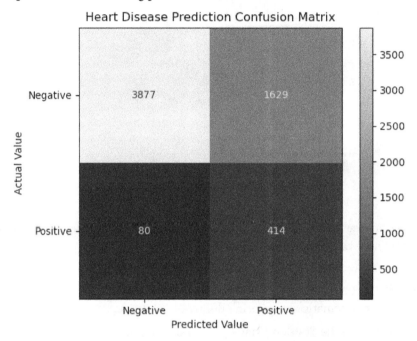

Figure 10.8 – Confusion matrix of heart disease prediction

12. This gives us a small improvement in sensitivity:

```
skmet.recall_score(y_test.values.ravel(), pred)
0.7935222672064778
skmet.recall_score(y_test.values.ravel(), pred2)
0.8380566801619433
```

The point here is not that we should change thresholds willy-nilly. This is often a bad idea. But we should keep two points in mind. First, when we have a highly imbalanced class, a 0.5 threshold may not make sense. Second, this is an important place to lean on domain knowledge. For some classification problems, a false positive is substantially less important than a false negative.

In this section, we focused on sensitivity, precision, and false positive rate as measures of model performance. That is partly because of space limitations, but also because of the issues with this particular target – imbalance classes and the likely preference for sensitivity. We will be emphasizing other measures, such as accuracy and specificity, in other models that we will be building in the next few chapters. In the rest of this chapter, we will look at a couple of extensions of logistic regression, regularization and multinomial logistic regression.

# Regularization with logistic regression

If you have already worked your way through *Chapter 7*, *Linear Regression Models*, and read the first section of this chapter, you already have a good idea of how regularization works. We add a penalty to the estimator that minimizes our parameter estimates. The size of that penalty is typically tuned based on a measure of model performance. We will work through that in this section. Follow these steps:

1. We will load the same modules that we worked with in the previous section, plus the modules we will need for the necessary hyperparameter tuning. We will use RandomizedSearchCV and uniform to find the best value for our penalty strength:

```
import pandas as pd
import numpy as np
from sklearn.model_selection import train_test_split
from sklearn.preprocessing import StandardScaler
from sklearn.preprocessing import OneHotEncoder
from sklearn.pipeline import make_pipeline
from sklearn.impute import SimpleImputer
from sklearn.compose import ColumnTransformer
from sklearn.model_selection import
RepeatedStratifiedKFold
from sklearn.linear_model import LogisticRegression
from sklearn.model_selection import RandomizedSearchCV
from scipy.stats import uniform
import os
import sys
sys.path.append(os.getcwd() + "/helperfunctions")
from preprocfunc import OutlierTrans,\
    MakeOrdinal, ReplaceVals
```

2. Next, we will load the heart disease data and do a little processing:

```
healthinfo = pd.read_csv("data/healthinfosample.csv")

healthinfo.set_index("personid", inplace=True)

healthinfo['heartdisease'] = \
  np.where(healthinfo.heartdisease=='No',0,1).\
  astype('int')
```

3. Next, we will organize our features to facilitate the column transformation we will do in a couple of steps:

```
num_cols = ['bmi','physicalhealthbaddays',
    'mentalhealthbaddays','sleeptimenightly']
binary_cols = ['smoking','alcoholdrinkingheavy',
  'stroke','walkingdifficult','physicalactivity',
  'asthma','kidneydisease','skincancer']
cat_cols = ['gender','ethnicity']
spec_cols1 = ['agecategory']
spec_cols2 = ['genhealth']
spec_cols3 = ['diabetic']

rep_dict = {
  'genhealth': {'Poor':0,'Fair':1,'Good':2,
    'Very good':3,'Excellent':4},
  'diabetic': {'No':0,
    'No, borderline diabetes':0,'Yes':1,
    'Yes (during pregnancy)':1}
}
```

4. Now, we must create testing and training DataFrames:

```
X_train, X_test, y_train, y_test = \
  train_test_split(healthinfo[num_cols +
    binary_cols + cat_cols + spec_cols1 +
    spec_cols2 + spec_cols3],\
  healthinfo[['heartdisease']], test_size=0.2,
    random_state=0)
```

5.  Then, we must set up the column transformations:

```
ohe = OneHotEncoder(drop='first', sparse=False)

standtrans = make_pipeline(OutlierTrans(3),
  SimpleImputer(strategy="median"),
  StandardScaler())
spectrans1 = make_pipeline(MakeOrdinal(),
  StandardScaler())
spectrans2 = make_pipeline(ReplaceVals(rep_dict),
  StandardScaler())
spectrans3 = make_pipeline(ReplaceVals(rep_dict))
bintrans = make_pipeline(ohe)
cattrans = make_pipeline(ohe)
coltrans = ColumnTransformer(
  transformers=[
    ("stand", standtrans, num_cols),
    ("spec1", spectrans1, spec_cols1),
    ("spec2", spectrans2, spec_cols2),
    ("spec3", spectrans3, spec_cols3),
    ("bin", bintrans, binary_cols),
    ("cat", cattrans, cat_cols),
  ]
)
```

6.  Now, we are ready to run our model. We will instantiate logistic regression and repeated stratified k-fold objects. Then, we will create a pipeline with our column transformation from the previous step and the logistic regression.

    After that, we will create a list of dictionaries for our hyperparameters, rather than just one dictionary, as we have done previously in this book. This is because not all hyperparameters work together. For example, we cannot use an L1 penalty with a newton-cg solver. The logisticregression__ (note the double underscore) prefix to the dictionary key names indicates that we want the values to be passed to the logistic regression step of our pipeline.

We will set the n_iter parameter to 20 for our randomized grid search to get it to sample hyperparameters 20 times. Each of those times, the grid search will select from the hyperparameters listed in one of the dictionaries. We will indicate that we want the grid search scoring to be based on the area under the ROC curve:

```
lr = LogisticRegression(random_state=1, class_
weight='balanced', max_iter=1000)


kf = RepeatedStratifiedKFold(n_splits=7, n_repeats=3,
random_state=0)


pipe1 = make_pipeline(coltrans, lr)


reg_params = [
    {
      'logisticregression__solver': ['liblinear'],
      'logisticregression__penalty': ['l1','l2'],
      'logisticregression__C': uniform(loc=0, scale=10)
    },
    {
      'logisticregression__solver': ['newton-cg'],
      'logisticregression__penalty': ['l2'],
      'logisticregression__C': uniform(loc=0, scale=10)
    },
    {
      'logisticregression__solver': ['saga'],
      'logisticregression__penalty': ['elasticnet'],
      'logisticregression__l1_ratio': uniform(loc=0,
scale=1),
      'logisticregression__C': uniform(loc=0, scale=10)
    }
]


rs = RandomizedSearchCV(pipe1, reg_params, cv=kf,
   n_iter=20, scoring='roc_auc')


rs.fit(X_train, y_train.values.ravel())
```

7.  After fitting the search, the `best_params` attribute gives us the parameters associated with the highest score. Elastic net regression, with an L1 ratio closer to L1 than to L2, performs the best:

```
rs.best_params_
{'logisticregression__C': 0.6918282397356423,
 'logisticregression__l1_ratio': 0.758705704020254,
 'logisticregression__penalty': 'elasticnet',
 'logisticregression__solver': 'saga'}


rs.best_score_
0.8410275986723489
```

8.  Let's look at some of the other top scores from the grid search. The best three models have pretty much the same score. One uses elastic net regression, another L1, and another L2.

The `cv_results_` dictionary of the grid search provides us with lots of information about the 20 models that were tried. The `params` list in that dictionary has a somewhat complicated structure because some keys are not present for some iterations, such as `L1_ratio`. We can use `json_normalize` to flatten the structure:

```
results = \
  pd.DataFrame(rs.cv_results_['mean_test_score'], \
    columns=['meanscore']).\
  join(pd.json_normalize(rs.cv_results_['params'])).\
  sort_values(['meanscore'], ascending=False)


results.head(3).T
```

|                                  | 15        | 4         | 12        |
| -------------------------------- | --------- | --------- | --------- |
| meanscore                        | 0.841     | 0.841     | 0.841     |
| logisticregression__C            | 0.692     | 1.235     | 0.914     |
| logisticregression__l1_ratio     | 0.759     | NaN       | NaN       |
| logisticregression__penalty      | elasticnet | l1       | l2        |
| logisticregression__solver       | saga      | liblinear | liblinear |

9. Let's take a look at the confusion matrix:

```
pred = rs.predict(X_test)

cm = skmet.confusion_matrix(y_test, pred)
cmplot = \
  skmet.ConfusionMatrixDisplay(confusion_matrix=cm,
  display_labels=['Negative', 'Positive'])
cmplot.plot()
cmplot.ax_.\
  set(title='Heart Disease Prediction Confusion Matrix',
  xlabel='Predicted Value', ylabel='Actual Value')
```

This generates the following plot:

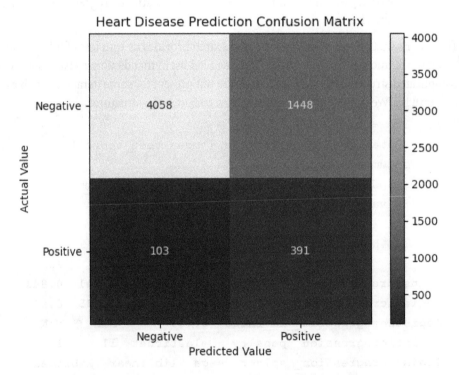

Figure 10.9 – Confusion matrix of heart disease prediction

10. Let's also look at some metrics. Our scores are largely unchanged from our model without regularization:

```
print("accuracy: %.2f, sensitivity: %.2f, specificity:
%.2f, precision: %.2f" %
  (skmet.accuracy_score(y_test.values.ravel(), pred),
  skmet.recall_score(y_test.values.ravel(), pred),
  skmet.recall_score(y_test.values.ravel(), pred,
    pos_label=0),
  skmet.precision_score(y_test.values.ravel(), pred)))
accuracy: 0.74, sensitivity: 0.79, specificity: 0.74,
precision: 0.21
```

Even though regularization provided no obvious improvement in the performance of our model, there are many times when it does. It is also not as necessary to worry about feature selection when using L1 regularization, as the weights for less important features will be 0.

We still haven't dealt with how to handle models where the target has more than two possible values, though almost all the discussion in the last two sections applies to multiclass models as well. In the next section, we will learn how to use multinomial logistic regression to model multiclass targets.

# Multinomial logistic regression

Logistic regression would not be as useful if it only worked for binary classification problems. Fortunately, we can use multinomial logistic regression when our target has more than two values.

In this section, we will work with data on machine failures as a function of air and process temperature, torque, and rotational speed.

> **Note**
>
> This dataset on machine failure is available for public use at https://www.kaggle.com/datasets/shivamb/machine-predictive-maintenance-classification. There are 10,000 observations, 12 features, and two possible targets. One is binary – that is, the machine failed or didn't. The other has types of failure. The instances in this dataset are synthetic, generated by a process designed to mimic machine failure rates and causes.

Let's learn how to use multinomial logistic regression to model machine failure:

1.  First, we will import the now-familiar libraries. We will also import `cross_validate`, which we first used in *Chapter 6, Preparing for Model Evaluation*, to help us evaluate our model:

```
import pandas as pd
import numpy as np
from sklearn.model_selection import train_test_split
from sklearn.preprocessing import StandardScaler
from sklearn.preprocessing import OneHotEncoder
from sklearn.pipeline import make_pipeline
from sklearn.impute import SimpleImputer
from sklearn.compose import ColumnTransformer
from sklearn.model_selection import
RepeatedStratifiedKFold
from sklearn.linear_model import LogisticRegression
from sklearn.model_selection import cross_validate
import os
import sys
sys.path.append(os.getcwd() + "/helperfunctions")
from preprocfunc import OutlierTrans
```

2.  We will load the machine failure data and take a look at its structure. We do not have any missing data. That's great news:

```
machinefailuretype = pd.read_csv("data/
machinefailuretype.csv")
machinefailuretype.info()
<class 'pandas.core.frame.DataFrame'>
RangeIndex: 10000 entries, 0 to 9999
Data columns (total 10 columns):
```

| #   | Column             | Non-Null Count   | Dtype   |
| --- | ------             | --------------   | ----    |
| 0   | udi                | 10000 non-null   | int64   |
| 1   | product            | 10000 non-null   | object  |
| 2   | machinetype        | 10000 non-null   | object  |
| 3   | airtemp            | 10000 non-null   | float64 |
| 4   | processtemperature | 10000 non-null   | float64 |

```
  5   rotationalspeed     10000 non-null      int64
  6   torque              10000 non-null      float64
  7   toolwear            10000 non-null      int64
  8   fail                10000 non-null      int64
  9   failtype            10000 non-null      object
dtypes: float64(3), int64(4), object(3)
memory usage: 781.4+ KB
```

3. Let's look at a few rows. `machinetype` has values of L, M, and H. These values are proxies for machines of low, medium, and high quality, respectively:

```
machinefailuretype.head()
    udi product machinetype airtemp processtemperature\
0   1   M14860   M           298     309
1   2   L47181   L           298     309
2   3   L47182   L           298     308
3   4   L47183   L           298     309
4   5   L47184   L           298     309


    Rotationalspeed   torque   toolwear   fail   failtype
0   1551              43       0          0      No Failure
1   1408              46       3          0      No Failure
2   1498              49       5          0      No Failure
3   1433              40       7          0      No Failure
4   1408              40       9          0      No Failure
```

4. We should also generate some frequencies:

```
machinefailuretype.failtype.value_counts(dropna=False).
sort_index()
Heat Dissipation Failure    112
No Failure                  9652
Overstrain Failure          78
Power Failure               95
Random Failures             18
Tool Wear Failure           45
Name: failtype, dtype: int64
```

```
machinefailuretype.machinetype.\
  value_counts(dropna=False).sort_index()
```
**H    1003**
**L    6000**
**M    2997**
**Name: machinetype, dtype: int64**

5. Let's collapse the `failtype` values and create numeric code for them. We will combine random failures and tool wear failures since the counts are so low for random failures:

```
def setcode(typetext):
  if (typetext=="No Failure"):
    typecode = 1
  elif (typetext=="Heat Dissipation Failure"):
    typecode = 2
  elif (typetext=="Power Failure"):
    typecode = 3
  elif (typetext=="Overstrain Failure"):
    typecode = 4
  else:
    typecode = 5
  return typecode

machinefailuretype["failtypecode"] = \
  machinefailuretype.apply(lambda x: setcode(x.failtype),
axis=1)
```

6. We should confirm that `failtypecode` does what we intended:

```
machinefailuretype.groupby(['failtypecode','failtype']).
size().\
  reset_index()
```

| | failtypecode | failtype | 0 |
|---|---|---|---|
| 0 | 1 | No Failure | 9652 |
| 1 | 2 | Heat Dissipation Failure | 112 |
| 2 | 3 | Power Failure | 95 |
| 3 | 4 | Overstrain Failure | 78 |
| 4 | 5 | Random Failures | 18 |
| 5 | 5 | Tool Wear Failure | 45 |

7.  Let's also get some descriptive statistics:

```
num_cols =
['airtemp','processtemperature','rotationalspeed',
   'torque','toolwear']
cat_cols = ['machinetype']


machinefailuretype[num_cols].agg(['min','median','max']).T
```

|                    | min   | median | max   |
|--------------------|-------|--------|-------|
| airtemp            | 295   | 300    | 304   |
| processtemperature | 306   | 310    | 314   |
| rotationalspeed    | 1,168 | 1,503  | 2,886 |
| torque             | 4     | 40     | 77    |
| toolwear           | 0     | 108    | 253   |

8.  Now, let's create the testing and training DataFrames. We will also set up the column transformations:

```
X_train, X_test, y_train, y_test =  \
   train_test_split(machinefailuretype[num_cols +
   cat_cols], machinefailuretype[['failtypecode']],
   test_size=0.2, random_state=0)


ohe = OneHotEncoder(drop='first', sparse=False)


standtrans = make_pipeline(OutlierTrans(3),
   SimpleImputer(strategy="median"),
   StandardScaler())
cattrans = make_pipeline(ohe)
coltrans = ColumnTransformer(
   transformers=[
      ("stand", standtrans, num_cols),
      ("cat", cattrans, cat_cols),
   ]
)
```

9. Now, let's set up a pipeline with our column transformations and our multinomial logistic regression model We just need to set the `multi_class` attribute to multinomial when we instantiate the logistic regression:

```
lr = LogisticRegression(random_state=0,
  multi_class='multinomial', solver='lbfgs',
  max_iter=1000)


kf = RepeatedStratifiedKFold(n_splits=10,
  n_repeats=5, random_state=0)


pipe1 = make_pipeline(coltrans, lr)
```

10. Now, we can generate a confusion matrix:

```
cm = skmet.confusion_matrix(y_test,
    pipe1.fit(X_train, y_train.values.ravel()).\
    predict(X_test))
cmplot = \
    skmet.ConfusionMatrixDisplay(confusion_matrix=cm,
    display_labels=['None',
'Heat','Power','Overstrain','Other'])
cmplot.plot()
cmplot.ax_.\
    set(title='Machine Failure Type Confusion Matrix',
    xlabel='Predicted Value', ylabel='Actual Value')
```

This produces the following plot:

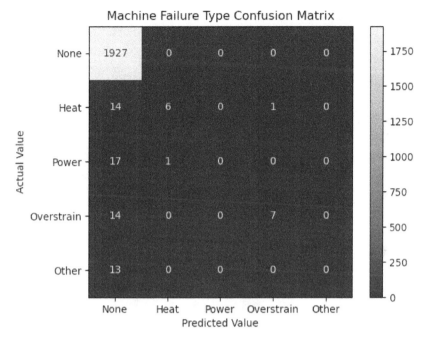

Figure 10.10 – Confusion matrix of predicted machine failure types

The confusion matrix shows that our model does not do a good job of predicting the failure type when there is a failure, particularly with power failures or other failures.

11.  We can use `cross_validate` to evaluate this model. We mainly get excellent scores for accuracy, precision, and sensitivity (recall). However, this is misleading. The weighted scores when the classes are so imbalanced (almost all of the instances have `no failure`) are very heavily influenced by the class that contains almost all of the values. Our model gets `no failure` correct reliably.

If we look at the `f1_macro` score (recall from *Chapter 6, Preparing for Model Evaluation*, that `f1` is the harmonic mean of precision and sensitivity), we will see that our model does not do very well for classes other than the `no failure` class. (The `macro` score is just a simple average.)

We could have just used a classification report here, as we did in *Chapter 6, Preparing for Model Evaluation*, but I sometimes find it helpful to generate the stats I need:

```
scores = cross_validate(
    pipe1, X_train, y_train.values.ravel(), \
    scoring=['accuracy', 'precision_weighted',
             'recall_weighted', 'f1_macro',
```

```
                'f1_weighted'],
   cv=kf, n_jobs=-1)


accuracy, precision, sensitivity, f1_macro, f1_weighted =
\
   np.mean(scores['test_accuracy']),\
   np.mean(scores['test_precision_weighted']),\
   np.mean(scores['test_recall_weighted']),\
   np.mean(scores['test_f1_macro']),\
   np.mean(scores['test_f1_weighted'])


accuracy, precision, sensitivity, f1_macro, f1_weighted
(0.9716499999999999,
 0.9541025493784612,
 0.9716499999999999,
 0.3820938909478524,
 0.9611411229222823)
```

In this section, we explored how to construct a multinomial logistic regression model. This approach works regardless of whether the target is nominal or ordinal. In this case, it was nominal. We also saw how we can extend the model evaluation approaches we used for logistic regression with a binary target. We reviewed how to interpret the confusion matrix and scoring metrics when we have more than two classes.

## Summary

Logistic regression has been a go-to tool for me for many, many years when I have needed to predict a categorical target. It is an efficient algorithm with low bias. Some of its disadvantages, such as high variance and difficulty handling highly correlated predictors, can be addressed with regularization and feature selection. We went over examples of doing that in this chapter. We also examined how to handle imbalanced classes in terms of what such targets mean for modeling and interpretation of results.

In the next chapter, we will look at a very popular alternative to logistic regression for classification – decision trees. We will see that decision trees have many advantages that make them a particularly good option if we need to model complexity, without having to worry as much about how our features are specified as we do with logistic regression.

# 11
# Decision Trees and Random Forest Classification

Decision trees and random forests are very popular classification models. This is partly because they are easy to train and interpret. They are also quite flexible. We can model complexity without necessarily having to increase the feature space or transform features. We do not even need to do anything special to apply the algorithm to multiclass problems, something we had to do with logistic regression.

On the other hand, decision trees can be less stable than other classification models, being fairly sensitive to small changes in the training data. Decision trees can also be biased when there is a significant class imbalance (when there are many more observations in one class than another). Fortunately, these issues can be addressed with techniques such as bagging to reduce variance and oversampling to deal with imbalance. We will examine these techniques in this chapter.

In this chapter, we will cover the following topics:

- Key concepts
- Decision tree models
- Implementing random forest
- Implementing gradient boosting

# Technical requirements

In addition to the scikit-learn modules we have been using so far, we will use SMOTENC from Imbalanced-learn. We will use SMOTENC to address class imbalance. The Imbalanced-learn library can be installed with `pip install -U imbalanced-learn`. All the code in this chapter was tested with scikit-learn versions 0.24.2 and 1.0.2.

# Key concepts

Decision trees are an exceptionally useful machine learning tool. They are non-parametric, easy to interpret, and can work with a wide range of data. No assumptions regarding linearity of relationships between features and targets, and normality of error terms, are made. It isn't even necessary to scale the data. Decision trees also often do a good job of capturing complex relationships between predictors and targets.

The flexibility of the decision tree algorithm, and its ability to model complicated and unanticipated relationships in the data, is due to the **recursive partitioning** procedure that's used to segment the data. Decision trees group observations based on the values of their features. This is done with a series of binary decisions, starting from an initial split at the root node, and ending with a leaf for each grouping. Each split is based on the feature, and feature values, that provide the most information about the target. More precisely, a split is chosen based on whether it produces the lowest Gini Impurity score. We will discuss Gini Impurity in more detail later.

All new observations with the same values, or the same range of values, along the branches from the root node to the leaf, get the same predicted value for the target. When the target is categorical, that is the most frequent value for the target for the training observations at that leaf.

The following diagram provides a fairly straightforward example of a decision tree, with made-up data and results for a model of college completion among those who attended college. For this decision tree, an initial split of high school GPA by those with a 3.0 or less and those with a GPA of greater than 3.0 was found to lead to the lowest impurity compared to other available features, as well as other thresholds. Therefore, high school GPA is our root node, also known as depth 0:

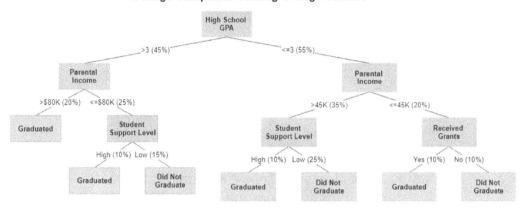

Figure 11.1 – A decision tree for college completion

The binary split at the root node results in 45% of the observations on the left-hand side of the tree and 55% on the right. At depth 1, there are binary splits on both sides based on parental income, though the threshold is different; $80k and $45k for the left and right-hand sides, respectively. For parental incomes above $80k, where the high school GPA is greater than 3, there is no more splitting. Here, we get to a prediction of **Graduated**. This is a leaf node.

We can navigate up the tree from each leaf to describe how the tree has segmented the data, just as we did for individuals with parental income above $80k and a high school GPA above 3. For example, the decision tree predicts not graduating for individuals who receive low levels of support from their schools, have parental incomes greater than $45k, and have a high school GPA less than or equal to 3.

So, how does the decision tree algorithm perform this magic? How does it select the features and the threshold amounts or class values? Why greater than $80k or $45k for parental income? Why received grants at depth 2 (split 3) for parental income less than or equal to $45k, but student support level for depth 2 for other leaves? One leaf does not even have a further split at depth 2.

One way of measuring the information about a class that a binary split provides is by how much it helps us distinguish between in-class and out-of-class membership. We frequently use Gini Impurity calculations for that evaluation, though entropy is sometimes used. The Gini Impurity statistic tells us how well class membership has been segmented at each node. This can be seen in the following formula:

$$Gini\ Impurity = 1 - \sum\_(k = 1)^{\wedge}m \text{▓} P\_k^{\wedge}2$$

Here, $P\_k$ is the probability of membership in the $k$ class and $m$ is the number of classes. If class membership is equal at a node, then Gini Impurity is 0.5. When completely pure, it is 0.

It might be helpful to try calculating Gini Impurity by hand to get a better sense of how it works. We can do that for the very simple decision tree shown in the following diagram. There are just two leaf nodes – one for individuals with a high school GPA greater than 3 and one for a high school GPA less than or equal to 3. (Again, these counts are made up for expository purposes. We assume the percentages used in *Figure 11.1* are counts out of 100 people.) Take a look at the following diagram:

## A Simplified Decision Tree With Gini Impurity
## College Completion Among College Starters

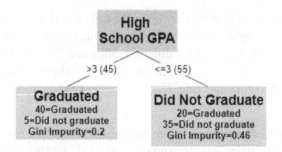

Figure 11.2 - A decision tree with one split and Gini Impurity calculations

Based on this model, **Graduated** would be predicted for folks with a high GPA since most of those individuals, 40 out of 45, did graduate. Gini Impurity is fairly low, which is good. We can calculate Gini Impurity for that node using the preceding formula:

$$1 - (40/45)^2 + (5/45)^2 = 0.2$$

Our model would predict **Did not graduate** for individuals with a high school GPA less than or equal to 3, since the majority of those individuals did not graduate from college. However, there is much less purity here. The Gini Impurity value for that node is as follows:

$$1 - (20/55)^2 + (35/55)^2 = 0.46$$

The decision tree algorithm calculates the weighted sum of the Gini Impurity values for all possible splits from a given point and chooses the split with the lowest score. The algorithm would follow a similar process if entropy was used rather than Gini Impurity. We will use scikit-learn's **classification and regression tree** (**CART**) algorithm to build decision trees in this chapter. That tool defaults to using Gini Impurity, though we could get it to use entropy instead.

A decision tree is what we call a greedy learner. The algorithm chooses a split that gives us the best Gini Impurity or entropy score at the current level. It does not examine how that choice affects the splits available subsequently, reconsidering the choice at the current level based on that information. This makes the algorithm much more efficient than it otherwise would be, but it may not provide the globally optimal solution.

The main disadvantage of decision trees is their high variance. They can overfit anomalous observations in the training data and not do well with new data as a result. Depending on the characteristics of our data, we can get a very different model each time we fit a decision tree. We can use ensemble methods, such as bagging or random forest, to address this issue.

# Using random forest for classification

Random forests, perhaps not surprisingly, are collections of decision trees. But this would not distinguish a random forest from bootstrap aggregating, commonly referred to as bagging. Bagging is often used to reduce the variance of machine learning algorithms, such as decision trees, that have high variances. With bagging, we generate random samples from our dataset, say 100. Then, we run our model, such as a decision tree classifier, on each of those samples, averaging the predictions.

However, the samples that are generated with bagging can be correlated, and the resulting decision trees may have many similarities. This is more likely to be the case when there are just a few features that explain much of the variation. Random forests address this issue by limiting the number of features that can be selected for each split. A good rule of thumb for a decision tree classification model is to take the square root of the number of features available to determine the number of features to use. For example, if there are 25 features, we would use five for each split.

Let's be a little more precise about the steps involved in constructing a random forest:

1. Randomly sample, with replacement, instances from the training data. The sample has the same number of observations as the original dataset.
2. Randomly select, with replacement, features from the sample. (A good number of features to select each time is the square root of the total number of features available.)

3.  Identify a feature, from those randomly selected in *Step 2*, where a split would result in nodes with the greatest purity.

4.  **Inner Loop**: Repeat *Steps 2* and *3* until a decision tree is constructed.

5.  **Outer Loop**: Repeat all steps, including the inner loop, until the desired number of trees is created. The result across all trees is determined by voting; that is, the class is predicted based on the most frequent class label across all trees for the given feature values.

One cool secondary effect of this process is that it generates test data for us, so to speak. The bootstrapping process – that is, sampling with replacement – will lead to many instances being left out of the sample for one or more trees, often as much as a third. These instances, known as **out-of-bag** samples, can be used to evaluate the model.

Basing our class predictions on many uncorrelated decision trees has the positive effect on variance (lowering it!) that you would expect. Random forest models are often more generalizable than decision tree models. They are less vulnerable to overfitting and less likely to be yanked around by anomalous data. But this comes with a price. Building a hundred or more decision trees draws more on system resources than just building one. We also lose the ease of interpretation of decision trees; it is harder to explain the importance of each feature.

# Using gradient-boosted decision trees

Conceptually, gradient-boosted decision trees are similar to a random forest. They rely on multiple decision trees to improve model performance. But they are performed sequentially, with each tree learning from the previous ones. Each new tree works from the residuals of the previous iteration.

The rate at which a gradient-boosted decision tree learns is determined by the $a$ hyperparameter. You might be wondering why would we not want our model to learn as fast as possible. A faster learning rate is more efficient and taxes system resources less. However, we can build a model that is more generalizable with a lower learning rate. There is less risk of overfitting. The optimal learning rate is ultimately an empirical question. We need to do some hyperparameter tuning to find that. We will do this in the last section of this chapter.

Just as was the case with random forests, we can improve our intuition about gradient boosting by going through the steps for a binary classification problem:

1.  Make an initial prediction for the target based on the mean value of the target across the sample. For a binary target, this is a proportion. Assign this prediction to all observations. (We use the log of the odds of class membership here.)

2.  Calculate the residual for each instance, which will be 1 minus the initial prediction for in-class instances, and 0 minus the prediction, or -prediction, for out-of-class instances.

3.  Construct a decision tree to predict the residuals.

4.  Generate new predictions based on the decision tree model.

5.  Adjust the previous prediction for each instance based on the new predictions scaled by the learning rate. As mentioned previously, we use a learning rate because we do not want predictions to move in the right direction too quickly.

6.  Loop back to *step 3* unless the maximum number of trees has been reached or the residuals are very small.

While this is a simplified explanation of how gradient boosting works, it does provide a good feel of what the algorithm is doing for us. Hopefully, it also helps you understand why gradient boosting has become so popular. The algorithm works repeatedly to adjust to previous errors but does so relatively efficiently and with less risk of overfitting than decision trees alone.

We will work through examples of decision trees, random forests, and gradient boosting in the rest of this chapter. We will discuss how to tune hyperparameters and how to evaluate these models. We will also go over the advantages and disadvantages of each approach.

# Decision tree models

We will work with the heart disease data again in this chapter. This will be a great way to compare our results from the logistic regression model to those of a non-parametric model such as a decision tree. Follow these steps:

1.  First, we load the same libraries that we have been using so far. The new modules are `DecisionTreeClassifier` from scikit-learn and `SMOTENC` from Imbalance Learn, which will help us deal with imbalanced data:

```
import pandas as pd
import numpy as np
from sklearn.model_selection import train_test_split
from sklearn.preprocessing import OneHotEncoder
from imblearn.pipeline import make_pipeline
from sklearn.compose import ColumnTransformer
from sklearn.model_selection import RandomizedSearchCV
from imblearn.over_sampling import SMOTENC
```

```
from sklearn.tree import DecisionTreeClassifier, plot_
tree
from scipy.stats import randint
import sklearn.metrics as skmet
import os
import sys
sys.path.append(os.getcwd() + "/helperfunctions")
from preprocfunc import MakeOrdinal,\
   ReplaceVals
```

2.  Let's load the heart disease data. We will convert our target, `heartdisease`, into a 0 and 1 integer. We will not repeat the frequencies and descriptive statistics that we generated from the data in the previous chapter. It is probably helpful to take a quick look at them if you did not work through *Chapter 10, Logistic Regression*:

```
healthinfo = pd.read_csv("data/healthinfosample.csv")

healthinfo.set_index("personid", inplace=True)

healthinfo.heartdisease.value_counts()
No      27467
Yes      2533
Name: heartdisease, dtype: int64

healthinfo['heartdisease'] = \
   np.where(healthinfo.heartdisease=='No',0,1).\
   astype('int')

healthinfo.heartdisease.value_counts()
0       27467
1        2533
Name: heartdisease, dtype: int64
```

Notice the class imbalance. Less than 10% of observations have heart disease. We will need to deal with that in our model.

3. Let's also look at values for the age category feature since it is a tad unusual. It contains character data for the age range. We will convert it into an ordinal feature using the MakeOrdinal class we loaded. For example, values of 18-24 will give us a value of 0, while values of 50-54 will be 6 after we do the transformation later:

```
healthinfo.agecategory.value_counts().\
   sort_index().reset_index()
```

|    | index       | agecategory |
|----|-------------|-------------|
| 0  | 18-24       | 1973        |
| 1  | 25-29       | 1637        |
| 2  | 30-34       | 1688        |
| 3  | 35-39       | 1938        |
| 4  | 40-44       | 2007        |
| 5  | 45-49       | 2109        |
| 6  | 50-54       | 2402        |
| 7  | 55-59       | 2789        |
| 8  | 60-64       | 3122        |
| 9  | 65-69       | 3191        |
| 10 | 70-74       | 2953        |
| 11 | 75-79       | 2004        |
| 12 | 80 or older | 2187        |

4. We should organize our features by data type as this will make some tasks easier later. We will also set up a dictionary to recode the genhealth and diabetic features:

```
num_cols = ['bmi','physicalhealthbaddays',
    'mentalhealthbaddays','sleeptimenightly']
binary_cols = ['smoking','alcoholdrinkingheavy',
    'stroke','walkingdifficult','physicalactivity',
    'asthma','kidneydisease','skincancer']
cat_cols = ['gender','ethnicity']
spec_cols1 = ['agecategory']
spec_cols2 = ['genhealth','diabetic']

rep_dict = {
   'genhealth': {'Poor':0,'Fair':1,'Good':2,
      'Very good':3,'Excellent':4},
   'diabetic': {'No':0,
```

```
      'No, borderline diabetes':0,'Yes':1,
      'Yes (during pregnancy)':1}
  }
```

5. Now, let's create training and testing DataFrames:

```
X_train, X_test, y_train, y_test =  \
  train_test_split(healthinfo[num_cols +
    binary_cols + cat_cols + spec_cols1 +
    spec_cols2],\
  healthinfo[['heartdisease']], test_size=0.2,
    random_state=0)
```

6. Next, we will set up the column transformations. We will use our custom classes to encode the agecategory feature as ordinal and replace character values with numeric ones for genhealth and diabetic.

   We will not transform the numerical columns since it is not typically necessary to scale those features when using a decision tree. We will not worry about outliers either as decision trees are less sensitive to them than in logistic regression. We will set remainder to passthrough to get the transformer to pass the remaining columns (the numerical columns) through as-is:

```
ohe = OneHotEncoder(drop='first', sparse=False)

spectrans1 = make_pipeline(MakeOrdinal())
spectrans2 = make_pipeline(ReplaceVals(rep_dict))
bintrans = make_pipeline(ohe)
cattrans = make_pipeline(ohe)
coltrans = ColumnTransformer(
  transformers=[
    ("bin", bintrans, binary_cols),
    ("cat", cattrans, cat_cols),
    ("spec1", spectrans1, spec_cols1),
    ("spec2", spectrans2, spec_cols2),
  ],
    remainder = 'passthrough'
)
```

7. We need to do a little work before we run our model. As you will see in the next step, we need to know how many features will be returned from the one-hot encoder. We should also grab the new feature names while we are at it. We will need them later as well. (We only need to fit the column transformer on a small random sample of the data for this purpose.) Take a look at the following code:

```
coltrans.fit(X_train.sample(1000))

new_binary_cols = \
  coltrans.\
  named_transformers_['bin'].\
  named_steps['onehotencoder'].\
  get_feature_names(binary_cols)
new_cat_cols = \
  coltrans.\
  named_transformers_['cat'].\
  named_steps['onehotencoder'].\
  get_feature_names(cat_cols)
```

8. Let's view the feature names:

```
new_cols = np.concatenate((new_binary_cols,
  new_cat_cols, np.array(spec_cols1 + spec_cols2 +
  num_cols)))

new_cols
array(['smoking_Yes', 'alcoholdrinkingheavy_Yes',
       'stroke_Yes', 'walkingdifficult_Yes',
       'physicalactivity_Yes', 'asthma_Yes',
       'kidneydisease_Yes', 'skincancer_Yes',
       'gender_Male', 'ethnicity_Asian',
       'ethnicity_Black', 'ethnicity_Hispanic',
       'ethnicity_Other', 'ethnicity_White',
       'agecategory', 'genhealth', 'diabetic', 'bmi',
       'physicalhealthbaddays', 'mentalhealthbaddays',
       'sleeptimenightly'], dtype=object)
```

9.  We need to deal with our imbalanced dataset before we fit our decision tree. We can use the SMOTENC module from Imbalanced-learn to oversample the heart disease class. This will generate enough representative instances of the heart disease class to balance the class memberships.

Next, we must instantiate a decision tree classifier and indicate that the leaf nodes need to have at least five observations and that the tree depth cannot exceed two. Then, we will use the column transformer to transform the training data and fit the model.

We will do some hyperparameter tuning later in this section. For now, we just want to produce a decision tree that is easy to interpret and visualize.

SMOTENC needs to know which columns are categorical. When we set up the column transformer, we encoded the binary columns first, and then the categorical columns. So, the number of binary columns plus the number of categorical columns gives us the endpoint of those column indexes. Then, we must pass a range, starting from 0 and ending at the number of categorical columns, to the categorical_features parameter of SMOTENC.

Now, we can create a pipeline containing the column transformations, the oversampling, and the decision tree classifier, and fit it:

```
catcolscnt = new_binary_cols.shape[0] + \
  new_cat_cols.shape[0]


smotenc = \
  SMOTENC(categorical_features=np.arange(0,catcolscnt),
  random_state=0)


dtc_example = DecisionTreeClassifier(
  min_samples_leaf=5, max_depth=2)


pipe0 = make_pipeline(coltrans, smotenc, dtc_example)


pipe0.fit(X_train, y_train.values.ravel())
```

> **Note**
>
> Oversampling can be a good option when we are concerned that our model is doing a poor job of capturing variation in a class because we have too few instances of that class, relative to one or more other classes. Oversampling duplicates instances of that class.
>
> **Synthetic Minority Oversampling Technique (SMOTE)** is an algorithm that uses KNN to duplicate instances. The implementation of SMOTE in this chapter is from Imbalanced-learn, specifically SMOTENC, which can handle categorical data.
>
> Oversampling is often done when the class imbalance is even worse than with this dataset, say 100 to 1. Nonetheless, I thought it would be helpful to demonstrate how to use SMOTE and similar tools in this chapter.

10. After running the fit, we can look at which features were identified as important. `agecategory`, `genhealth`, and `diabetic` are important features in this simple model:

```
feature_imp = \
  pipe0.named_steps['decisiontreeclassifier'].\
  tree_.compute_feature_importances(normalize=False)


feature_impgt0 = feature_imp>0


feature_implabs = np.column_stack((feature_imp.\
  ravel(), new_cols))


feature_implabs[feature_impgt0]
array([[0.10241844433036575, 'agecategory'],
       [0.04956947743193013, 'genhealth'],
       [0.012777650193266089, 'diabetic']],
      dtype=object)
```

11. Next, we can generate a graph of the decision tree:

```
plot_tree(pipe0.named_steps['decisiontreeclassifier'],
  feature_names=new_cols,
  class_names=['No Disease','Disease'], fontsize=10)
```

This produces the following plot:

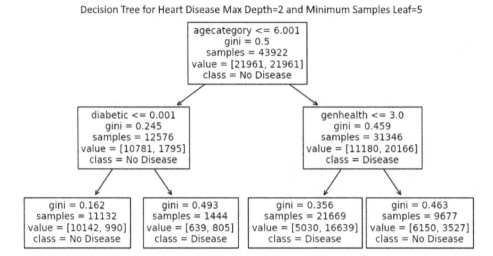

Decision Tree for Heart Disease Max Depth=2 and Minimum Samples Leaf=5

Figure 11.3 – Decision tree example with a heart disease target

The initial binary split, at the root node (also known as depth 0), is based on whether agecategory is less than or equal to 6. (Recall that agecategory was originally a character feature. Initial values of 50-54 get a value of 6 after our encoding.) If the root node statement is true, it leads to the node down one level and to the left. If the statement is false, it leads to the node down one level and to the right. The number of samples is the number of observations that made it to that node. So, the sample value of 12576 at the diabetic<=0.001 (that is, not diabetic) node reflects the number of instances for which the statement from the parent node was true; that is, 12576 instances had values for the age category that were less than or equal to 6.

The value list within each node gives us the instances in each class in the training data. In this case, the first value is for the count of no disease observations. The second value is the count of the observations with heart disease. For example, there are 10781 no disease observations and 1795 disease observations at the diabetic<=0.001 node.

Decision trees predict the most frequent class at the leaf mode. So, this model would predict no disease for individuals less than or equal to 54 (agecategory<=6) and who are not diabetic (diabetic<=0.001). There were 10142 no disease observations for that group in the training data, and 990 disease observations. This gives us a leaf node with a very good Gini Impurity of 0.162.

If the person has diabetes, even if they are 54 or younger, our model predicts heart disease. It predicts this less definitively, however. Gini Impurity is 0.493. By comparison, the prediction of disease for individuals over 54 (agecate-gory<=6.001 is false) that have less than excellent health (genhealth<=3.0) has a significantly lower Gini Impurity.

Our model predicts no disease for individuals over 54 and with general health equal to 4. (We coded general health values of *Excellent* as 4 when we did the column transformations.) However, the poor Gini Impurity score indicates that our model does not exactly make that prediction with confidence.

12.  Let's look at some metrics for this model. The model has okay, though not great, sensitivity, predicting about 70% of the time that there is heart disease. Precision, the rate at which we are correct when we make a positive prediction, is quite low. Only 19% of the time that we make a positive prediction are we correct:

```
pred = pipe0.predict(X_test)


print("accuracy: %.2f, sensitivity: %.2f, specificity:
%.2f, precision: %.2f"  %
  (skmet.accuracy_score(y_test.values.ravel(), pred),
  skmet.recall_score(y_test.values.ravel(), pred),
  skmet.recall_score(y_test.values.ravel(), pred,
    pos_label=0),
  skmet.precision_score(y_test.values.ravel(), pred)))
accuracy: 0.72, sensitivity: 0.70, specificity: 0.73,
precision: 0.19
```

This model makes some fairly arbitrary decisions regarding hyperparameters, setting maximum depth at 2 and the minimum number of samples for a leaf at 5. We should explore alternative values for these hyperparameters that will yield a better-performing model. Let's do a randomized grid search to find those values.

13.  Let's set up a pipeline with the column transformations, the oversampling, and a decision tree classifier. We will also create a dictionary with ranges for the minimum leaf size and maximum tree depth hyperparameters. Note that there are two underscores after decisiontreeclassifier for each dictionary key:

```
dtc = DecisionTreeClassifier(random_state=0)


pipe1 = make_pipeline(coltrans, smotenc, dtc)
```

```
dtc_params = {
 'decisiontreeclassifier__min_samples_leaf': randint(100,
1200),
 'decisiontreeclassifier__max_depth': randint(2, 11)
}
```

14. Now, we can run the randomized grid search. We will run 20 iterations to test a good number of values for our hyperparameters:

```
rs = RandomizedSearchCV(pipe1, dtc_params, cv=5,
  n_iter=20, scoring="roc_auc")


rs.fit(X_train, y_train.values.ravel())


rs.best_params_
{'decisiontreeclassifier__max_depth': 9,
 'decisiontreeclassifier__min_samples_leaf': 954}


rs.best_score_
0.7964540832005679
```

15. Let's look at the score from each iteration:

```
results = \
  pd.DataFrame(rs.cv_results_['mean_test_score'], \
    columns=['meanscore']).\
  join(pd.DataFrame(rs.cv_results_['params'])).\
  sort_values(['meanscore'], ascending=False).\
  rename(columns=\
    {'decisiontreeclassifier__max_depth':'maxdepth',
     'decisiontreeclassifier__min_samples_leaf':\
     'samples'})
```

This produces the following output. The best-performing models have substantially greater max_depth than the model we constructed earlier. Our model also does best with much higher minimum instances for each leaf:

|    | meanscore | maxdepth | samples |
|----|-----------|----------|---------|
| 15 | 0.796     | 9        | 954     |
| 13 | 0.796     | 8        | 988     |

| 4 | 0.795 | 7 | 439 |
| 19 | 0.795 | 9 | 919 |
| 12 | 0.794 | 9 | 856 |
| 3 | 0.794 | 9 | 510 |
| 2 | 0.794 | 9 | 1038 |
| 5 | 0.793 | 8 | 575 |
| 0 | 0.793 | 10 | 1152 |
| 10 | 0.793 | 7 | 1080 |
| 6 | 0.793 | 6 | 1013 |
| 8 | 0.793 | 10 | 431 |
| 17 | 0.793 | 6 | 896 |
| 14 | 0.792 | 6 | 545 |
| 16 | 0.784 | 5 | 180 |
| 1 | 0.778 | 4 | 366 |
| 11 | 0.775 | 4 | 286 |
| 9 | 0.773 | 4 | 138 |
| 18 | 0.768 | 3 | 358 |
| 7 | 0.765 | 3 | 907 |

16. Let's generate a confusion matrix:

```
pred2 = rs.predict(X_test)

cm = skmet.confusion_matrix(y_test, pred2)
cmplot = \
  skmet.ConfusionMatrixDisplay(confusion_matrix=cm,
  display_labels=['Negative', 'Positive'])
cmplot.plot()
cmplot.ax_.\
  set(title='Heart Disease Prediction Confusion Matrix',
  xlabel='Predicted Value', ylabel='Actual Value')
```

This produces the following plot:

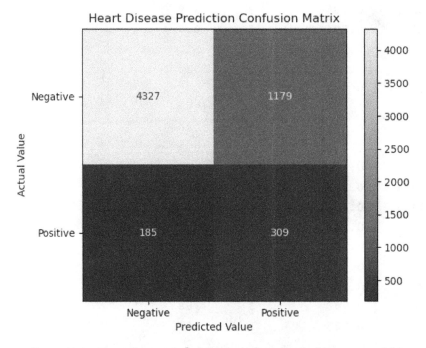

Figure 11.4 – Heart disease confusion matrix from the decision tree model

One thing about this plot that you may have noticed right away is how low our precision is. The vast majority of the time we predict positive, we are wrong.

17. Let's look at the accuracy, sensitivity, specificity, and precision scores to see if there is much improvement in the metrics from the model without hyperparameter tuning. We do noticeably worse with sensitivity, which is 63% now compared to 70% earlier. However, we do a little better with specificity. We correctly predict negative on the testing data 79% of the time now, compared to 73% with the earlier model:

```
print("accuracy: %.2f, sensitivity: %.2f, specificity:
%.2f, precision: %.2f"  %
    (skmet.accuracy_score(y_test.values.ravel(), pred2),
    skmet.recall_score(y_test.values.ravel(), pred2),
    skmet.recall_score(y_test.values.ravel(), pred2,
        pos_label=0),
    skmet.precision_score(y_test.values.ravel(),
        pred2)))
accuracy: 0.77, sensitivity: 0.63, specificity: 0.79,
precision: 0.21
```

Decision trees are a good starting point for classification models. They make very few assumptions about the underlying data and do not require much preprocessing. Notice that we did not do any scaling or outlier detection in this example, as that is often not essential with decision trees. We also get a model that is fairly easy to understand or explain, so long as we limit the number of depths.

We can often improve the performance of our decision tree models with random forests, for reasons we discussed at the beginning of this chapter. A key reason for this is that variance is reduced when using random forests instead of decision trees.

In the next section, we'll take a look at random forests.

# Implementing random forest

Let's try to improve our heart disease model with a random forest:

1. First, let's load the same libraries that we used in the previous section, except we will import the random forest classifier this time:

```
import pandas as pd
import numpy as np
from imblearn.pipeline import make_pipeline
from sklearn.ensemble import RandomForestClassifier
from sklearn.model_selection import RandomizedSearchCV
from scipy.stats import randint
import sklearn.metrics as skmet
import os
import sys
sys.path.append(os.getcwd() + "/helperfunctions")
import healthinfo as hi
```

We also load the healthinfo module; it loads the health information data and does our preprocessing. There is nothing fancy here. The preprocessing code we stepped through earlier was just copied to the helperfunctions subfolder of the current working directory.

2.  Now, let's grab the data that's been processed by the `healthinfo` module so that we can use it for our random forest classifier:

```
X_train = hi.X_train
X_test = hi.X_test
y_train = hi.y_train
y_test = hi.y_test
```

3.  Let's instantiate a random forest classifier and create a pipeline for the grid search. We will also create a dictionary for the hyperparameters to search. In addition to the `max_depth` and `min_samples_leaf` hyperparameters that we used for the decision tree, an important hyperparameter for a random forest is `n_estimators`. This indicates the number of trees to use for that iteration of the search. We will also add `entropy` as a criterion, in addition to `gini`, which we have been using so far:

```
rfc = RandomForestClassifier(random_state=0)

pipe1 = make_pipeline(hi.coltrans, hi.smotenc, rfc)

rfc_params = {
  'randomforestclassifier__min_samples_leaf':
    randint(100, 1200),
  'randomforestclassifier__max_depth':
    randint(2, 11),
  'randomforestclassifier__n_estimators':
    randint(100, 3000),
  'randomforestclassifier__criterion':
    ['gini','entropy']
}
rs = RandomizedSearchCV(pipe1, rfc_params, cv=5,
  n_iter=20, scoring="roc_auc")

rs.fit(X_train, y_train.values.ravel())
```

4.  We can use the `best_params_` and `best_score_` attributes of the randomized grid search object to find the best parameters and the associated score, respectively. The best model has `1023` trees and a maximum depth of `9`.

    Here, we can see some improvements in the `roc_auc` score over the decision tree model from the previous section:

    ```
    rs.best_params_
    {'randomforestclassifier__criterion': 'gini',
     'randomforestclassifier__max_depth': 9,
     'randomforestclassifier__min_samples_leaf': 667,
     'randomforestclassifier__n_estimators': 1023}


    rs.best_score_
    0.8210934290375318
    ```

5.  The random forest's results are more difficult to interpret than those for a single decision tree, but a good place to start is with the feature importance. The top three features are the same as the ones we saw with the decision tree – that is, `agecategory`, `genhealth`, and `diabetic`:

    ```
    feature_imp = \
        rs.best_estimator_['randomforestclassifier'].\
        feature_importances_

    feature_implabs = np.column_stack((feature_imp.\
        ravel(), hi.new_cols))

    pd.DataFrame(feature_implabs,
        columns=['importance','feature']).\
        sort_values(['importance'], ascending=False)
    ```

    |    | importance | feature             |
    |----|------------|---------------------|
    | 14 | 0.321      | agecategory         |
    | 15 | 0.269      | genhealth           |
    | 16 | 0.159      | diabetic            |
    | 13 | 0.058      | ethnicity_White     |
    | 0  | 0.053      | smoking_Yes         |
    | 18 | 0.033      | physicalhealthbaddays |
    | 8  | 0.027      | gender_Male         |

| 3  | 0.024 | walkingdifficult_Yes      |
|----|-------|---------------------------|
| 20 | 0.019 | sleeptimenightly          |
| 11 | 0.010 | ethnicity_Hispanic        |
| 17 | 0.007 | bmi                       |
| 19 | 0.007 | mentalhealthbaddays       |
| 1  | 0.007 | alcoholdrinkingheavy_Yes  |
| 5  | 0.003 | asthma_Yes                |
| 10 | 0.002 | ethnicity_Black           |
| 4  | 0.001 | physicalactivity_Yes      |
| 7  | 0.001 | skincancer_Yes            |
| 2  | 0.000 | stroke_Yes                |
| 6  | 0.000 | kidneydisease_Yes         |
| 9  | 0.000 | ethnicity_Asian           |
| 12 | 0.000 | ethnicity_Other           |

6.  Let's look at some metrics. There is some improvement in sensitivity over the previous model, though not much in any other measure. We keep the same relatively decent specificity score. Overall, this is our best model so far:

```
print("accuracy: %.2f, sensitivity: %.2f, specificity:
%.2f, precision: %.2f"  %
  (skmet.accuracy_score(y_test.values.ravel(), pred),
  skmet.recall_score(y_test.values.ravel(), pred),
  skmet.recall_score(y_test.values.ravel(), pred,
    pos_label=0),
  skmet.precision_score(y_test.values.ravel(), pred)))
accuracy: 0.77, sensitivity: 0.69, specificity: 0.78,
precision: 0.22
```

We may still be able to improve on our model's performance metrics. We should at least attempt some gradient boosting. As we discussed in the *Using gradient-boositng decision trees* section of this chapter, gradient-boosted decision trees can sometimes result in a better model than a random forest. This is because each tree learns from the errors of previous ones.

# Implementing gradient boosting

In this section, we will try to improve our random forest model using gradient boosting. One thing we will have to watch out for is overfitting, which can be more of an issue with gradient boosting decision trees than with random forests. This is because the trees for random forests do not learn from other trees, whereas with gradient boosting, each tree builds on the learning of previous trees. Our choice of hyperparameters here is key. Let's get started:

1.  We will start by importing the necessary libraries. We will use the same modules we used for random forests, except we will import GradientBoostingClassifier from ensemble rather than RandomForestClassifier:

    ```
    import pandas as pd
    import numpy as np
    from imblearn.pipeline import make_pipeline
    from sklearn.model_selection import RandomizedSearchCV
    from sklearn.ensemble import GradientBoostingClassifier
    import sklearn.metrics as skmet
    from scipy.stats import uniform
    from scipy.stats import randint
    import matplotlib.pyplot as plt
    import os
    import sys
    sys.path.append(os.getcwd() + "/helperfunctions")
    import healthinfo as hi
    ```

2.  Now, let's grab the data that's been processed by the healthinfo module for our random forest classifier:

    ```
    X_train = hi.X_train
    X_test = hi.X_test
    y_train = hi.y_train
    y_test = hi.y_test
    ```

3.  Next, we will instantiate a gradient-boosting classifier instance and add it to a pipeline, along with our steps for preprocessing the health data (which are in the module we imported).

We will create a dictionary that contains the hyperparameters for the gradient-boosting classifier. These include the familiar minimum samples per leaf, maximum depth, and the number of estimators hyperparameters. We will also add values to check for the learning rate:

```
gbc = GradientBoostingClassifier(random_state=0)

pipe1 = make_pipeline(hi.coltrans, hi.smotenc, gbc)

gbc_params = {
 'gradientboostingclassifier__min_samples_leaf':
     randint(100, 1200),
 'gradientboostingclassifier__max_depth':
     randint(2, 20),
 'gradientboostingclassifier__learning_rate':
     uniform(loc=0.02, scale=0.25),
 'gradientboostingclassifier__n_estimators':
     randint(100, 1200)
}
```

4.  Now, we are ready to do our grid search. (Note that this may take some time to run on your machine.) The best model out of the seven iterations we ran had a learning rate of 0.25 and 308 estimators, or trees. It had a decent roc_auc score of 0.82:

```
rs = RandomizedSearchCV(pipe1, gbc_params, cv=5,
  n_iter=7, scoring="roc_auc")

rs.fit(X_train, y_train.values.ravel())

rs.best_params_
{'gradientboostingclassifier__learning_rate': 0.2528,
 'gradientboostingclassifier__max_depth': 3,
 'gradientboostingclassifier__min_samples_leaf': 565,
 'gradientboostingclassifier__n_estimators': 308}

rs.best_score_
0.8162378796382679
```

5. Let's look at the feature importance:

```
feature_imp = pd.Series(rs.\
    best_estimator_['gradientboostingclassifier'].\
    feature_importances_, index=hi.new_cols)
feature_imp.loc[feature_imp>0.01].\
        plot(kind='barh')
plt.tight_layout()
plt.title('Gradient Boosting Feature Importance')
```

This produces the following plot:

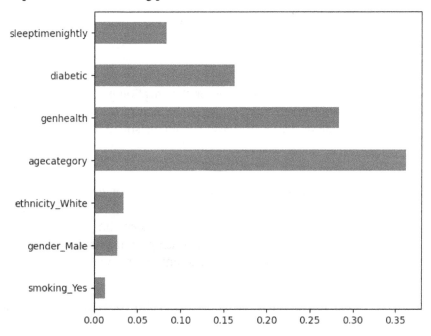

Figure 11.5 – Gradient boosting feature importance

6. Let's look at some metrics. Interestingly, we get outstanding accuracy and specificity but abysmal sensitivity. This may be due to overfitting:

```
pred = rs.predict(X_test)

print("accuracy: %.2f, sensitivity: %.2f, specificity:
%.2f, precision: %.2f" %
    (skmet.accuracy_score(y_test.values.ravel(), pred),
    skmet.recall_score(y_test.values.ravel(), pred),
```

```
skmet.recall_score(y_test.values.ravel(), pred,
    pos_label=0),
skmet.precision_score(y_test.values.ravel(), pred)))
accuracy: 0.91, sensitivity: 0.19, specificity: 0.97,
precision: 0.40
```

Although our results were mixed, gradient-boosted decision trees are often a great choice for a classification model. This is particularly true if we are modeling complicated relationships between features and the target. When a decision tree is an appropriate choice, but we are concerned about high variance with our decision tree model, gradient-boosted decision trees are at least as good an option as random forests.

Now, let's summarize what we've learned in this chapter.

# Summary

This chapter explored how to use decision trees for classification problems. Although the examples in this chapter all involved a binary target, the algorithms we worked with can also handle multiclass problems. Unlike the switch from logistic to multinomial logistic regression, few changes need to be made to use the algorithms well when our target has more than two values.

We looked at two approaches to dealing with the high variance of decision trees. One approach is to use a random forest, which is a form of bagging. This will reduce the variance in our predictions. Another approach is to use gradient-boosted decision trees. Boosting can help us capture very complicated relationships in the data, but there is a non-trivial risk of overfitting. It is particularly important to tune our hyperparameters with that in mind.

In the next chapter, we explore another well-known algorithm for classification: K-nearest neighbors.

# 12
# K-Nearest Neighbors for Classification

**K-nearest neighbors** (**KNN**) is a good choice for a classification model when there are not many observations or features and predicting class membership does not need to be very efficient. It is a lazy learner, so it is quicker to fit than other classification algorithms but considerably slower at classifying new observations. It can also yield less accurate predictions at the extremes, though this can be improved by adjusting $k$ appropriately. We will consider these choices carefully in the model we will develop in this chapter.

KNN is perhaps the most straightforward non-parametric algorithm we could select, making it a good diagnostic tool. No assumptions need to be made about the distribution of features or the relationship that features have with the target. There are not many hyperparameters to tune, and the two key hyperparameters – nearest neighbors and the distance metric – are quite easy to interpret.

KNN can be used successfully for both binary and multiclass problems without any extensions to the algorithm.

In this chapter, we will cover the following topic:

- Key concepts of KNN
- KNN for binary classification
- KNN for multiclass classification

# Technical requirements

In addition to the usual scikit-learn libraries, we will need the `imblearn` (Imbalanced Learn) library to run the code in this chapter. This library helps us handle significant class imbalance. `imblearn` can be installed with `pip install imbalanced-learn`, or `conda install -c conda-forge imbalanced-learn` if you are using Anaconda. All the code has been tested using scikit-learn versions 0.24.2 and 1.0.2.

# Key concepts of KNN

KNN might be the most intuitive algorithm that we will discuss in this book. The idea is to find $k$ instances whose attributes are most similar, where that similarity matters for the target. That last clause is an important, though perhaps obvious, qualification. We care about similarity among those attributes associated with the target's value.

For each observation where we need to predict the target, KNN finds the $k$ training observations whose features are most similar to those of that observation. When the target is categorical, KNN selects the most frequent value of the target for the $k$ training observations. (We often select an odd value for $k$ for classification problems to avoid ties.)

By *training* observations, I mean those observations that have known target values. No real training is done with KNN since it's a lazy learner. I will discuss that in more detail in this section.

The following diagram illustrates the use of KNN for classification with values of 1 and 3 for $k$. When **k=1**, we would predict that our new observation, **X**, would be in the circle class. When **k=3**, it would be assigned to the square class:

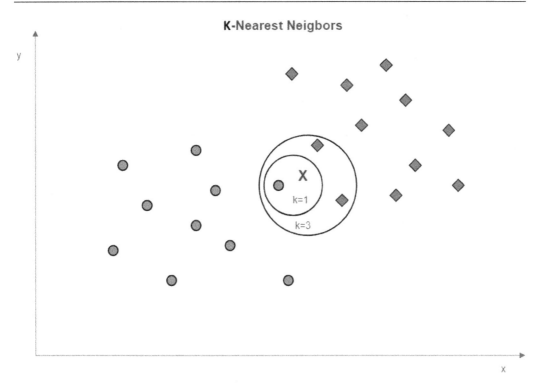

Figure 12.1 – KNN with values of 1 and 3 for k

But what do we mean by similar, or nearest, instances? There are several ways to measure similarity, but the most common measure is the Euclidean distance. The Euclidean distance is the sum of the squared difference between two points. This may remind you of the Pythagorean theorem. The Euclidean distance from point $a$ to point $b$ is as follows:

$$D(a, b) = \sqrt{\sum_{i=1}^{n} (a_i - b_i)^2}$$

A reasonable alternative to the Euclidean distance is the Manhattan distance. The Manhattan distance from point $a$ to point $b$ is as follows:

$$D(a, b) = \sum_{i=1}^{n} |a_i - b_i|$$

The default distance measure in scikit-learn is Minkowski. The Minkowski distance from point $a$ to point $b$ is as follows:

$$D(a, b) = \left( \sum_{i=1}^{n} |a_i - b_i|^p \right)^{\frac{1}{p}}$$

Notice that when $p$ is 1, it is the same as the Manhattan distance. When $p$ is 2, it is the same as the Euclidean distance.

The Manhattan distance is sometimes called the taxicab distance. This is because it reflects the distance between two points along a path on a grid. The following diagram illustrates the Manhattan distance and compares it to the Euclidean distance:

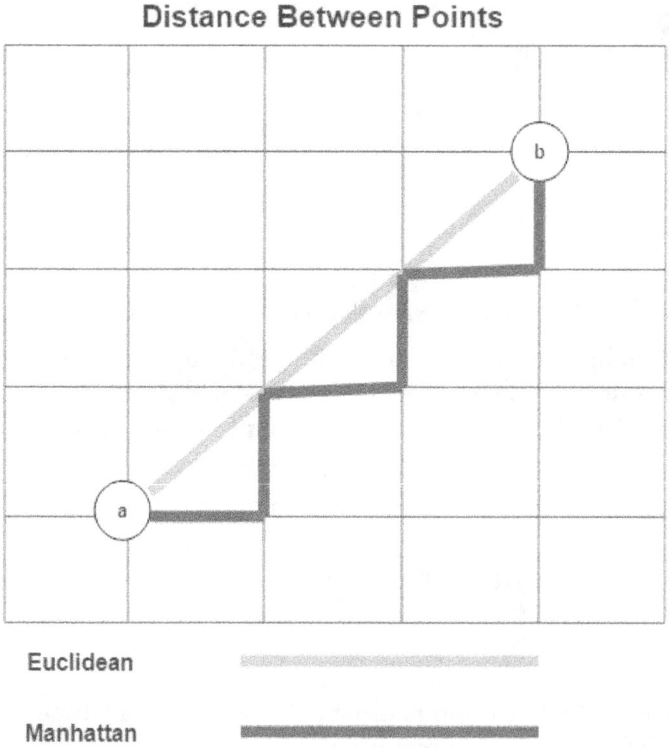

Figure 12.2 – The Euclidean and Manhattan measures of distance

Using the Manhattan distance can yield better results when features are very different in terms of type or scale. However, we can treat the choice of distance measure as an empirical question; that is, we can try both (or other distance measures) and see which gives us the best-performing model. We will demonstrate this with a grid search in the next section.

As you likely suspect, KNN models are sensitive to the choice of $k$. Lower values of $k$ will result in a model that attempts to identify subtle distinctions between observations. There is a substantial risk of overfitting at very low values of $k$. But at high values of $k$, our model may not be flexible enough. We are once again confronted with the variance-bias trade-off. Lower $k$ values result in less bias and more variance, while higher values result in the opposite.

There is no definitive answer to the choice of $k$. But a good rule of thumb is to use the square root of the number of observations. However, just as we would do for the distance measure, we should test a model's performance at different values of $k$. KNN is a non-parametric algorithm. No assumptions are made about the attributes of the underlying data, such as linearity or normally distributed features. This makes KNN quite flexible. It can be used to model a variety of relationships between features and the target.

As mentioned previously, KNN is a lazy learner algorithm. No calculations are performed at training time. The learning happens only during testing. This has its advantages and disadvantages. It may not be a good choice when there are many instances or dimensions in the data, and the speed of predictions matters. KNN also tends not to perform well when we have sparse data, such as datasets that contain many 0 values.

We will use KNN in the next section to build a binary classification model, before constructing a couple of multiclass models in the following section.

# KNN for binary classification

The KNN algorithm has some of the same advantages as the decision tree algorithm. No prior assumptions about the distribution of features or residuals have to be met. It is a suitable algorithm for the heart disease model we tried to build in the last two chapters. The dataset is not very large (30,000 observations) and does not have too many features.

> **Note**
>
> The heart disease dataset is available for public download at `https://www.kaggle.com/datasets/kamilpytlak/personal-key-indicators-of-heart-disease`. It is derived from the United States Center for Disease Control survey data on more than 400,000 individuals from 2020. I have randomly sampled 30,000 observations from this dataset for the analysis in this section. Data columns include whether respondents ever had heart disease, body mass index, smoking history, heavy alcohol drinking, age, diabetes, and kidney disease.

Let's get started with our model:

1.  First, we must load some of the same libraries we have used over the last couple of chapters. We will also load `KneighborsClassifier`:

```
import pandas as pd
import numpy as np
from imblearn.pipeline import make_pipeline
from sklearn.model_selection import RandomizedSearchCV,\
    RepeatedStratifiedKFold
from sklearn.neighbors import KNeighborsClassifier
from sklearn.feature_selection import SelectKBest, chi2
from scipy.stats import randint
import sklearn.metrics as skmet
from sklearn.model_selection import cross_validate
import os
import sys
sys.path.append(os.getcwd() + "/helperfunctions")
import healthinfo as hi
```

The `healthinfo` module contains all of the code we used in *Chapter 10, Logistic Regression*, to load the health information data and do the preprocessing. There is no need to repeat those steps here. If you have not read *Chapter 10, Logistic Regression*, it might be helpful to at least scan the code in the second section of that chapter. That will give you a better sense of the features.

2.  Now, let's grab the data that's been processed by the `healthinfo` module and display the feature names:

```
X_train = hi.X_train
X_test = hi.X_test
y_train = hi.y_train
y_test = hi.y_test
new_cols = hi.new_cols

new_cols
array(['smoking_Yes', 'alcoholdrinkingheavy_Yes',
        'stroke_Yes', 'walkingdifficult_Yes',
        'physicalactivity_Yes', 'asthma_Yes',
        'kidneydisease_Yes', 'skincancer_Yes',
```

```
        'gender_Male', 'ethnicity_Asian',
        'ethnicity_Black', 'ethnicity_Hispanic',
        'ethnicity_Other', 'ethnicity_White',
        'agecategory', 'genhealth', 'diabetic', 'bmi',
        'physicalhealthbaddays', 'mentalhealthbaddays',
        'sleeptimenightly'], dtype=object)
```

3.  We can use K-fold cross-validation to assess this model. We discussed K-fold cross-validation in *Chapter 6, Preparing for Model Evaluation*. We will indicate that we want 10 splits repeated 10 times. The defaults are 5 and 10, respectively.

    The precision of our model, the rate at which we are correct when we predict heart disease, is exceptionally poor at 0.17. Sensitivity, the rate at which we predict heart disease when there is heart disease, is also low at 0.56:

    ```
    knn_example = KNeighborsClassifier(n_neighbors=5, n_
    jobs=-1)


    kf = RepeatedStratifiedKFold(n_splits=10, n_repeats=10,
    random_state=0)


    pipe0 = make_pipeline(hi.coltrans, hi.smotenc, knn_
    example)


    scores = cross_validate(pipe0, X_train,
      y_train.values.ravel(), \
      scoring=['accuracy','precision','recall','f1'], \
      cv=kf, n_jobs=-1)


    print("accuracy: %.2f, sensitivity: %.2f, precision:
    %.2f, f1: %.2f" %
      (np.mean(scores['test_accuracy']),\
      np.mean(scores['test_recall']),\
      np.mean(scores['test_precision']),\
      np.mean(scores['test_f1'])))
    accuracy: 0.73, sensitivity: 0.56, precision: 0.17, f1:
    0.26
    ```

4.  We can improve the performance of our model with some hyperparameter tuning. Let's create a dictionary for several neighbors and distance metrics. We will also try different values for the number of features selected with our `filter` method:

```
knn = KNeighborsClassifier(n_jobs=-1)


pipe1 = make_pipeline(hi.coltrans, hi.smotenc,
    SelectKBest(score_func=chi2), knn)


knn_params = {
 'selectkbest__k':
    randint(1, len(new_cols)),
 'kneighborsclassifier__n_neighbors':
    randint(5, 300),
 'kneighborsclassifier__metric':
    ['euclidean','manhattan','minkowski']
}
```

5.  We will base the scoring in the grid search on the area under the **receiver operating characteristic curve (ROC curve)**. We covered ROC curves in *Chapter 6, Preparing for Model Evaluation*:

```
rs = RandomizedSearchCV(pipe1, knn_params, cv=5,
scoring="roc_auc")


rs.fit(X_train, y_train.values.ravel())
```

6.  We can use the best estimator attribute of the randomized grid search to get the selected features from `selectkbest`:

```
selected = rs.best_estimator_['selectkbest'].\
  get_support()


selected.sum()
11
new_cols[selected]
array(['smoking_Yes', 'alcoholdrinkingheavy_Yes',
        'walkingdifficult_Yes', 'ethnicity_Black',
        'ethnicity_Hispanic', 'agecategory',
```

```
            'genhealth', 'diabetic', 'bmi',
            'physicalhealthbaddays','mentalhealthbaddays'],
         dtype=object)
```

7. We can also take a look at the best parameters and the best score. 11 features (out of 17) were selected, as we saw in the previous step. A *k* (n_neighbors) of 254 and the Manhattan distance metric were the other hyperparameters of the highest scoring model:

```
rs.best_params_
{'kneighborsclassifier__metric': 'manhattan',
 'kneighborsclassifier__n_neighbors': 251,
 'selectkbest__k': 11}


rs.best_score_
0.8030553205304845
```

8. Let's look at some more metrics for this model. We are doing much better with sensitivity but not with any of the other metrics:

```
pred = rs.predict(X_test)


print("accuracy: %.2f, sensitivity: %.2f, specificity:
%.2f, precision: %.2f" %
   (skmet.accuracy_score(y_test.values.ravel(), pred),
   skmet.recall_score(y_test.values.ravel(), pred),
   skmet.recall_score(y_test.values.ravel(), pred,
     pos_label=0),
   skmet.precision_score(y_test.values.ravel(), pred)))
accuracy: 0.67, sensitivity: 0.82, specificity: 0.66,
precision: 0.18
```

9. We should also plot the confusion matrix. To do this, we can look at the relatively decent sensitivity. Here, we correctly identify most of the actual positives as positive. However, this comes at the expense of many false positives. We can see this in the precision score from the previous step. Most of the time we predict positive, we are wrong:

```
cm = skmet.confusion_matrix(y_test, pred)
cmplot = skmet.ConfusionMatrixDisplay(
   confusion_matrix=cm,
```

```
        display_labels=['Negative', 'Positive'])
cmplot.plot()
cmplot.ax_.set(title='Heart Disease Prediction Confusion
Matrix',
    xlabel='Predicted Value', ylabel='Actual Value')
```

This produces the following plot:

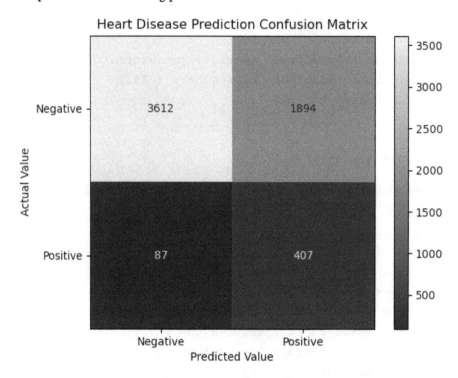

Figure 12.3 – Heart disease confusion matrix after hyperparameter tuning

In this section, you learned how to use KNN with a binary target. We can follow very similar steps to use KNN for a multiclass classification problem.

# KNN for multiclass classification

Constructing a KNN multiclass model is quite straightforward since it involves no special extensions to the algorithm, such as those needed to adapt logistic regression to targets with more than two values. We can see this by working with the same machine failure data that we worked with in the *Multinomial logistic regression* section of *Chapter 10, Logistic Regression*.

> **Note**
>
> This dataset on machine failure is available for public use at `https://www.`
> `kaggle.com/datasets/shivamb/machine-predictive-`
> `maintenance-classification`. There are 10,000 observations,
> 12 features, and two possible targets. One is binary and specifies whether the
> machine failed or did not. The other contains types of failure. The instances in
> this dataset are synthetic, generated by a process designed to mimic machine
> failure rates and causes.

Let's build our machine failure type model:

1.  First, let's load the now-familiar modules:

    ```
    import pandas as pd
    import numpy as np
    from sklearn.model_selection import train_test_split
    from sklearn.preprocessing import OneHotEncoder,
    MinMaxScaler
    from imblearn.pipeline import make_pipeline
    from sklearn.impute import SimpleImputer
    from sklearn.compose import ColumnTransformer
    from sklearn.model_selection import RandomizedSearchCV
    from sklearn.neighbors import KNeighborsClassifier
    from imblearn.over_sampling import SMOTENC
    from sklearn.feature_selection import SelectKBest, chi2
    import sklearn.metrics as skmet
    import os
    import sys
    sys.path.append(os.getcwd() + "/helperfunctions")
    from preprocfunc import OutlierTrans
    ```

2.  Let's load the machine failure data and take a look at its structure. There are 10,000
    observations and no missing data. There is a combination of categorical and
    numerical data:

    ```
    machinefailuretype = pd.read_csv("data/
    machinefailuretype.csv")


    machinefailuretype.info()
    <class 'pandas.core.frame.DataFrame'>
    ```

```
RangeIndex: 10000 entries, 0 to 9999
Data columns (total 10 columns):
 #   Column              Non-Null Count  Dtype
---  ------              --------------  -----
 0   udi                 10000 non-null  int64
 1   product             10000 non-null  object
 2   machinetype         10000 non-null  object
 3   airtemp             10000 non-null  float64
 4   processstemperature 10000 non-null  float64
 5   rotationalspeed     10000 non-null  int64
 6   torque              10000 non-null  float64
 7   toolwear            10000 non-null  int64
 8   fail                10000 non-null  int64
 9   failtype            10000 non-null  object
dtypes: float64(3), int64(4), object(3)
memory usage: 781.4+ KB
```

3.  Let's also look at a few observations:

```
machinefailuretype.head()
   udi product machinetype airtemp processstemperature\
0   1   M14860      M         298          309
1   2   L47181      L         298          309
2   3   L47182      L         298          308
3   4   L47183      L         298          309
4   5   L47184      L         298          309


   Rotationalspeed  Torque  toolwear  fail   failtype
0        1551          43       0       0    No Failure
1        1408          46       3       0    No Failure
2        1498          49       5       0    No Failure
3        1433          40       7       0    No Failure
4        1408          40       9       0    No Failure
```

4. We should also do some frequencies on the categorical features. The overwhelming majority of observations, 97%, have no failures. This rather stark class imbalance will likely be difficult to model. There are three machine types – high quality, low quality, and medium quality:

```
machinefailuretype.failtype.value_counts(dropna=False).
sort_index()
Heat Dissipation Failure  112
No Failure                9652
Overstrain Failure        78
Power Failure             95
Random Failures           18
Tool Wear Failure         45
Name: failtype, dtype: int64


machinefailuretype.machinetype.\
  value_counts(dropna=False).sort_index()
H       1003
L       6000
M       2997
Name: machinetype, dtype: int64
```

5. Let's collapse some of the `failtype` values and check our work. First, we will define a function, `setcode`, to map the failure type text to a failure type code. We will assign random failures and tool wear failures to code 5 for other failures:

```
def setcode(typetext):
  if (typetext=="No Failure"):
    typecode = 1
  elif (typetext=="Heat Dissipation Failure"):
    typecode = 2
  elif (typetext=="Power Failure"):
    typecode = 3
  elif (typetext=="Overstrain Failure"):
    typecode = 4
  else:
    typecode = 5
  return typecode
```

```
machinefailuretype["failtypecode"] = \
  machinefailuretype.apply(lambda x: setcode(x.failtype),
axis=1)

machinefailuretype.groupby(['failtypecode','failtype']).
size().\
  reset_index()
```

|   | failtypecode | failtype | 0 |
|---|---|---|---|
| 0 | 1 | No Failure | 9652 |
| 1 | 2 | Heat Dissipation Failure | 112 |
| 2 | 3 | Power Failure | 95 |
| 3 | 4 | Overstrain Failure | 78 |
| 4 | 5 | Random Failures | 18 |
| 5 | 5 | Tool Wear Failure | 45 |

6.  We should look at some descriptive statistics for our numeric features:

```
num_cols = ['airtemp', 'processtemperature',
  'rotationalspeed', 'torque', 'toolwear']
cat_cols = ['machinetype']

machinefailuretype[num_cols].agg(['min','median','max']).T
```

|  | min | median | max |
|---|---|---|---|
| airtemp | 295 | 300 | 304 |
| processtemperature | 306 | 310 | 314 |
| rotationalspeed | 1,168 | 1,503 | 2,886 |
| torque | 4 | 40 | 77 |
| toolwear | 0 | 108 | 253 |

7.  Now, we are ready to create training and testing DataFrames. We will use the failure type code we just created for our target:

```
X_train, X_test, y_train, y_test =  \
  train_test_split(machinefailuretype[num_cols + cat_
cols],\
  machinefailuretype[['failtypecode']], test_size=0.2,
random_state=0)
```

8.  Now, let's set up the column transformation. For the numerical features, we will set outliers to the median and then scale the data. We will use min-max scaling, which will return values from 0 to 1 (the default for MinMaxScaler). We are using this scaler, rather than the standard scaler, to avoid negative values. The feature selection method we will use later, selectkbest, does not work with negative values:

```
ohe = OneHotEncoder(drop='first', sparse=False)

cattrans = make_pipeline(ohe)
standtrans = make_pipeline(
  OutlierTrans(3),SimpleImputer(strategy="median"),
  MinMaxScaler())
coltrans = ColumnTransformer(
  transformers=[
      ("cat", cattrans, cat_cols),
      ("stand", standtrans, num_cols),
  ]
)
```

9.  Let's also take a peek at the columns that we will have after the encoding. We need this in advance of the oversampling we will do since the SMOTENC module needs the column indexes of the categorical features. We are oversampling to handle the significant class imbalance. We discussed this in more detail in *Chapter 11, Decision Trees and Random Forest Classification*:

```
coltrans.fit(X_train.sample(1000))

new_cat_cols = \
  coltrans.\
  named_transformers_['cat'].\
  named_steps['onehotencoder'].\
  get_feature_names(cat_cols)

new_cols = np.concatenate((new_cat_cols, np.array(num_cols)))

print(new_cols)
```

```
['machinetype_L' 'machinetype_M' 'airtemp'
'processtemperature' 'rotationalspeed' 'torque'
'toolwear']
```

10. Next, we will set up a pipeline for our model. The pipeline will do a column transformation, oversampling using SMOTENC, feature selection with selectkbest, and then run a KNN model. Remember that we have to pass the column indexes of the categorical features to SMOTENC for it to run correctly:

```
catcolscnt = new_cat_cols.shape[0]
smotenc = SMOTENC(categorical_features=np.
arange(0,catcolscnt), random_state=0)


knn = KNeighborsClassifier(n_jobs=-1)


pipe1 = make_pipeline(coltrans, smotenc,
SelectKBest(score_func=chi2), knn)
```

11. Now, we are ready to fit our model. We will do a randomized grid search to identify the best value for *k* and the distance metric for the KNN. We will also search for the best *k* value for the feature selection:

```
knn_params = {
  'selectkbest__k': np.arange(1, len(new_cols)),
  'kneighborsclassifier__n_neighbors': np.arange(5, 175,
2),
  'kneighborsclassifier__metric':
['euclidean','manhattan','minkowski']
}


rs = RandomizedSearchCV(pipe1, knn_params, cv=5,
scoring="roc_auc_ovr_weighted")


rs.fit(X_train, y_train.values.ravel())
```

12. Let's take a look at what our grid search found. All the features except for processtemperature were worth keeping in the model. The best values for *k* and the distance metric for the KNN were 125 and minkowski, respectively. The best score, based on the area under the ROC curve, was 0.9:

```
selected = rs.best_estimator_['selectkbest'].get_
support()
```

```
selected.sum()
6
```

```
new_cols[selected]
array(['machinetype_L', 'machinetype_M', 'airtemp',
       'rotationalspeed', 'torque', 'toolwear'],
      dtype=object)
rs.best_params_
{'selectkbest__k': 6,
 'kneighborsclassifier__n_neighbors': 125,
 'kneighborsclassifier__metric': 'minkowski'}
```

```
rs.best_score_
0.899426752716227
```

13. Let's look at a confusion matrix. Looking at the first row, we can see that a sizeable number of failures were found when no failure happened. However, our model does correctly identify most of the actual heat, power, and overstrain failures. This may not be a horrible precision and sensitivity trade-off. Depending on the problem, we may accept a large number of false positives to get an acceptable level of sensitivity in our model:

```
pred = rs.predict(X_test)

cm = skmet.confusion_matrix(y_test, pred)
cmplot = skmet.ConfusionMatrixDisplay(confusion_
matrix=cm,
    display_labels=['None',
'Heat','Power','Overstrain','Other'])
cmplot.plot()
cmplot.ax_.set(title='Machine Failure Type Confusion
Matrix',
  xlabel='Predicted Value', ylabel='Actual Value')
```

This produces the following plot:

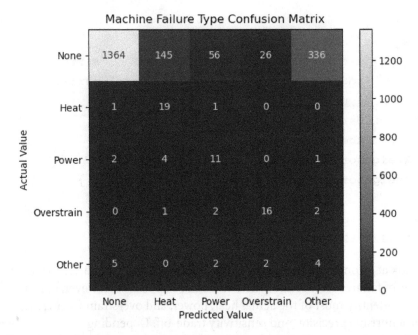

Figure 12.4 – Confusion matrix for machine failure type after hyperparameter tuning

14. We should also look at a classification report. You may remember from *Chapter 6, Preparing for Model Evaluation*, that the macro average is a simple average across classes. Here, we are more interested in the weighted average. The weighted F1-score is not bad at 0.81. Recall that F1 is the harmonic mean of precision and sensitivity:

```
print(skmet.classification_report(y_test, pred,
   target_names=\
   ['None', 'Heat','Power','Overstrain','Other']))
```

|  | Precision | recall | f1-score | support |
|---|---|---|---|---|
| None | 0.99 | 0.71 | 0.83 | 1927 |
| Heat | 0.11 | 0.90 | 0.20 | 21 |
| Power | 0.15 | 0.61 | 0.24 | 18 |
| Overstrain | 0.36 | 0.76 | 0.49 | 21 |
| Other | 0.01 | 0.31 | 0.02 | 13 |
|  |  |  |  |  |
| accuracy |  |  | 0.71 | 2000 |
| macro avg | 0.33 | 0.66 | 0.36 | 2000 |
| weighted avg | 0.96 | 0.71 | 0.81 | 2000 |

The class imbalance for the machine failure type makes it particularly difficult to model. Still, our KNN model does relatively well, assuming that the large number of false positives is not problematic. In this case, a false positive may not be nearly as much of an issue as a false negative. It may just involve doing more checks on machines that seem to be in danger of failing. If we compare that with being surprised by an actual machine failure, a bias toward sensitivity rather than precision may be appropriate.

Let's try KNN on another multiclass problem.

# KNN for letter recognition

We can take pretty much the same approach we just took for predicting machine failure with letter recognition. So long as we have features that do a good job of discriminating between the letters, KNN is a reasonable choice for that model. We will try this in this section.

> **Note**
>
> We will work with letter recognition data in this section. It is available for public use at `https://archive-beta.ics.uci.edu/ml/datasets/letter+recognition`. There are 26 letters (all capitals) and 20 different fonts. 16 different features capture different attributes of each letter.

Let's build the model:

1.  First, we will load the same libraries we have already been using:

    ```
    import pandas as pd
    import numpy as np
    from sklearn.model_selection import train_test_split
    from sklearn.model_selection import StratifiedKFold, \
        GridSearchCV
    from sklearn.neighbors import KNeighborsClassifier
    import sklearn.metrics as skmet
    from scipy.stats import randint
    ```

2.  Now, we will load the data and look at the first few instances. There are 20,000 observations and 17 columns. `letter` is our target:

    ```
    letterrecognition = pd.read_csv("data/letterrecognition.
    csv")


    letterrecognition.shape
    ```

```
(20000, 17)
```

```
letterrecognition.head().T
```

|          | 0  | 1  | 2  | 3  | 4  |
|----------|----|----|----|----|----|
| letter   | T  | I  | D  | N  | G  |
| xbox     | 2  | 5  | 4  | 7  | 2  |
| ybox     | 8  | 12 | 11 | 11 | 1  |
| width    | 3  | 3  | 6  | 6  | 3  |
| height   | 5  | 7  | 8  | 6  | 1  |
| onpixels | 1  | 2  | 6  | 3  | 1  |
| xbar     | 8  | 10 | 10 | 5  | 8  |
| ybar     | 13 | 5  | 6  | 9  | 6  |
| x2bar    | 0  | 5  | 2  | 4  | 6  |
| y2bar    | 6  | 4  | 6  | 6  | 6  |
| xybar    | 6  | 13 | 10 | 4  | 6  |
| x2ybar   | 10 | 3  | 3  | 4  | 5  |
| xy2bar   | 8  | 9  | 7  | 10 | 9  |
| x-ege    | 0  | 2  | 3  | 6  | 1  |
| xegvy    | 8  | 8  | 7  | 10 | 7  |
| y-ege    | 0  | 4  | 3  | 2  | 5  |
| yegvx    | 8  | 10 | 9  | 8  | 10 |

3.  Now, let's create training and testing DataFrames:

```
X_train, X_test, y_train, y_test =  \
  train_test_split(letterrecognition.iloc[:,1:],\
  letterrecognition.iloc[:,0:1], test_size=0.2,
  random_state=0)
```

4.  Next, let's instantiate a KNN instance. We will also set up stratified K-fold cross-validation and a dictionary for the hyperparameters. We will search for the best hyperparameter for *k* (n_neighbors) and the distance metric:

```
knn = KNeighborsClassifier(n_jobs=-1)


kf = StratifiedKFold(n_splits=5, shuffle=True, random_
state=0)


knn_params = {
```

```
    'n_neighbors': np.arange(3, 41, 2),
    'metric': ['euclidean','manhattan','minkowski']
}
```

5.  Now, we are ready to do an exhaustive grid search. We are doing an exhaustive search here since we do not have a lot of hyperparameters to check. The best-performing distance metric is Euclidean. The best value for *k* for nearest neighbors is 3. This model gets us nearly 95% accuracy:

```
gs = GridSearchCV(knn, knn_params, cv=kf,
scoring='accuracy')


gs.fit(X_train, y_train.values.ravel())


gs.best_params_
{'metric': 'euclidean', 'n_neighbors': 3}


gs.best_score_
0.9470625
```

6.  Let's generate predictions and plot a confusion matrix:

```
pred = gs.best_estimator_.predict(X_test)


letters = np.sort(letterrecognition.letter.unique())


cm = skmet.confusion_matrix(y_test, pred)
cmplot = \
  skmet.ConfusionMatrixDisplay(confusion_matrix=cm,
  display_labels=letters)
cmplot.plot()
cmplot.ax_.set(title='Letters',
  xlabel='Predicted Value', ylabel='Actual Value')
```

This produces the following plot:

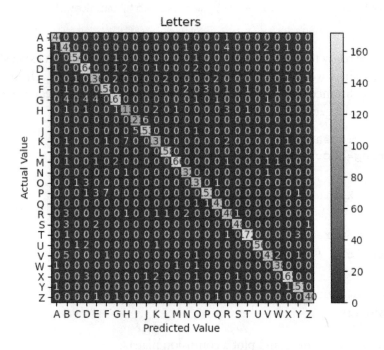

Figure 12.5 – Confusion matrix for letter predictions

Let's quickly summarize what we've learned in this chapter.

## Summary

This chapter demonstrated how easy it is to use KNN for binary or multiclass classification. Since KNN does not make assumptions about normality or linearity, it can be used in cases where logistic regression may not yield the best results. This flexibility does bring with it a real risk of overfitting, so care has to be taken with the choice of *k*. We also explored how to hyperparameter tune both binary and multiclass models in this chapter. Finally, KNN is not a great option when we care about the speed of our predictions or if we are working with a large dataset. Decision tree or random forest classification, which we explored in the previous chapter, is often a better choice in those cases.

Another really good choice is support vector classification. We will explore support vector classification in the next chapter.

# 13
# Support Vector Machine Classification

There are some similarities between support vector classification models and k-nearest neighbors models. They are both intuitive and flexible. However, support vector classification, due to the nature of the algorithm, scales better than k-nearest neighbor. Unlike logistic regression, it can handle nonlinear models rather easily. The strategies and issues with using support vector machines for classification are similar to those we discussed in *Chapter 8*, *Support Vector Regression*, when we used support vector machines for regression.

One of the key advantages of **support vector classification** (**SVC**) is the ability it gives us to reduce model complexity without increasing our feature space. But it also provides multiple levers we can adjust to limit the possibility of overfitting. We can choose a linear model or select from several nonlinear kernels. We can use a regularization parameter, much as we did for logistic regression. With extensions, we can also use these same techniques to construct multiclass models.

We will explore the following topics in this chapter:

- Key concepts for SVC
- Linear SVC models
- Nonlinear SVM classification models
- SVMs for multiclass classification

# Technical requirements

We will stick to the pandas, NumPy, and scikit-learn libraries in this chapter. All code in this chapter was tested with scikit-learn versions 0.24.2 and 1.0.2. The code that displays the decision boundaries needs scikit-learn version 1.1.1 or later.

# Key concepts for SVC

We can use **support vector machines** (**SVMs**) to find a line or curve to separate instances by class. When classes can be discriminated by a line, they are said to be **linearly separable**.

There may, however, be many possible linear classifiers, as we can see in *Figure 13.1*. Each line successfully discriminates between the two classes, represented by dots and squares, using the two features x1 and x2. The key difference is in how the lines would classify new instances, represented by the transparent rectangle. Using the line closest to the squares would cause the transparent rectanglez to be classified as a dot. Using either of the other two lines would classify it as a square.

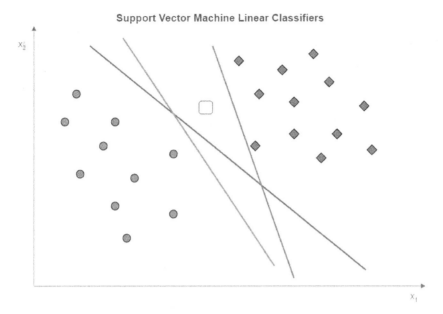

Figure 13.1 – Three possible linear classifiers

When a linear discriminant is very close to training instances, as is the case with two of the lines in *Figure 13.2*, there is a greater risk of misclassifying new instances. We want a line that gives us the maximum margin between classes; one that is furthest away from border data points for each class. That is the middle line in *Figure 13.1*, but it can be seen more clearly in *Figure 13.2*:

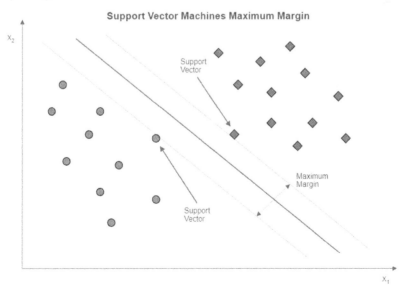

Figure 13.2 – SVM classification and maximum margin

The bold line splits the maximum margin and is referred to as the decision boundary. The border data points for each class are known as the support vectors.

We use SVM to find the linear discriminant with the maximum margin between classes. It does this by finding an equation representing a margin that can be maximized, where the margin is the distance between a data point and the separating hyperplane. With two features, as in *Figure 13.2*, that hyperplane is just a line. However, this can be generalized to feature spaces with more dimensions.

With data points such as those in *Figure 13.2*, we can use what is known as **hard margin classification** without problems; that is, we can be strict about all observations for each class being on the correct side of the decision boundary. But what if our data points look like those in *Figure 13.3*? Here, there is a square very close to the dots. The hard margin classifier is the left line, giving us quite tiny margins.

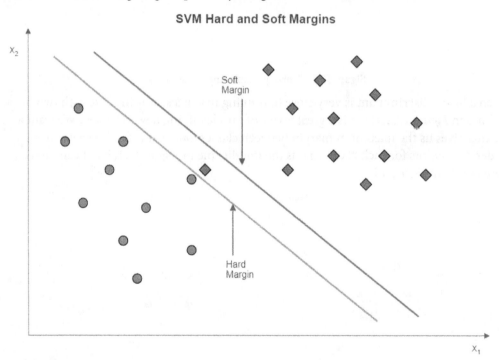

Figure 13.3 – SVMs with hard and soft margins

If we use **soft margin classification** instead, we get the line to the right. Soft margin classification relaxes the constraint that all instances have to be correctly separated. As is the case with the data in *Figure 13.3*, allowing for a small number of misclassifications in the training data can give us a larger margin. We ignore the wayward square and get a decision boundary represented by the soft margin line.

The amount of relaxation of the constraint is determined by the *C* hyperparameter. The larger the value of *C*, the greater the penalty for margin violations. Not surprisingly, models with larger *C* values are more prone to overfitting. *Figure 13.4* illustrates how the margin changes with values of *C*. At *C = 1*, the penalty for misclassification is low, giving us a much greater margin than when *C* is 100. Even at a *C* of 100, however, some margin violation still happens.

**SVC Soft Margins**

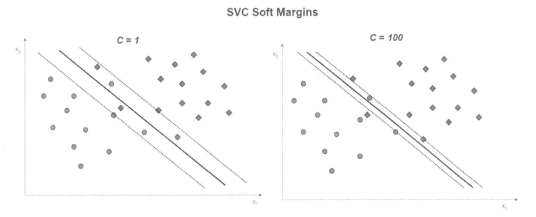

Figure 13.4 – Soft margins at different C values

As a practical matter, we almost always build our SVC models with soft margins. The default value for *C* in scikit-learn is 1.

# Nonlinear SVM and the kernel trick

We have not yet fully addressed the issue of linear separability with SVC. For simplicity, it is helpful to return to a classification problem involving two features. Let's say a plot of two features against a categorical target looks like the illustration in *Figure 13.5*. The target has two possible values, represented by the dots and squares. x1 and x2 are numeric and have negative values.

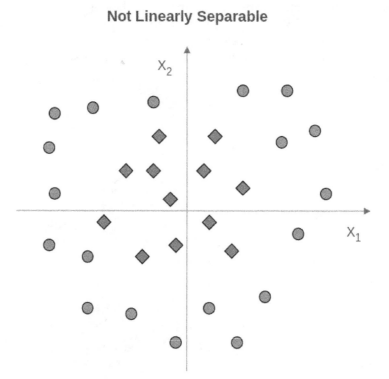

Figure 13.5 – Class labels not linearly separable with two features

What can we do in a case like this to identify a margin between the classes? It is often the case that a margin can be identified at a higher dimension. In this example, we can use a polynomial transformation, as illustrated in *Figure 13.6*:

## Nonlinear Transformation

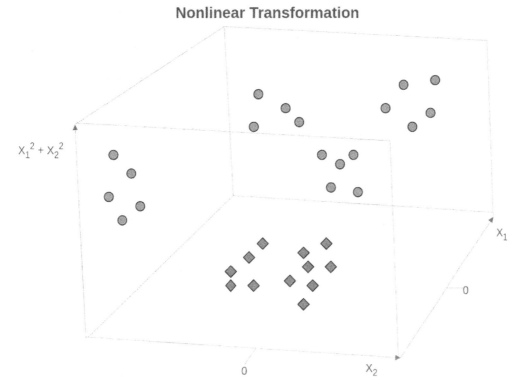

Figure 13.6 – Using polynomial transformation to establish the margin

There is now a third dimension, which is the sum of the squares of x1 and x2. The dots are all higher than the squares. This is similar to how we used polynomial transformation with linear regression.

One drawback of this approach is that we can quickly end up with too many features for our model to perform well. This is where the **kernel trick** comes in very handy. SVC can use a kernel function to expand the feature space implicitly without actually creating more features. This is done by creating a vector of values that can be used to fit a nonlinear margin.

While this allows us to fit a polynomial transformation like the hypothetical one illustrated in *Figure 13.6*, the most frequently used kernel function with SVC is the **radial basis function** (**RBF**). RBF is popular because it is faster than other common kernel functions and because it can be used with the gamma hyperparameter for additional flexibility. The equation for the RBF kernel is as follows:

$$k(x_i, x_j) = \exp\left(-\gamma \|x_i - x_j\|^2\right)$$

Here, $x_i$ and $x_j$ are data points. Gamma, $\gamma$, determines the amount of influence of each point. With high values of gamma, points have to be very close to each other to be grouped together. At very high values of gamma, we start to see islands of points.

Of course, what is a high value for gamma, or of $C$, depends partly on our data. A good approach is to create visualizations of decision boundaries at different values for gamma and $C$ before doing much modeling. This will give us a sense of whether or not we are underfitting or overfitting at different hyperparameter values. We will plot decision boundaries at different values of gamma and $C$ in this chapter.

## Multiclass classification with SVC

All of our discussion about SVC so far has centered on binary classification. Fortunately, all of the key concepts that apply to SVMs for binary classification also apply to classification when our target has more than two possible values. We transform the multiclass problem into a binary classification problem by modeling it as a **one-versus-one**, or a **one-versus-rest** problem.

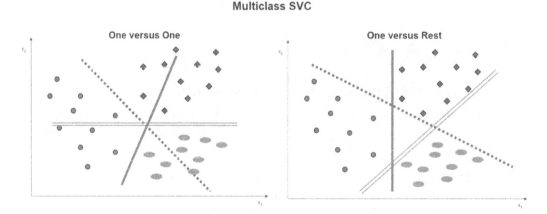

Figure 13.7 – Multiclass SVC options

One-versus-one classification is easy to illustrate in a three-class example, as shown on the left side of *Figure 13.7*. A decision boundary is estimated between each class and each of the other classes. For example, the dotted line is the decision boundary for the dot class versus the square class. The solid line is the decision boundary between the dots and the ovals.

With one-versus-rest classification, a decision boundary is constructed between each class and those instances that are not of that class. This is illustrated on the right side of *Figure 13.7*. The solid line is the decision boundary between the dots and the instances that are not dots (those that are squares or ovals). The dotted and double lines are the decision boundaries for the squares versus the rest and the ovals versus the rest of the instances, respectively.

We can construct both linear and nonlinear SVC models using either one-versus-one or one-versus-rest classification. We can also specify values for $C$ to construct soft margins. However, the construction of more decision boundaries with each of these techniques claims greater computational resources than SVC for binary classification. If we have a large number of observations, many features, and more than a couple of parameters to tune, we will likely need very good system resources to get timely results.

The three-class example hides one thing that is different about one-versus-one and one-versus-rest classifiers. With three classes, they use the same number of classifiers (three), but the number of classifiers increases relatively rapidly with one-versus-one. The number of classifiers will always be equal to the number of class values with one-versus-rest, whereas, with one-versus-one, it is equal to the following:

$$S = N * \frac{N - 1}{2}$$

Here, $S$ is the number of classifiers and $N$ is the cardinality (the number of class values) of the target. So, with a cardinality of 4, one-versus-rest needs 4 classifiers, and one-versus-one uses 6.

We explore multiclass SVC models in the last section of this chapter, but let's start with a relatively straightforward linear model to see SVC in action. There are two things to keep in mind when doing the preprocessing for an SVC model. First, SVC is sensitive to the scale of features, so we will need to address that before fitting our model. Second, if we are using hard margins or high values for $C$, outliers might have a large effect on our model.

# Linear SVC models

We can often get good results by using a linear SVC model. When we have more than two features, there is no easy way to visualize whether our data is linearly separable or not. We often decide on linear or nonlinear based on hyperparameter tuning. For this section, we will assume we can get good performance with a linear model and soft margins.

We will work with data on **National Basketball Association** (**NBA**) games in this section. The dataset has statistics from each NBA game from the 2017/2018 season through the 2020/2021 season. This includes the home team, whether the home team won, the visiting team, shooting percentages for visiting and home teams, turnovers, rebounds, and assists by both teams, and a number of other measures.

> **Note**
>
> NBA game data is available for download for the public at `https://www.kaggle.com/datasets/wyattowalsh/basketball`. This dataset has game data starting with the 1946/1947 NBA season. It uses nba_api to pull stats from `nba.com`. That API is available at `https://github.com/swar/nba_api`.

Let's build a linear SVC model:

1.  We start by loading the familiar libraries. The only new modules are `LinearSVC` and `DecisionBoundaryDisplay`. We will use `DecisionBoundaryDisplay` to show the boundaries of a linear model:

    ```python
    import pandas as pd
    import numpy as np
    from sklearn.model_selection import train_test_split
    from sklearn.preprocessing import OneHotEncoder,
    StandardScaler
    from sklearn.svm import LinearSVC
    from scipy.stats import uniform
    from sklearn.impute import SimpleImputer
    from sklearn.pipeline import make_pipeline
    from sklearn.compose import ColumnTransformer
    from sklearn.feature_selection import RFECV
    from sklearn.inspection import DecisionBoundaryDisplay
    from sklearn.model_selection import cross_validate, \
        RandomizedSearchCV, RepeatedStratifiedKFold
    import sklearn.metrics as skmet
    import seaborn as sns
    import os
    import sys
    sys.path.append(os.getcwd() + "/helperfunctions")
    from preprocfunc import OutlierTrans
    ```

2.  We are ready to load the NBA game data. We just have a little cleaning to do. A small number of observations have missing values for our target, `WL_HOME`, whether the home team won. We remove those observations. We convert the `WL_HOME` feature to a `0` and `1` feature.

There is not much of a problem with a class imbalance here. This will save us some time later:

```
nbagames = pd.read_csv("data/nbagames2017plus.csv",
parse_dates=['GAME_DATE'])

nbagames = \
  nbagames.loc[nbagames.WL_HOME.isin(['W','L'])]

nbagames.shape
(4568, 149)

nbagames['WL_HOME'] = \
  np.where(nbagames.WL_HOME=='L',0,1).astype('int')

nbagames.WL_HOME.value_counts(dropna=False)
1    2586
0    1982
Name: WL_HOME, dtype: int64
```

3. Let's organize our features by data type:

```
num_cols = ['FG_PCT_HOME','FTA_HOME','FG3_PCT_HOME',
  'FTM_HOME','FT_PCT_HOME','OREB_HOME','DREB_HOME',
  'REB_HOME','AST_HOME','STL_HOME','BLK_HOME',
  'TOV_HOME','FG_PCT_AWAY','FTA_AWAY','FG3_PCT_AWAY',
  'FT_PCT_AWAY','OREB_AWAY','DREB_AWAY','REB_AWAY',
  'AST_AWAY','STL_AWAY','BLK_AWAY','TOV_AWAY']
cat_cols = ['SEASON']
```

4. Let's look at some descriptive statistics. (I have omitted some features from the printout to save space.) We will need to scale these features since they have very different ranges. There are no missing values but we will generate some when we assign missings to extreme values:

```
nbagames[['WL_HOME'] + num_cols].
agg(['count','min','median','max']).T
              count    min    median    max
WL_HOME       4,568    0.00   1.00      1.00
FG_PCT_HOME   4,568    0.27   0.47      0.65
```

| | | | | |
|---|---|---|---|---|
| FTA_HOME | 4,568 | 1.00 | 22.00 | 64.00 |
| FG3_PCT_HOME | 4,568 | 0.06 | 0.36 | 0.84 |
| FTM_HOME | 4,568 | 1.00 | 17.00 | 44.00 |
| FT_PCT_HOME | 4,568 | 0.14 | 0.78 | 1.00 |
| OREB_HOME | 4,568 | 1.00 | 10.00 | 25.00 |
| DREB_HOME | 4,568 | 18.00 | 35.00 | 55.00 |
| REB_HOME | 4,568 | 22.00 | 45.00 | 70.00 |
| AST_HOME | 4,568 | 10.00 | 24.00 | 50.00 |
| ......... | | | | |
| FT_PCT_AWAY | 4,568 | 0.26 | 0.78 | 1.00 |
| OREB_AWAY | 4,568 | 0.00 | 10.00 | 26.00 |
| DREB_AWAY | 4,568 | 18.00 | 34.00 | 56.00 |
| REB_AWAY | 4,568 | 22.00 | 44.00 | 71.00 |
| AST_AWAY | 4,568 | 9.00 | 24.00 | 46.00 |
| STL_AWAY | 4,568 | 0.00 | 8.00 | 19.00 |
| BLK_AWAY | 4,568 | 0.00 | 5.00 | 15.00 |
| TOV_AWAY | 4,568 | 3.00 | 14.00 | 30.00 |

5. We should also review the correlations of the features:

```
corrmatrix = nbagames[['WL_HOME'] + \
  num_cols].corr(method="pearson")

sns.heatmap(corrmatrix,
  xticklabels=corrmatrix.columns,
  yticklabels=corrmatrix.columns, cmap="coolwarm")
plt.title('Heat Map of Correlation Matrix')
plt.tight_layout()
plt.show()
```

This produces the following plot:

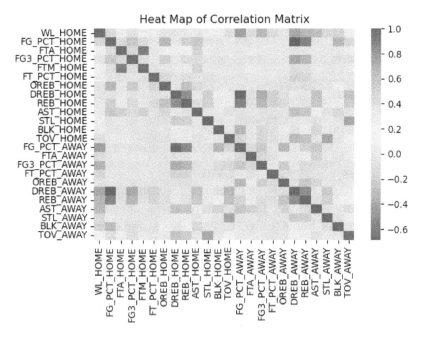

Figure 13.8 – Heat map of NBA game statistics correlations

Several features are correlated with the target, including the field goal percentage of the home team (FG_PCT_HOME) and defensive rebounds of the home team (DREB_HOME).

There is also correlation among the features. For example, the field goal percentage of the home team (FG_PCT_HOME) and the 3-point field goal percentage of the home team (FG3_PCT_HOME) are positively correlated, not surprisingly. Also, rebounds of the home team (REB_HOME) and defensive rebounds of the home team (DREB_HOME) are likely too closely correlated for any model to disentangle their impact.

6. Next, we create training and testing DataFrames:

```
X_train, X_test, y_train, y_test =  \
  train_test_split(nbagames[num_cols + cat_cols],\
  nbagames[['WL_HOME']], test_size=0.2, random_state=0)
```

7. We need to set up our column transformations. For the numeric columns, we check for outliers and scale the data. We one-hot encode the one categorical feature, SEASON. We will use these transformations later with the pipeline for our grid search:

```
ohe = OneHotEncoder(drop='first', sparse=False)


cattrans = make_pipeline(ohe)
```

```
standtrans = make_pipeline(OutlierTrans(2),
  SimpleImputer(strategy="median"), StandardScaler())
coltrans = ColumnTransformer(
  transformers=[
    ("cat", cattrans, cat_cols),
    ("stand", standtrans, num_cols)
  ]
)
```

8.   Before constructing our model, let's look at a decision boundary from a linear SVC model. We base the boundary on two features correlated with the target: the field goal percentage of the home team (FG_PCT_HOME) and defensive rebounds of the home team (DREB_HOME).

We create a function, dispbound, which will use the DecisionBoundaryDisplay module to show the boundary. This module is available with scikit-learn versions 1.1.1 or later. DecisionBoundaryDisplay needs a model to fit, two features, and target values:

```
pipe0 = make_pipeline(OutlierTrans(2),
  SimpleImputer(strategy="median"), StandardScaler())

X_train_enc = pipe0.\
  fit_transform(X_train[['FG_PCT_HOME','DREB_HOME']])

def dispbound(model, X, xvarnames, y, title):
  dispfit = model.fit(X,y)
  disp = DecisionBoundaryDisplay.from_estimator(
    dispfit, X, response_method="predict",
    xlabel=xvarnames[0], ylabel=xvarnames[1],
    alpha=0.5,
  )
  scatter = disp.ax_.scatter(X[:,0], X[:,1],
    c=y, edgecolor="k")

  disp.ax_.set_title(title)
  legend1 = disp.ax_.legend(*scatter.legend_elements(),
    loc="lower left", title="Home Win")
  disp.ax_.add_artist(legend1)
```

```
dispbound(LinearSVC(max_iter=1000000,loss='hinge'),
    X_train_enc, ['FG_PCT_HOME','DREB_HOME'],
    y_train.values.ravel(),
    'Linear SVC Decision Boundary')
```

This produces the following plot:

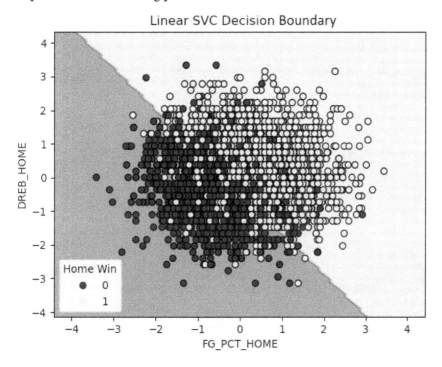

Figure 13.9 – Decision boundary for a two-feature linear SVC model

We get a pretty decent linear boundary with just the two features. That is great news, but let's do a more carefully constructed model.

9. To build our model, we first instantiate a linear SVC object and set up recursive feature elimination. We then add the column transformation, the feature selection, and the linear SVC to a pipeline and fit it:

```
svc = LinearSVC(max_iter=1000000, loss='hinge',
    random_state=0)

rfecv = RFECV(estimator=svc, cv=5)
```

```
pipe1 = make_pipeline(coltrans, rfecv, svc)

pipe1.fit(X_train, y_train.values.ravel())
```

10. Let's see what features were selected from our recursive feature elimination. We need to first get the column names after the one-hot encoding. We can then use the `get_support` method of the `rfecv` object to get the features that were selected. (You will get a deprecated warning regarding `get_feature_names` if you are using scikit-learn versions 1 or later. You can use `get_feature_names_out` instead, though that will not work with earlier versions of scikit-learn.):

```
new_cat_cols = \
  pipe1.named_steps['columntransformer'].\
  named_transformers_['cat'].\
  named_steps['onehotencoder'].\
  get_feature_names(cat_cols)

new_cols = np.concatenate((new_cat_cols, np.array(num_
cols)))
sel_cols = new_cols[pipe1['rfecv'].get_support()]
np.set_printoptions(linewidth=55)
sel_cols
array(['SEASON_2018', 'SEASON_2019', 'SEASON_2020',
       'FG_PCT_HOME', 'FTA_HOME', 'FG3_PCT_HOME',
       'FTM_HOME', 'FT_PCT_HOME', 'OREB_HOME',
       'DREB_HOME', 'REB_HOME', 'AST_HOME',
       'TOV_HOME', 'FG_PCT_AWAY', 'FTA_AWAY',
       'FG3_PCT_AWAY', 'FT_PCT_AWAY', 'OREB_AWAY',
       'DREB_AWAY', 'REB_AWAY', 'AST_AWAY',
       'BLK_AWAY', 'TOV_AWAY'], dtype=object)
```

11. We should look at the coefficients. Coefficients for each of the selected columns can be accessed with the `coef_` attribute of the `linearsvc` object. Perhaps not surprisingly, the shooting percentages of the home team (`FG_PCT_HOME`) and the away team (`FG_PCT_AWAY`) are the most important positive and negative predictors of the home team winning. The next most important features are the number of turnovers of the away and home teams:

```
pd.Series(pipe1['linearsvc'].\
  coef_[0], index=sel_cols).\
```

```
sort_values(ascending=False)
```

```
FG_PCT_HOME        2.21
TOV_AWAY           1.20
REB_HOME           1.19
FTM_HOME           0.95
FG3_PCT_HOME       0.94
FT_PCT_HOME        0.31
AST_HOME           0.25
OREB_HOME          0.18
DREB_AWAY          0.11
SEASON_2018        0.10
FTA_HOME          -0.05
BLK_AWAY          -0.07
SEASON_2019       -0.11
SEASON_2020       -0.19
AST_AWAY          -0.44
OREB_AWAY         -0.47
DREB_HOME         -0.49
FT_PCT_AWAY       -0.53
REB_AWAY          -0.63
FG3_PCT_AWAY      -0.80
FTA_AWAY          -0.81
TOV_HOME          -1.19
FG_PCT_AWAY       -1.91
dtype: float64
```

12. Let's take a look at the predictions. Our model predicts the home team winning very well:

```
pred = pipe1.predict(X_test)

print("accuracy: %.2f, sensitivity: %.2f, specificity:
%.2f, precision: %.2f"   %
    (skmet.accuracy_score(y_test.values.ravel(), pred),
    skmet.recall_score(y_test.values.ravel(), pred),
    skmet.recall_score(y_test.values.ravel(), pred, pos_
label=0),
```

```
         skmet.precision_score(y_test.values.ravel(), pred)))
accuracy: 0.93, sensitivity: 0.95, specificity: 0.92,
precision: 0.93
```

13. We should confirm that these metrics are not a fluke by doing some cross-validation. We use repeated stratified k folds for our validation, indicating that we want seven folds and 10 iterations. We get pretty much the same results as we did during the previous step:

```
kf = RepeatedStratifiedKFold(n_splits=7,n_repeats=10,\
    random_state=0)

scores = cross_validate(pipe1, X_train, \
    y_train.values.ravel(), \
    scoring=['accuracy','precision','recall','f1'], \
    cv=kf, n_jobs=-1)

print("accuracy: %.2f, precision: %.2f, sensitivity: \
%.2f, f1: %.2f"  %
    (np.mean(scores['test_accuracy']),\
    np.mean(scores['test_precision']),\
    np.mean(scores['test_recall']),\
    np.mean(scores['test_f1'])))
accuracy: 0.93, precision: 0.93, sensitivity: 0.95, f1:
0.94
```

14. We have been using the default value of C of 1 so far. We can try to identify a better value for C with a randomized grid search:

```
svc_params = {
 'linearsvc__C': uniform(loc=0, scale=100)
}

rs = RandomizedSearchCV(pipe1, svc_params, cv=10,
    scoring='accuracy', n_iter=20, random_state=0)

rs.fit(X_train, y_train.values.ravel())

rs.best_params_
```

```
{'linearsvc__C': 54.88135039273247}
```

```
rs.best_score_
0.9315809566584325
```

The best C value is 2.02 and the best accuracy score is 0.9316.

15. Let's take a closer look at the scores for each of the 20 times we ran the grid search. Each score is the average accuracy score across 10 folds. We actually get pretty much the same score regardless of the C value:

```
results = \
  pd.DataFrame(rs.cv_results_['mean_test_score'], \
    columns=['meanscore']).\
  join(pd.DataFrame(rs.cv_results_['params'])).\
  sort_values(['meanscore'], ascending=False)
```

```
results
```

|    | meanscore | linearsvc__C |
|----|-----------|--------------|
| 0  | 0.93      | 54.88        |
| 8  | 0.93      | 96.37        |
| 18 | 0.93      | 77.82        |
| 17 | 0.93      | 83.26        |
| 13 | 0.93      | 92.56        |
| 12 | 0.93      | 56.80        |
| 11 | 0.93      | 52.89        |
| 1  | 0.93      | 71.52        |
| 10 | 0.93      | 79.17        |
| 7  | 0.93      | 89.18        |
| 6  | 0.93      | 43.76        |
| 5  | 0.93      | 64.59        |
| 3  | 0.93      | 54.49        |
| 2  | 0.93      | 60.28        |
| 19 | 0.93      | 87.00        |
| 9  | 0.93      | 38.34        |
| 4  | 0.93      | 42.37        |
| 14 | 0.93      | 7.10         |
| 15 | 0.93      | 8.71         |
| 16 | 0.93      | 2.02         |

16. Let's now look at some of the predictions. Our model does well across the board, but not any better than the initial model:

```
pred = rs.predict(X_test)

print("accuracy: %.2f, sensitivity: %.2f, specificity:
%.2f, precision: %.2f"   %
    (skmet.accuracy_score(y_test.values.ravel(), pred),
    skmet.recall_score(y_test.values.ravel(), pred),
    skmet.recall_score(y_test.values.ravel(), pred, pos_
label=0),
    skmet.precision_score(y_test.values.ravel(), pred)))
accuracy: 0.93, sensitivity: 0.95, specificity: 0.92,
precision: 0.93
```

17. Let's also look at a confusion matrix:

```
cm = skmet.confusion_matrix(y_test, pred)
cmplot = \
    skmet.ConfusionMatrixDisplay(confusion_matrix=cm,
display_labels=['Loss', 'Won'])
cmplot.plot()
cmplot.ax_.set(title='Home Team Win Confusion Matrix',
    xlabel='Predicted Value', ylabel='Actual Value')
```

This produces the following plot:

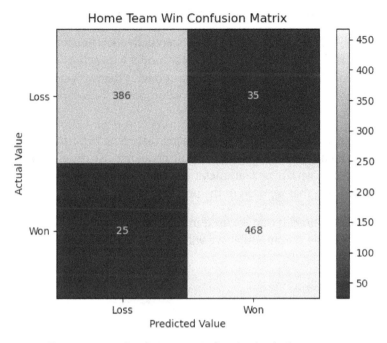

Figure 13.10 – Confusion matrix for wins by the home team

Our model largely predicts home team wins and losses correctly. Tuning the value of C did not make much of a difference, as we get pretty much the same accuracy regardless of the C value.

> **Note**
>
> You may have noticed that we are using the accuracy metric more often with the NBA games data than with the heart disease and machine failure data that we have worked with in previous chapters. We focused more on sensitivity with that data. There are two reasons for that. First, accuracy is a more compelling measure when classes are closer to being balanced for reasons we discussed in detail in *Chapter 6, Preparing for Model Evaluation*. Second, in predicting heart disease and machine power failure, we are biased towards sensitivity, as the cost of a false negative is higher than that of a false positive in those domains. For predicting NBA games, there is no such bias.

One advantage of linear SVC models is how easy they are to interpret. We are able to look at coefficients, which helps us make sense of the model and communicate the basis of our predictions to others. It nonetheless can be helpful to confirm that we do not get better results with a nonlinear model. We will do that in the next section.

# Nonlinear SVM classification models

Although nonlinear SVC is more complicated conceptually than linear SVC, as we saw in the first section of this chapter, running a nonlinear model with scikit-learn is relatively straightforward. The main difference from a linear model is that we need to do a fair bit more hyperparameter tuning. We have to specify values for C, for gamma, and for the kernel we want to use.

While there are theoretical reasons for hypothesizing that some hyperparameter values might work better than others for a given modeling challenge, we usually resolve those values empirically, that is, with hyperparameter tuning. We try that in this section with the same NBA games data that we used in the previous section:

1.  We load the same libraries that we used in the previous section. We also import the LogisticRegression module. We will use that with a feature selection wrapper method later:

    ```
    import pandas as pd
    import numpy as np
    from sklearn.preprocessing import MinMaxScaler
    from sklearn.pipeline import make_pipeline
    from sklearn.svm import SVC
    from sklearn.linear_model import LogisticRegression
    from scipy.stats import uniform
    from sklearn.feature_selection import RFECV
    from sklearn.impute import SimpleImputer
    from scipy.stats import randint
    from sklearn.model_selection import RandomizedSearchCV
    import sklearn.metrics as skmet
    import os
    import sys
    sys.path.append(os.getcwd() + "/helperfunctions")
    from preprocfunc import OutlierTrans
    ```

2.  We import the `nbagames` module, which has the code that loads and preprocesses the NBA games data. This is just a copy of the code that we ran in the previous section to prepare the data for modeling. There is no need to repeat those steps here.

    We also import the `dispbound` function we used in the previous section to display decision boundaries. We copied that code into a file called `displayfunc.py` in the `helperfunctions` subfolder of the current directory:

    ```
    import nbagames as ng
    from displayfunc import dispbound
    ```

3.  We use the `nbagames` module to get the training and testing data:

    ```
    X_train = ng.X_train
    X_test = ng.X_test
    y_train = ng.y_train
    y_test = ng.y_test
    ```

4.  Before constructing a model, let's look at the decision boundaries for a couple of different kernels with two features: the field goal percentage of the home team (FG_PCT_HOME) and defensive rebounds of the home team (DREB_HOME). We start with the `rbf` kernel, using different values for gamma and C:

    ```
    pipe0 = make_pipeline(OutlierTrans(2),
      SimpleImputer(strategy="median"),
      StandardScaler())
    X_train_enc = \
      pipe0.fit_transform(X_train[['FG_PCT_HOME',
        'DREB_HOME']])

    dispbound(SVC(kernel='rbf', gamma=30, C=1),
      X_train_enc, ['FG_PCT_HOME','DREB_HOME'],
      y_train.values.ravel(),
      "SVC with rbf kernel-gamma=30, C=1")
    ```

Running this a few different ways produces the following plots:

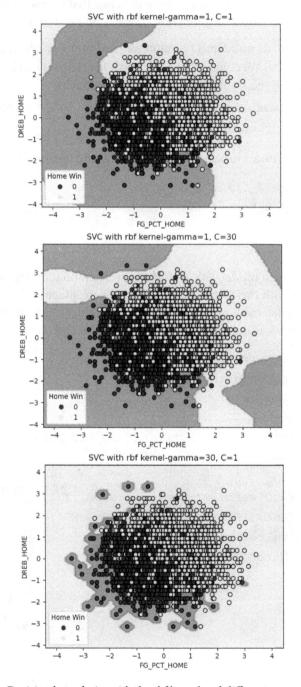

Figure 13.11 – Decision boundaries with the rbf kernel and different gamma and C values

At values for gamma and C near the default, we see some bending of the decision boundary to accommodate a few wayward points in the loss class. These are instances where the home team lost despite having very high defensive rebound totals. With the rbf kernel, two of these instances are now correctly classified. There are also a couple of instances with a high home team field goal percentage but low home team defensive rebounds, which are now correctly classified. However, there is not much change overall in our predictions compared with the linear model from the previous section.

But this changes significantly if we increase values for C or gamma. Recall that higher values of C increase the penalty for misclassification. This leads to boundaries that wind around instances more.

Increasing gamma to 30 causes substantial overfitting. High values of gamma mean that data points have to be very close to each other to be grouped together. This results in decision boundaries closely tied to small numbers of instances, sometimes just one instance.

5. We can also show the boundaries for a polynomial kernel. We will keep the C value at the default to focus on the effect of changing the number of degrees:

```
dispbound(SVC(kernel='poly', degree=7),
  X_train_enc, ['FG_PCT_HOME','DREB_HOME'],
  y_train.values.ravel(),
  "SVC with polynomial kernel - degree=7")
```

Running this a couple of different ways produces the following plots:

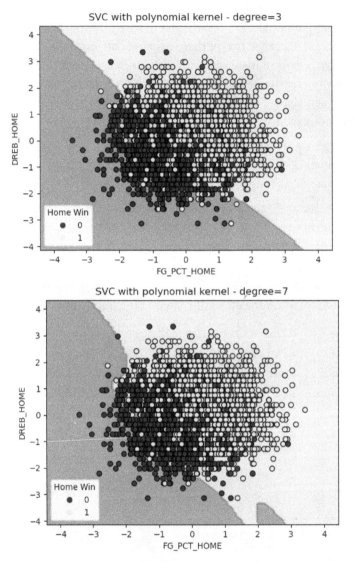

Figure 13.12 – Decision boundaries with polynomial kernel and different degrees

We can see some bending of the decision boundary at higher degree levels to handle a couple of unusual instances. There is not much overfitting here, but not really much improvement in our predictions either.

This at least hints at what to expect when we construct the model. We should try some nonlinear models but there is a good chance that they will not lead to much improvement over the linear model we used in the previous section.

6.  Now, we are ready to set up the pipeline that we will use for our nonlinear SVC. Our pipeline will do the column transformation and a recursive feature elimination. We use logistic regression for the feature selection:

```
rfecv = RFECV(estimator=LogisticRegression())


svc = SVC()


pipel = make_pipeline(ng.coltrans, rfecv, svc)
```

7.  We create a dictionary to use for our hyperparameter tuning. This dictionary is structured somewhat differently from other dictionaries we have used for this purpose. That is because certain hyperparameters only work with certain other hyperparameters. For example, gamma does not work with a linear kernel:

```
svc_params = [
    {
        'svc__kernel': ['rbf'],
        'svc__C': uniform(loc=0, scale=20),
        'svc__gamma': uniform(loc=0, scale=100)
    },
    {
        'svc__kernel': ['poly'],
        'svc__degree': randint(1, 5),
        'svc__C': uniform(loc=0, scale=20),
        'svc__gamma': uniform(loc=0, scale=100)
    },
    {
        'svc__kernel': ['linear','sigmoid'],
        'svc__C': uniform(loc=0, scale=20)
    }
]
```

> **Note**
>
> You may have noticed that one of the kernels we will be using is linear, and wonder how this is different from the linear SVC module we used in the previous section. `LinearSVC` will often converge faster, particularly with large datasets. It does not use the kernel trick. We will also likely get different results as the optimization is different in several ways.

8.  Now we are ready to fit an SVC model. The best model is actually one with a linear kernel:

```
rs = RandomizedSearchCV(pipe1, svc_params, cv=5,
    scoring='accuracy', n_iter=10, n_jobs=-1,
    verbose=5, random_state=0)


rs.fit(X_train, y_train.values.ravel())


rs.best_params_
{'svc__C': 1.1342595463488636, 'svc__kernel': 'linear'}


rs.best_score_
0.9299405955437289
```

9.  Let's take a closer look at the hyperparameters selected and the associated accuracy scores. We can get the 20 randomly chosen hyperparameter combinations from the params list from the grid object's cv_results_ dictionary. We can get the mean test score from that same dictionary.

    We sort by accuracy score in descending order. Linear kernels outperform polynomial and rbf kernels, though not substantially better than polynomial at 3, 4, and 5 degrees. rbf kernels perform particularly poorly:

```
results = \
  pd.DataFrame(rs.cv_results_['mean_test_score'], \
    columns=['meanscore']).\
  join(pd.json_normalize(rs.cv_results_['params'])).\
  sort_values(['meanscore'], ascending=False)


results
```

|  | C | gamma | kernel | degree |
|---|---|---|---|---|
| meanscore | | | | |
| 0.93 | 1.13 | NaN | linear | NaN |
| 0.89 | 1.42 | 64.82 | poly | 3.00 |
| 0.89 | 9.55 | NaN | sigmoid | NaN |
| 0.89 | 11.36 | NaN | sigmoid | NaN |
| 0.89 | 2.87 | 75.86 | poly | 5.00 |
| 0.64 | 12.47 | 43.76 | poly | 4.00 |

| 0.64 | 15.61 | 72.06 | poly | 4.00 |
| 0.57 | 11.86 | 84.43 | rbf | NaN |
| 0.57 | 16.65 | 77.82 | rbf | NaN |
| 0.57 | 19.57 | 79.92 | rbf | NaN |

> **Note**
>
> We use the pandas `json_nomalize` method to handle the somewhat messy hyperparameter combinations we pull from the `params` list. It is messy because different hyperparameters are available depending on the kernel used. This means that the 20 dictionaries in the `params` list will have different keys. For example, the polynomial kernels will have values for degrees. The linear and `rbf` kernels will not.

10. We can access the support vectors via the `best_estimator_` attribute. There are 625 support vectors *holding up* the decision boundary:

```
rs.best_estimator_['svc'].\
  support_vectors_.shape
(625, 18)
```

11. Finally, we can take a look at the predictions. Not surprisingly, we do not get better results than we got with the linear SVC model that we ran in the last section. I say not surprisingly because the best model was found to be a model with a linear kernel:

```
pred = rs.predict(X_test)

print("accuracy: %.2f, sensitivity: %.2f, specificity:
%.2f, precision: %.2f" %
  (skmet.accuracy_score(y_test.values.ravel(), pred),
  skmet.recall_score(y_test.values.ravel(), pred),
  skmet.recall_score(y_test.values.ravel(), pred,
    pos_label=0),
  skmet.precision_score(y_test.values.ravel(), pred)))
accuracy: 0.93, sensitivity: 0.94, specificity: 0.91,
precision: 0.93
```

Although we have not improved upon our model from the previous section, it was still a worthwhile exercise to experiment with some nonlinear models. Indeed, this is often how we discover whether we have data that can be successfully separated linearly. This is typically difficult to visualize and so we rely on hyperparameter tuning to tell us which kernel classifies our data best.

This section and the previous one demonstrate the key techniques for using SVMs for binary classification. Much of what we have done so far applies to multiclass classification as well. We will take a look at SVC modeling strategies when our target has more than two values in the next section.

# SVMs for multiclass classification

All of the same concerns that we had when we used SVC for binary classification apply when we are doing multiclass classification. We need to determine whether the classes are linearly separable, and if not, which kernel will yield the best results. As discussed in the first section of this chapter, we also need to decide whether that classification is best modeled as one-versus-one or one-versus-rest. One-versus-one finds decision boundaries that separate each class from each of the other classes. One-versus-rest finds decision boundaries that distinguish each class from all other instances. We try both approaches in this section.

We will work with the machine failure data that we worked with in previous chapters.

> **Note**
>
> This dataset on machine failure is available for public use at https://www.kaggle.com/datasets/shivamb/machine-predictive-maintenance-classification. There are 10,000 observations, 12 features, and two possible targets. One is binary: the machine failed or did not. The other has types of failure. The instances in this dataset are synthetic, generated by a process designed to mimic machine failure rates and causes.

Let's build a multiclass SVC model:

1.  We start by loading the same libraries that we have been using in this chapter:

    ```
    import pandas as pd
    from sklearn.model_selection import train_test_split
    from sklearn.preprocessing import OneHotEncoder,
    MinMaxScaler
    from sklearn.pipeline import make_pipeline
    from sklearn.svm import SVC
    from scipy.stats import uniform
    from sklearn.impute import SimpleImputer
    from sklearn.compose import ColumnTransformer
    from sklearn.model_selection import RandomizedSearchCV
    ```

```
import sklearn.metrics as skmet
import os
import sys
sys.path.append(os.getcwd() + "/helperfunctions")
from preprocfunc import OutlierTrans
```

2.  We will load the machine failure type dataset and take a look at its structure. There is a mixture of character and numeric data. There are no missing values:

```
machinefailuretype = pd.read_csv("data/
machinefailuretype.csv")
machinefailuretype.info()
<class 'pandas.core.frame.DataFrame'>
RangeIndex: 10000 entries, 0 to 9999
Data columns (total 10 columns):
 #   Column                Non-Null Count      Dtype
---  ------                --------------      -----
 0   udi                   10000 non-null      int64
 1   product               10000 non-null      object
 2   machinetype           10000 non-null      object
 3   airtemp               10000 non-null      float64
 4   processtemperature    10000 non-null      float64
 5   rotationalspeed       10000 non-null      int64
 6   torque                10000 non-null      float64
 7   toolwear              10000 non-null      int64
 8   fail                  10000 non-null      int64
 9   failtype              10000 non-null      object
dtypes: float64(3), int64(4), object(3)
memory usage: 781.4+ KB
```

3.  Let's look at a few observations:

```
machinefailuretype.head()
   udi product machinetype airtemp processtemperature\
0   1   M14860          M      298           309
1   2   L47181          L      298           309
2   3   L47182          L      298           308
3   4   L47183          L      298           309
```

| | | | | | |
|---|---|---|---|---|---|
| 4 | 5 | L47184 | L | 298 | 309 |

| | rotationalspeed | torque | toolwear | fail | failtype |
|---|---|---|---|---|---|
| 0 | 1551 | 43 | 0 | 0 | No Failure |
| 1 | 1408 | 46 | 3 | 0 | No Failure |
| 2 | 1498 | 49 | 5 | 0 | No Failure |
| 3 | 1433 | 40 | 7 | 0 | No Failure |
| 4 | 1408 | 40 | 9 | 0 | No Failure |

4. Let's also look at the distribution of the target. We have a significant class imbalance, so we will need to deal with that in some way:

```
machinefailuretype.failtype.\
  value_counts(dropna=False).sort_index()
Heat Dissipation Failure     112
No Failure                   9652
Overstrain Failure           78
Power Failure                95
Random Failures              18
Tool Wear Failure            45
Name: failtype, dtype: int64
```

5. We can save ourselves some trouble later by creating a numeric code for failure type, which we will use rather than the character value. We do not need to put this into a pipeline since we are not introducing any data leakage in the conversion:

```
def setcode(typetext):
  if (typetext=="No Failure"):
    typecode = 1
  elif (typetext=="Heat Dissipation Failure"):
    typecode = 2
  elif (typetext=="Power Failure"):
    typecode = 3
  elif (typetext=="Overstrain Failure"):
    typecode = 4
  else:
    typecode = 5
  return typecode
```

```
machinefailuretype["failtypecode"] = \
  machinefailuretype.apply(lambda x: setcode(x.failtype),
axis=1)
```

6.  We should also look at some descriptive statistics. We will need to scale the features:

```
num_cols = ['airtemp','processtemperature',
  'rotationalspeed','torque','toolwear']
cat_cols = ['machinetype']

machinefailuretype[num_cols].agg(['min','median','max']).T
                   min        median     max
airtemp            295.30     300.10     304.50
processtemperature 305.70     310.10     313.80
rotationalspeed    1,168.00   1,503.00   2,886.00
torque             3.80       40.10      76.60
toolwear           0.00       108.00     253.00
```

7.  Let's now create training and testing DataFrames. We should also use the stratify parameter to ensure an equal distribution of target values in our training and testing data:

```
X_train, X_test, y_train, y_test =  \
  train_test_split(machinefailuretype[num_cols + cat_
cols],\
  machinefailuretype[['failtypecode']],\
  stratify=machinefailuretype[['failtypecode']], \
  test_size=0.2, random_state=0)
```

8.  We set up the column transformations we need to run. For the numeric columns, we set outliers to the median and then scale the values. We do one-hot-encoding of the one categorical feature, machinetype. It has H, M, and L values for high, medium, and low quality:

```
ohe = OneHotEncoder(drop='first', sparse=False)

cattrans = make_pipeline(ohe)
standtrans = make_pipeline(OutlierTrans(3),
  SimpleImputer(strategy="median"),
```

```
   MinMaxScaler())
coltrans = ColumnTransformer(
  transformers=[
    ("cat", cattrans, cat_cols),
    ("stand", standtrans, num_cols),
  ]
)
```

9.  Next, we set up a pipeline with the column transformation and the SVC instance. We set the `class_weight` parameter to `balanced` to deal with class imbalance. This applies a weight that is inversely related to the frequency of the target class:

```
svc = SVC(class_weight='balanced', probability=True)
pipel = make_pipeline(coltrans, svc)
```

We only have a handful of features in this case so we will not worry about feature selection. (We might still be concerned about features that are highly correlated, but that is not an issue with this dataset.)

10. We create a dictionary with the hyperparameter combinations to use with a grid search. This is largely the same as the dictionary we used in the previous section, except we have added a decision function shape key. This will cause the grid search to try both one-versus-one (`ovo`) and one-versus-rest (`ovr`) classification:

```
svc_params = [
  {
    'svc__kernel': ['rbf'],
    'svc__C': uniform(loc=0, scale=20),
    'svc__gamma': uniform(loc=0, scale=100),
    'svc__decision_function_shape': ['ovr','ovo']
  },
  {
    'svc__kernel': ['poly'],
    'svc__degree': np.arange(0,6),
    'svc__C': uniform(loc=0, scale=20),
    'svc__gamma': uniform(loc=0, scale=100),
    'svc__decision_function_shape': ['ovr','ovo']
  },
  {
    'svc__kernel': ['linear','sigmoid'],
```

```
    'svc__C': uniform(loc=0, scale=20),
    'svc__decision_function_shape': ['ovr','ovo']
  }
]
```

11. Now we are ready to run the randomized grid search. We will base our scoring on the area under the ROC curve. The best hyperparameters include the one-versus-one decision function and the `rbf` kernel:

```
rs = RandomizedSearchCV(pipe1, svc_params, cv=7,
scoring="roc_auc_ovr", n_iter=10)
rs.fit(X_train, y_train.values.ravel())
rs.best_params_
{'svc__C': 5.609789456747942,
 'svc__decision_function_shape': 'ovo',
 'svc__gamma': 27.73459801111866,
 'svc__kernel': 'rbf'}


rs.best_score_
0.9187636814475847
```

12. Let's see the score for each iteration. In addition to the best model that we saw in the previous step, there are several other hyperparameter combinations that have scores that are nearly as high. One-versus-rest with a linear kernel does nearly as well as the best-performing model:

```
results = \
  pd.DataFrame(rs.cv_results_['mean_test_score'], \
    columns=['meanscore']).\
  join(pd.json_normalize(rs.cv_results_['params'])).\
  sort_values(['meanscore'], ascending=False)


results
```

| | meanscore | svc__C | svc__decision_function_shape | svc__gamma | svc__kernel |
|---|---|---|---|---|---|
| 7 | 0.92 | 5.61 | ovo | 27.73 | rbf |
| 5 | 0.91 | 9.43 | ovr | NaN | linear |
| 3 | 0.91 | 5.40 | ovr | NaN | linear |
| 0 | 0.90 | 19.84 | ovr | 28.70 | rbf |

| 8 | 0.87 | 5.34 | ovo | 93.87 | rbf |
| 6 | 0.86 | 8.05 | ovr | 80.57 | rbf |
| 9 | 0.86 | 4.41 | ovo | 66.66 | rbf |
| 1 | 0.86 | 3.21 | ovr | 85.35 | rbf |
| 4 | 0.85 | 0.01 | ovo | 38.24 | rbf |
| 2 | 0.66 | 7.61 | ovr | NaN | sigmoid |

13. We should take a look at the confusion matrix:

```
pred = rs.predict(X_test)


cm = skmet.confusion_matrix(y_test, pred)
cmplot = skmet.ConfusionMatrixDisplay(confusion_
matrix=cm,
    display_labels=['None',
'Heat','Power','Overstrain','Other'])
cmplot.plot()
cmplot.ax_.set(title='Machine Failure Type Confusion
Matrix',
    xlabel='Predicted Value', ylabel='Actual Value')
```

This produces the following plot:

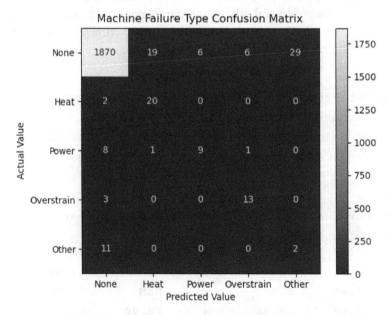

Figure 13.13 – Confusion matrix for machine failure type prediction

14. Let's also do a classification report. We do not get great scores for sensitivity for most classes, though our model does predict heat and overstrain failures pretty well:

```
print(skmet.classification_report(y_test, pred,
  target_names=['None', 'Heat','Power', 'Overstrain',
'Other']))
```

|  | precision | recall | f1-score | support |
|---|---|---|---|---|
| None | 0.99 | 0.97 | 0.98 | 1930 |
| Heat | 0.50 | 0.91 | 0.65 | 22 |
| Power | 0.60 | 0.47 | 0.53 | 19 |
| Overstrain | 0.65 | 0.81 | 0.72 | 16 |
| Other | 0.06 | 0.15 | 0.09 | 13 |
|  |  |  |  |  |
| accuracy |  |  | 0.96 | 2000 |
| macro avg | 0.56 | 0.66 | 0.59 | 2000 |
| weighted avg | 0.97 | 0.96 | 0.96 | 2000 |

When modeling targets such as machine failure types that have a high class imbalance, we are often more concerned with metrics other than accuracy. This is partly determined by our domain knowledge. Avoiding false negatives may be more important than avoiding false positives. Doing a thorough check on a machine too early is definitely preferable to doing it too late.

The 96 to 97 percent weighted precision, recall (sensitivity), and f1 scores do not provide a good sense of the performance of our model. They mainly reflect the large class imbalance and the fact that it is very easy to predict no machine failure. The much lower macro averages (which are just simple averages across classes) indicate that our model struggles to predict some types of machine failure.

This example illustrates that it is relatively easy to extend SVC to models that have targets with more than two values. We can specify whether we want to use one-versus-one or one-versus-rest classification. The one-versus-rest approach can be faster when the number of classes is above three since there will be fewer classifiers trained.

# Summary

In this chapter, we explored the different strategies for implementing SVC. We used linear SVC (which does not use kernels), which can perform very well when our classes are linearly separable. We then examined how to use the kernel trick to extend SVC to cases where the classes are not linearly separable. Finally, we used one-versus-one and one-versus-rest classification to handle targets with more than two values.

SVC is an exceptionally useful technique for binary and multiclass classification. It can handle both straightforward and complicated relationships between features and the target. There are few supervised learning problems for which SVMs should not at least be considered. However, it is not very efficient with very large datasets.

In the next chapter, we will explore another popular and flexible classification algorithm, Naive Bayes.

# 14
# Naïve Bayes Classification

In this chapter, we will examine situations where naïve Bayes might be a more efficient classifier than the ones we have examined so far. Naïve Bayes is a very intuitive and easy-to-implement classifier. Assuming our features are independent, we may even get improved performance over logistic regession, particularly if we are not using regularization with the latter.

In this chapter, we will discuss the fundamental assumptions of naïve Bayes and how the algorithm is used to tackle some of the modeling challenges we have already explored, as well as some new ones, such as text classification. We will consider when naïve Bayes is a good option and when it is not. We will also examine the interpretation of naïve Bayes models.

We will cover the following topics in this chapter:

- Key concepts
- Naïve Bayes classification models
- Naïve Bayes for text classification

# Technical requirements

We will mainly stick to the pandas, NumPy, and scikit-learn libraries in this chapter. The only exception is the imbalanced-learn library, which can be installed with `pip install imbalanced-learn`. All the code in this chapter was tested with scikit-learn versions 0.24.2 and 1.0.2.

# Key concepts

The naïve Bayes classifier uses Bayes' theorem to predict class membership. Bayes' theorem describes the relationship between the probability of an event and the probability of an event given new, relevant data. The probability of an event given new data is called the **posterior probability**. The probability of an event occurring before the new data is appropriately referred to as the **prior probability**.

Bayes' theorem gives us the following equation:

$$posterior\ probability = \frac{probability\ of\ data\ given\ event * prior\ probability}{probability\ of\ data}$$

The posterior probability (the probability of an event given new data) is equal to the probability of the data given the event, times the prior probability of the event, divided by the probability of the new data.

Somewhat less colloquially, this is typically written as follows:

$$P(A|B) = \frac{P(B|A)P(A)}{P(B)}$$

Here, $A$ is an event, such as class membership, and $B$ is new information. When applied to classification, we get the following equation:

$$P(y|x_1, ..., x_j) = \frac{P(x_1, ..., x_j|y)P(y)}{P(x_1, ..., x_j)}$$

Here, $P(y \mid x_1, ..., x_j)$ is the probability of class membership for an instance given the features for the instance, and $P(x_1, ..., x_j \mid y)$ is the probability of features given class membership. $P(y)$ is the probability of class membership, while $P(x_1, ..., x_j)$ is the probability of feature values. Therefore, the posterior probability, $P(y \mid x_1, ..., x_j)$, is equal to the probability of the feature values given class membership, times the probability of class membership, divided by the probability of the feature values.

The assumption here is that the features are independent of one another. This is what gives this method the *naïve* adjective. As a practical matter, though, feature independence is not necessary to get good results with naïve Bayes.

Naïve Bayes can work with numeric or categorical features. We typically use Gaussian naïve Bayes when we have mainly numeric features. As its name implies, Gaussian naïve Bayes assumes that the conditional probability of feature values, $P(x_1, ..., x_j \mid y)$, follows a normal distribution. $P(x_1, ..., x_j \mid y)$ can then be calculated relatively straightforwardly using the standard deviations and means of features within each class.

When our features are discrete or are counts, we can use multinomial naïve Bayes instead. More generally, it works well when the conditional probability of feature values follows a multinomial distribution. A common application of multinomial naïve Bayes is with text classification using the **bag-of-words** approach. With bag-of-words, the features are the counts of each word in a document. We can apply Bayes' theorem to estimate the probability of class membership:

$$P(C_k \mid W) = \frac{P(C_k)P(W \mid C_k)}{P(W)}$$

Here, $P(C_k \mid W)$ is the probability of membership in the $k$ class, given a vector of word counts, $W$. We will put this to good use in the last section of this chapter.

There is a fairly wide range of text classification tasks for which naïve Bayes is applicable. It is used in sentiment analysis, spam detection, and news story categorization, to name just a few examples.

Naïve Bayes is an efficient algorithm for both training and prediction and often performs very well. It is quite scalable, working well with large numbers of instances and features. It is also very easy to interpret. The algorithm does best when model complexity is not necessary for good predictions. Even when naïve Bayes is not likely to be the approach that will yield the least bias, it is often useful for diagnostic purposes, or to check the results from a different algorithm.

The NBA data we worked with in the previous chapter might be a good candidate for modeling with naïve Bayes. We will explore that in the next section.

# Naïve Bayes classification models

One of the attractions of naïve Bayes is that you can get decent results quickly, even when you have lots of data. Both fitting and predicting are fairly easy on system resources. Another advantage is that relatively complex relationships can be captured without having to transform the feature space or doing much hyperparameter tuning. We can demonstrate this with the NBA data we worked with in the previous chapter.

We will work with data on **National Basketball Association** (NBA) games in this section. The dataset contains statistics from each NBA game from the 2017/2018 season through the 2020/2021 season. This includes the home team; whether the home team won; the visiting team; shooting percentages for visiting and home teams; turnovers, rebounds, and assists by both teams; and several other measures.

---

**Note**

The NBA game data can be downloaded by the public at `https://www.kaggle.com/datasets/wyattowalsh/basketball`. This dataset contains game data starting with the 1946/1947 NBA season. It uses `nba_api` to pull stats from `nba.com`. That API is available at `https://github.com/swar/nba_api`.

---

Let's build a classification model using naïve Bayes:

1. We will load the same libraries we have been using in the last few chapters:

```
import pandas as pd
import numpy as np
from sklearn.model_selection import train_test_split
from sklearn.preprocessing import OneHotEncoder,
StandardScaler
from sklearn.impute import SimpleImputer
from sklearn.pipeline import make_pipeline
from sklearn.compose import ColumnTransformer
from sklearn.feature_selection import RFE
from sklearn.naive_bayes import GaussianNB
from sklearn.linear_model import LogisticRegression
from sklearn.model_selection import cross_validate, \
    RandomizedSearchCV, RepeatedStratifiedKFold
import sklearn.metrics as skmet
import os
import sys
sys.path.append(os.getcwd() + "/helperfunctions")
from preprocfunc import OutlierTrans
```

2. Next, we will load the NBA games data. We need to do a little data cleaning here. A handful of observations have missing values for whether the home team won or not, `WL_HOME`. We will remove those as that will be our target. We will also convert `WL_HOME` into an integer. Notice that there is not so much class imbalance that we need to take aggressive steps to deal with it:

```
nbagames = pd.read_csv("data/nbagames2017plus.csv",
parse_dates=['GAME_DATE'])
nbagames = nbagames.loc[nbagames.WL_HOME.isin(['W','L'])]
nbagames.shape
(4568, 149)
nbagames['WL_HOME'] = \
  np.where(nbagames.WL_HOME=='L',0,1).astype('int')
nbagames.WL_HOME.value_counts(dropna=False)
1    2586
0    1982
Name: WL_HOME, dtype: int64
```

3. Now, let's create training and testing DataFrames, organizing them by numeric and categorical features. We should also generate some descriptive statistics. Since we did that in the previous chapter, we will not repeat that here; however, it might be helpful to go back and take a look at those numbers to get ready for the modeling stage:

```
num_cols = ['FG_PCT_HOME','FTA_HOME','FG3_PCT_HOME',
  'FTM_HOME','FT_PCT_HOME','OREB_HOME','DREB_HOME',
  'REB_HOME','AST_HOME','STL_HOME','BLK_HOME',
  'TOV_HOME', 'FG_PCT_AWAY','FTA_AWAY','FG3_PCT_AWAY',
  'FT_PCT_AWAY','OREB_AWAY','DREB_AWAY','REB_AWAY',
  'AST_AWAY','STL_AWAY','BLK_AWAY','TOV_AWAY']
cat_cols = ['TEAM_ABBREVIATION_HOME','SEASON']

X_train, X_test, y_train, y_test = \
  train_test_split(nbagames[num_cols + cat_cols],\
  nbagames[['WL_HOME']], test_size=0.2,random_state=0)
```

4. Now, we need to set up the column transformations. We will deal with some outliers for the numeric features, assigning those values and any missing values to the median. Then, we will use the standard scaler. We will set up one-hot encoding for the categorical features:

```
ohe = OneHotEncoder(drop='first', sparse=False)

cattrans = make_pipeline(ohe)
standtrans = make_pipeline(OutlierTrans(2),
  SimpleImputer(strategy="median"), StandardScaler())
coltrans = ColumnTransformer(
  transformers=[
    ("cat", cattrans, cat_cols),
    ("stand", standtrans, num_cols)
  ]
)
```

5. We are now ready to run a naïve Bayes classifier. We will add a Gaussian naïve Bayes instance to a pipeline that will be run after the column transformation and some recursive feature elimination:

```
nb = GaussianNB()

rfe = RFE(estimator=LogisticRegression(),
  n_features_to_select=15)

pipe1 = make_pipeline(coltrans, rfe, nb)
```

6. Let's evaluate this model with K-fold cross-validation. We get okay scores, though not as good as those with support vector classification in the previous chapter:

```
kf = RepeatedStratifiedKFold(n_splits=7,n_repeats=10,\
  random_state=0)

scores = cross_validate(pipe1, X_train, \
  y_train.values.ravel(), \
  scoring=['accuracy','precision','recall','f1'], \
  cv=kf, n_jobs=-1)
```

```
print("accuracy: %.2f, precision: %.2f,
  sensitivity: %.2f, f1: %.2f" %
  (np.mean(scores['test_accuracy']),\
  np.mean(scores['test_precision']),\
  np.mean(scores['test_recall']),\
  np.mean(scores['test_f1'])))
```
**accuracy: 0.81, precision: 0.84, sensitivity: 0.83, f1: 0.83**

7.  With Gaussian naïve Bayes, there is only one hyperparameter we must worry about tuning. We can determine how much smoothing to use with the `var_smoothing` hyperparameter. We can do a randomized grid search to figure out the best value.

    The `var_smoothing` hyperparameter determines how much is added to variances, which will cause models to be less dependent on instances close to mean values:

    ```
    nb_params = {
        'gaussiannb__var_smoothing': np.logspace(0,-9,
    num=100)
    }


    rs = RandomizedSearchCV(pipe1, nb_params, cv=kf, \
      scoring='accuracy')
    rs.fit(X_train, y_train.values.ravel())
    ```

8.  We get somewhat better accuracy:

    ```
    rs.best_params_
    ```
    **{'gaussiannb__var_smoothing': 0.657933224657568}**

    ```
    rs.best_score_
    ```
    **0.8608648056923919**

9.  We should also take a look at the results from the different iterations. As we can see, larger smoothing values do better:

    ```
    results = \
      pd.DataFrame(rs.cv_results_['mean_test_score'], \
        columns=['meanscore']).\
      join(pd.DataFrame(rs.cv_results_['params'])).\
    ```

```
sort_values(['meanscore'], ascending=False)

results
        meanscore     gaussiannb__var_smoothing
2       0.86086       0.65793
1       0.85118       0.03511
9       0.81341       0.00152
5       0.81212       0.00043
7       0.81180       0.00019
8       0.81169       0.00002
3       0.81152       0.00000
6       0.81152       0.00000
0       0.81149       0.00000
4       0.81149       0.00000
```

10. We can also look at the average fit and score times for each iteration:

```
print("fit time: %.3f, score time: %.3f"   %
    (np.mean(rs.cv_results_['mean_fit_time']),\
    np.mean(rs.cv_results_['mean_score_time'])))
fit time: 0.660, score time: 0.029
```

11. Let's look at the predictions for the best model. In addition to improving accuracy, there is an improvement in sensitivity, from 0.83 to 0.92:

```
pred = rs.predict(X_test)

print("accuracy: %.2f, sensitivity: %.2f, \
    specificity: %.2f, precision: %.2f"   %
    (skmet.accuracy_score(y_test.values.ravel(), pred),
    skmet.recall_score(y_test.values.ravel(), pred),
    skmet.recall_score(y_test.values.ravel(), pred, \
        pos_label=0),
    skmet.precision_score(y_test.values.ravel(), pred)))
accuracy: 0.86, sensitivity: 0.92, specificity: 0.79,
precision: 0.83
```

12. It is a good idea to also look at a confusion matrix to get a better sense of how the model does:

```
cm = skmet.confusion_matrix(y_test, pred)
cmplot = skmet.ConfusionMatrixDisplay(
    confusion_matrix=cm, display_labels=['Loss', 'Won'])
cmplot.plot()
cmplot.ax_.set(title='Home Team Win Confusion Matrix',
    xlabel='Predicted Value', ylabel='Actual Value')
```

This produces the following plot:

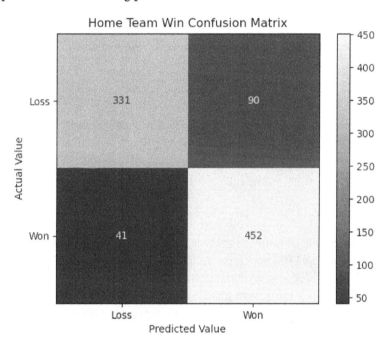

Figure 14.1 – Confusion matrix for home team wins based on the Gaussian naïve Bayes model

This is not bad, though still not as good as our support vector model in the previous chapter. In particular, we would like to do a little better at predicting losses. This is also reflected in the relatively low specificity score of **0.79** that we saw in the previous step. Recall that specificity is the rate at which we correctly predict negative values out of the actual negatives.

On the other hand, the fitting and scoring ran quite swiftly. We also did not need to do much hyperparameter tuning. Naïve Bayes can often be a good place to start when modeling either a binary or multiclass target.

Naïve Bayes has turned out to be an even more popular option for text classification. We will use it for that purpose in the next section.

# Naïve Bayes for text classification

It is perhaps surprising that an algorithm based on calculating conditional probabilities could be useful for text classification. But this follows fairly straightforwardly with a key simplifying assumption. Let's assume that our documents can be well represented by the counts of each word in the document, without regard for word order or grammar. This is known as a bag-of-words. The relationship that a bag-of-words has to a categorical target – say, spam/not spam or positive/negative – can be modeled successfully with multinomial naïve Bayes.

We will work with text message data in this section. The dataset we will use contains labels for spam and not spam messages.

> **Note**
>
> This dataset on text messages can be downloaded by the public at `https://www.kaggle.com/datasets/team-ai/spam-text-message-classification`. It contains two columns: the text message and the spam or not spam (ham) label.

Let's do some text classification with naïve Bayes:

1.  We will need a couple of modules that we have not used so far in this book. We will import `MultinomialNB`, which we will need to construct a multinomial naïve Bayes model. We will also need `CountVectorizer` to create a bag-of-words. We will import the `SMOTE` module to handle class imbalance. Note that we will use an *imbalanced-learn* pipeline rather than a *scikit-learn* one. This is because we will be using `SMOTE` in our pipeline:

    ```
    import pandas as pd
    import numpy as np
    from sklearn.model_selection import train_test_split
    from imblearn.pipeline import make_pipeline
    from imblearn.over_sampling import SMOTE
    from sklearn.naive_bayes import MultinomialNB
    from sklearn.feature_extraction.text import
    CountVectorizer
    import sklearn.metrics as skmet
    ```

> **Note**
>
> We are using SMOTE in this section to do oversampling; that is, we will be duplicating instances in underrepresented classes. Oversampling can be a good option when we are concerned that our model is doing a poor job of capturing variation in a class because we have too few instances of that class, relative to one or more other classes. Oversampling duplicates instances of that class.

2. Next, we will load the text message dataset. We will convert our target into an integer variable and confirm that it worked as expected. Note the significant class imbalance. Let's look at the first few rows to get a better feel for the data:

```
spamtext = pd.read_csv("data/spamtext.csv")
spamtext['spam'] = np.where(spamtext.
category=='spam',1,0)
```

```
spamtext.groupby(['spam','category']).size()
spam   category
0      ham            4824
1      spam            747
dtype: int64
```

```
spamtext.head()
   category   message                                    spam
0   ham        Go until jurong point, crazy..             0
1   ham        Ok lar... Joking wif u oni...              0
2   spam       Free entry in 2 a wkly comp to win...      1
3   ham        U dun say so early hor... U c already..0
4   ham        Nah I don't think he goes to usf, ..       0
```

3. Now, we create training and testing DataFrames. We will use the `stratify` parameter to ensure equal distributions of target values in the training and testing data.

   We will also instantiate a `CountVectorizer` object to create our bag-of-words later. We indicate that we want some words to be ignored because they do not provide useful information. We could have created a stop word list, but here, we will take advantage of scikit-learn's list of stop words in English:

```
X_train, X_test, y_train, y_test =  \
  train_test_split(spamtext[['message']],\
```

```
      spamtext[['spam']], test_size=0.2,\
      stratify=spamtext[['spam']], random_state=0)
   countvectorizer = CountVectorizer(analyzer='word', \
      stop_words='english')
```

4. Let's look at how the vectorizer works with a couple of observations from our data. To make it easier to view, we will only pull from messages that contain fewer than 50 characters.

   Using the vectorizer, we get counts for all non stop words used for each observation. For example, like is used once in the first message and not at all in the second. This gives like a value of 1 for the first observation in the transformed data and a value of 0 for the second observation.

   We won't use anything from this step in our model. We are only doing this for illustrative purposes:

```
smallsample = \
   X_train.loc[X_train.message.str.len()<50].\
      sample(2, random_state=35)


smallsample
```

|      | message |
|------|---------|
| 2079 | I can take you at like noon |
| 5393 | I dont know exactly could you ask chechi. |

```
ourvec = \
   pd.DataFrame(countvectorizer.\
   fit_transform(smallsample.values.ravel()).\
   toarray(),\
   columns=countvectorizer.get_feature_names())


ourvec
```

|   | ask | chechi | dont | exactly | know | like | noon |
|---|-----|--------|------|---------|------|------|------|
| 0 | 0   | 0      | 0    | 0       | 0    | 1    | 1    |
| 1 | 1   | 1      | 1    | 1       | 1    | 0    | 0    |

5.  Now, let's instantiate a `MultinomialNB` object and add it to a pipeline. We will add oversampling using `SMOTE` to handle the class imbalance:

```
nb = MultinomialNB()

smote = SMOTE(random_state=0)

pipe1 = make_pipeline(countvectorizer, smote, nb)

pipe1.fit(X_train.values.ravel(),
   y_train.values.ravel())
```

6.  Now, let's look at some predictions. We get an impressive **0.97** accuracy rate and equally good specificity. This excellent specificity suggests that we do not have many false positives. The somewhat lower sensitivity indicates that we are not catching some of the positives (the spam messages), though we are still doing quite well:

```
pred = pipe1.predict(X_test.values.ravel())

print("accuracy: %.2f, sensitivity: %.2f, specificity:
%.2f, precision: %.2f"  %
   (skmet.accuracy_score(y_test.values.ravel(), pred),
   skmet.recall_score(y_test.values.ravel(), pred),
   skmet.recall_score(y_test.values.ravel(), pred, pos_
label=0),
   skmet.precision_score(y_test.values.ravel(), pred)))
accuracy: 0.97, sensitivity: 0.87, specificity: 0.98,
precision: 0.87
```

7.  It is helpful to visualize our model's performance with a confusion matrix:

```
cm = skmet.confusion_matrix(y_test, pred)
cmplot = skmet.ConfusionMatrixDisplay(
   confusion_matrix=cm, \
   display_labels=['Not Spam', 'Spam'])
cmplot.plot()
cmplot.ax_.set(
   title='Spam Prediction Confusion Matrix',
   xlabel='Predicted Value', ylabel='Actual Value')
```

This produces the following plot:

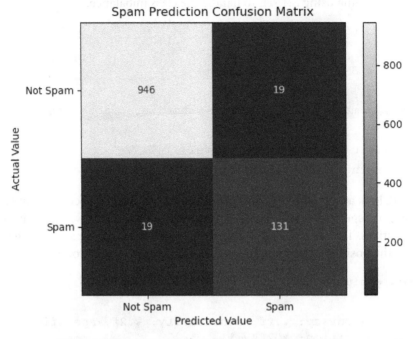

Figure 14.2 – Spam prediction using multinomial naïve Bayes

Naïve Bayes can yield excellent results when constructing a text classification model. The metrics are often quite good and quite efficient. This was a very straightforward binary classification problem. However, naïve Bayes can also be effective with multiclass text classification problems. The algorithm can be applied in pretty much the same way as we did here with multiclass targets.

## Summary

Naïve Bayes is a great algorithm to add to our regular toolkit for solving classification problems. It is not often the approach that will produce predictions with the least bias. However, the flip side is also true. There is less risk of overfitting, particularly when working with continuous features. It is also quite efficient, scaling well to a large number of observations and a large feature space.

The next two chapters of this book will explore unsupervised learning algorithms – those where we do not have a target to predict. In the next chapter, we will examine principal component analysis, and then K-means clustering in the chapter after that.

# Section 5 – Clustering and Dimensionality Reduction with Unsupervised Learning

The last two chapters of this book examines unsupervised learning models. These are models where there is no target to predict. Even without a target there are many insights that can be gleaned from our data. Dimension reduction with principal component analysis (PCA) allows us to capture the variance of our features with fewer components than the original number of features.

The components created with PCA can be used for visualizations, or to identify processes that are important but cannot really be captured well by each feature. PCA can also be used when we need to reduce the feature space in a supervised learning model. We will demonstrate how to create and evaluate a PCA in the next chapter.

Clustering helps us group instances by those which have more in common with each other than with those in any other group. This often reveals relationships that are not otherwise obvious. We look at two popular clustering algorithms, K-means and DBSCAN, in this chapter. Clustering works well when we are able to find the right hyperparameter values for our model -- the number of clusters (k) for k-means, and the value of epsilon for DBSCAN, which determines the size of the radius around core instances in a cluster. We will go over choosing the best hyperparameter values for these clustering algorithms in the final chapter of this book.

This section comprises the following chapters:

- *Chapter 15, Principal Component Analysis*
- *Chapter 16, K-Means and DBSCAN Clustering*

# 15
# Principal Component Analysis

Dimension reduction is one of the more important concepts/strategies in machine learning. It is sometimes equated with feature selection, but that is too narrow a view of dimension reduction. Our models often have to deal with an excess of features, some of which are capturing the same information. Not addressing the issue substantially increases the risk of overfitting or of unstable results. But dropping some of our features is not the only tool in our toolbox here. Feature extraction strategies, such as **principal component analysis (PCA)**, can often yield good results.

We can use PCA to reduce the dimensions (the number of features) of our dataset without losing significant predictive power. The number of principal components necessary to capture most of the variance in the data is typically less than the number of features, often much less.

These components can be used in our regression or classification models rather than the initial features. Not only can this speed up how quickly our model learns, but it may decrease the variance of our estimates. The key disadvantage of this feature extraction strategy is that the new features will usually be more difficult to interpret. PCA is also not a good choice when we have categorical features.

We will develop our understanding of how PCA works by first examining how each component is constructed. We will construct a PCA, interpret the results, and then use those results in a classification model. Finally, we will use kernels to improve PCA when our components might not be linearly separable.

Specifically, we will explore the following topics in this chapter:

- Key concepts of PCA
- Feature extraction with PCA
- Using kernels with PCA

# Technical requirements

We will mainly stick to the pandas, NumPy, and scikit-learn libraries in this chapter. All code was tested with scikit-learn versions 0.24.2 and 1.0.2.

The code for this chapter can be downloaded from the GitHub repository: `https://github.com/PacktPublishing/Data-Cleaning-and-Exploration-with-Machine-Learning`.

# Key concepts of PCA

PCA produces multiple linear combinations of features and each linear combination is a component. It identifies a component that captures the largest amount of variance, and a second component that captures the largest amount of remaining variance, and then a third component, and so on until a stopping point we specify is reached. The stopping point can be based on the number of components, the percent of the variation explained, or domain knowledge.

One very useful characteristic of principal components is that they are mutually orthogonal. This means that they are uncorrelated, which is really good news for modeling. *Figure 15.1* shows two components constructed from the features $x_1$ and $x_2$. The maximum variance is captured with *PC1*, the maximum remaining variance with *PC2*. (The data points in the figure are made up.)  Notice that the two vectors are orthogonal (perpendicular).

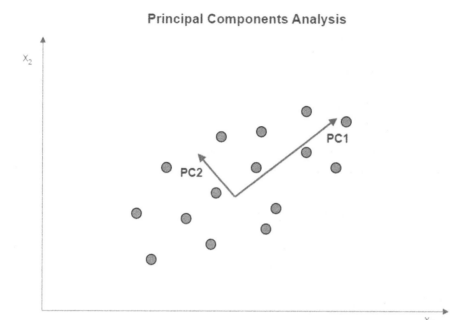

Figure 15.1 – An illustration of PCA with two features

Those of you who have done a lot of factor analysis probably get the general idea, even if this is your first time exploring PCA. Principal components are not very different from factors, though there are some conceptual and mathematical differences. With PCA, all of the variance is analyzed. Only the shared variance between variables is analyzed with factor analysis. In factor analysis, the unobserved factors are understood as having *caused* the observed variables. No assumptions about underlying, unobserved forces need to be made with PCA.

So, how is this bit of computational magic done? Principal components can be calculated by following these steps:

1. Standardize your data.

2. Calculate the covariance matrix of your variables.

3. Calculate the eigenvectors and eigenvalues of the covariance matrix.

4. Sort the eigenvectors by eigenvalues in descending order. The first eigenvector is principal component 1, the second is principal component 2, and so on.

It is not necessary to completely understand these steps to understand the discussion in the rest of this chapter. We will get scikit-learn to do this work for us. Still, it might improve your intuition if you computed the covariance matrix of a very small subset of data (two or three columns and just a few rows) and then did the eigendecomposition of that matrix. A somewhat easier way to experiment with constructing components, while still being illustrative, is to use the NumPy linear algebra functions (`numpy.linalg`). The key point here is how computationally straightforward it is to derive the principal components.

PCA is used for many machine learning tasks. It can be used to resize images, to analyze financial data, or for recommender systems. Essentially, it may be a good choice for any application where there are a large number of features and many of them are correlated.

Some of you no doubt noticed that I slipped in, without remarking on it, that PCA constructs *linear combinations* of features. What do we do when linear separability is not feasible, as we encountered with support vector machines? Well, it turns out that the kernel trick that we relied on with support vector machines also works with PCA. We will explore how to implement kernel PCA in this chapter. However, we will start with a relatively straightforward PCA example.

# Feature extraction with PCA

PCA can be used for dimension reduction in preparation for a model we will run subsequently. Although PCA is not, strictly speaking, a feature selection tool, we can run it in pretty much the same way we ran the wrapper feature selection methods in *Chapter 5, Feature Selection*. After some preprocessing (such as handling outliers), we generate the components, which we can then use as our new features. Sometimes we do not actually use these components in a model. Rather, we generate them mainly to help us visualize our data better.

To illustrate the use of PCA, we will work with data on **National Basketball Association (NBA)** games. The dataset has statistics from each NBA game from the 2017/2018 season through the 2020/2021 season. This includes the home team; whether the home team won; the visiting team; shooting percentages for visiting and home teams; turnovers, rebounds, and assists by both teams; and a number of other measures.

> **Note**
> NBA game data is available for download for the public at `https://www.kaggle.com/datasets/wyattowalsh/basketball`. This dataset has game data starting with the 1946/1947 NBA season. It uses the `nba_api` to pull stats from nba.com. That API is available at `https://github.com/swar/nba_api`.

Let's use PCA in a model:

1. We start by loading the required libraries. You have seen all of these in the previous chapters except for scikit-learn's PCA module:

```
import pandas as pd
import numpy as np
from sklearn.model_selection import train_test_split
from sklearn.preprocessing import StandardScaler
from sklearn.pipeline import make_pipeline
from sklearn.impute import SimpleImputer
from sklearn.linear_model import LogisticRegression
from sklearn.decomposition import PCA
from sklearn.model_selection import RandomizedSearchCV
from scipy.stats import uniform
from scipy.stats import randint
import os
import sys
sys.path.append(os.getcwd() + "/helperfunctions")
from preprocfunc import OutlierTrans
```

2. Next, we load the NBA data and do a little cleaning. A few instances do not have values for whether the home team won or lost, WL_HOME, so we remove them. WL_HOME will be our target. We will try to model it later, after we have constructed our components. Notice that the home team wins a majority of the time, but the class imbalance is not bad:

```
nbagames = pd.read_csv("data/nbagames2017plus.csv",
parse_dates=['GAME_DATE'])

nbagames = nbagames.loc[nbagames.WL_HOME.isin(['W','L'])]
nbagames.shape
(4568, 149)

nbagames['WL_HOME'] = \
   np.where(nbagames.WL_HOME=='L',0,1).astype('int')

nbagames.WL_HOME.value_counts(dropna=False)
1    2586
```

```
0    1982
Name: WL_HOME, dtype: int64
```

3.  We should look at some descriptive statistics:

```
num_cols = ['FG_PCT_HOME','FTA_HOME','FG3_PCT_HOME',
    'FTM_HOME','FT_PCT_HOME','OREB_HOME','DREB_HOME',
    'REB_HOME','AST_HOME','STL_HOME','BLK_HOME',
    'TOV_HOME', 'FG_PCT_AWAY','FTA_AWAY','FG3_PCT_AWAY',
    'FT_PCT_AWAY','OREB_AWAY','DREB_AWAY','REB_AWAY',
    'AST_AWAY','STL_AWAY','BLK_AWAY','TOV_AWAY']


nbagames[['WL_HOME'] + num_cols].
agg(['count','min','median','max']).T
```

This produces the following output. There are no missing values, but our features have very different ranges. We will need to do some scaling:

|  | count | min | median | max |
|---|---|---|---|---|
| WL_HOME | 4,568.00 | 0.00 | 1.00 | 1.00 |
| FG_PCT_HOME | 4,568.00 | 0.27 | 0.47 | 0.65 |
| FTA_HOME | 4,568.00 | 1.00 | 22.00 | 64.00 |
| FG3_PCT_HOME | 4,568.00 | 0.06 | 0.36 | 0.84 |
| FTM_HOME | 4,568.00 | 1.00 | 17.00 | 44.00 |
| FT_PCT_HOME | 4,568.00 | 0.14 | 0.78 | 1.00 |
| OREB_HOME | 4,568.00 | 1.00 | 10.00 | 25.00 |
| DREB_HOME | 4,568.00 | 18.00 | 35.00 | 55.00 |
| REB_HOME | 4,568.00 | 22.00 | 45.00 | 70.00 |
| AST_HOME | 4,568.00 | 10.00 | 24.00 | 50.00 |
| STL_HOME | 4,568.00 | 0.00 | 7.00 | 22.00 |
| BLK_HOME | 4,568.00 | 0.00 | 5.00 | 20.00 |
| TOV_HOME | 4,568.00 | 1.00 | 14.00 | 29.00 |
| FG_PCT_AWAY | 4,568.00 | 0.28 | 0.46 | 0.67 |
| FTA_AWAY | 4,568.00 | 3.00 | 22.00 | 54.00 |
| FG3_PCT_AWAY | 4,568.00 | 0.08 | 0.36 | 0.78 |
| FT_PCT_AWAY | 4,568.00 | 0.26 | 0.78 | 1.00 |
| OREB_AWAY | 4,568.00 | 0.00 | 10.00 | 26.00 |
| DREB_AWAY | 4,568.00 | 18.00 | 34.00 | 56.00 |
| REB_AWAY | 4,568.00 | 22.00 | 44.00 | 71.00 |

| AST_AWAY | 4,568.00 | 9.00 | 24.00 | 46.00 |
| STL_AWAY | 4,568.00 | 0.00 | 8.00 | 19.00 |
| BLK_AWAY | 4,568.00 | 0.00 | 5.00 | 15.00 |
| TOV_AWAY | 4,568.00 | 3.00 | 14.00 | 30.00 |

4.  Let's also examine how our features are correlated:

```
corrmatrix = nbagames[['WL_HOME'] + num_cols].\
    corr(method="pearson")
```

```
sns.heatmap(corrmatrix, xticklabels=corrmatrix.columns,
    yticklabels=corrmatrix.columns, cmap="coolwarm")
plt.title('Heat Map of Correlation Matrix')
plt.tight_layout()
plt.show()
```

This produces the following plot:

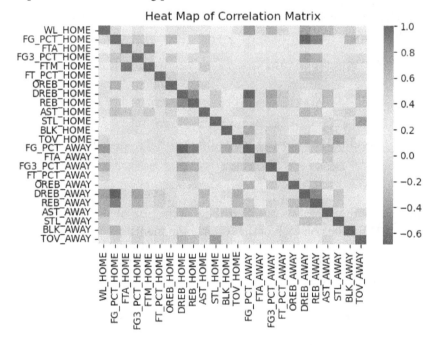

Figure 15.2 – Heat map of NBA features

A number of features are significantly positively or negatively correlated. For example, the field goal (shooting) percentage of the home team (FG_PCT_HOME) and three-point field goal percentage of the home team (FG3_PCT_HOME) are positively correlated, not surprisingly. Also, rebounds of the home team (REB_HOME) and defensive rebounds of the home team (DREB_HOME) are likely too closely correlated for any model to disentangle their impact.

This dataset might be a good candidate for PCA. Although some features are highly correlated, we will still lose information by dropping some. PCA at least offers the possibility of dealing with the correlation without losing that information.

5. Now we create training and testing DataFrames:

```
X_train, X_test, y_train, y_test = \
    train_test_split(nbagames[num_cols],\
    nbagames[['WL_HOME']],test_size=0.2, random_state=0)
```

6. We are ready now to create the components. We, somewhat arbitrarily, indicate that we want seven components. (Later, we will use hyperparameter tuning to choose the number of components.) We set up our pipeline to do some preprocessing before running the PCA:

```
pca = PCA(n_components=7)

pipe1 = make_pipeline(OutlierTrans(2),
        SimpleImputer(strategy="median"),
        StandardScaler(), pca)


pipe1.fit(X_train)
```

7. We can now use the components_ attribute of the pca object. This returns the scores of all 23 features on each of the seven components:

```
components = pd.DataFrame(pipe1['pca'].components_,
    columns=num_cols)

components.T.to_excel('views/components.xlsx')
```

This produces the following spreadsheet:

| | pc1 | pc2 | pc3 | pc4 | pc5 | pc6 | pc7 |
|---|---|---|---|---|---|---|---|
| FG_PCT_HOME | -0.38 | 0.29 | -0.01 | 0.05 | 0.21 | 0.03 | 0.12 |
| FTA_HOME | 0.09 | -0.04 | 0.53 | -0.25 | 0.21 | 0.06 | 0.09 |
| FG3_PCT_HOME | -0.27 | 0.23 | -0.07 | 0.04 | 0.17 | 0.06 | 0.23 |
| FTM_HOME | 0.06 | -0.02 | 0.55 | -0.29 | 0.25 | 0.04 | 0.01 |
| FT_PCT_HOME | -0.06 | 0.03 | 0.15 | -0.15 | 0.14 | -0.03 | -0.23 |
| OREB_HOME | 0.18 | 0.01 | 0.07 | -0.15 | -0.35 | 0.29 | 0.26 |
| DREB_HOME | 0.28 | 0.38 | -0.06 | -0.12 | -0.04 | 0.06 | 0.07 |
| REB_HOME | 0.33 | 0.32 | -0.01 | -0.18 | -0.23 | 0.20 | 0.20 |
| AST_HOME | -0.22 | 0.22 | -0.12 | 0.11 | 0.12 | 0.19 | 0.44 |
| STL_HOME | -0.02 | -0.08 | 0.22 | 0.48 | -0.02 | 0.42 | 0.01 |
| BLK_HOME | 0.13 | 0.15 | 0.05 | 0.15 | 0.14 | -0.10 | 0.18 |
| TOV_HOME | -0.05 | 0.20 | 0.30 | 0.22 | -0.38 | -0.35 | -0.02 |
| FG_PCT_AWAY | -0.30 | -0.37 | 0.05 | -0.13 | -0.13 | -0.03 | 0.16 |
| FTA_AWAY | 0.06 | 0.05 | 0.22 | 0.02 | 0.20 | -0.06 | 0.20 |
| FG3_PCT_AWAY | -0.21 | -0.27 | 0.00 | -0.13 | -0.11 | -0.02 | 0.20 |
| FT_PCT_AWAY | -0.08 | -0.05 | 0.00 | -0.09 | 0.01 | -0.03 | -0.20 |
| OREB_AWAY | 0.15 | 0.03 | 0.01 | 0.25 | 0.34 | -0.38 | 0.22 |
| DREB_AWAY | 0.34 | -0.32 | -0.13 | 0.07 | 0.05 | -0.01 | 0.09 |
| REB_AWAY | 0.37 | -0.25 | -0.10 | 0.19 | 0.23 | -0.21 | 0.19 |
| AST_AWAY | -0.19 | -0.23 | 0.03 | -0.13 | -0.12 | -0.14 | 0.45 |
| STL_AWAY | -0.05 | 0.14 | 0.27 | 0.21 | -0.39 | -0.42 | 0.02 |
| BLK_AWAY | 0.14 | -0.12 | 0.01 | -0.05 | -0.21 | 0.04 | 0.29 |
| TOV_AWAY | -0.01 | -0.15 | 0.26 | 0.49 | -0.04 | 0.35 | -0.04 |

Figure 15.3 – Principal components of NBA features

Each feature accounts for some portion of the variance with each component. (If for each component, you square each of the 23 scores and then sum the squares, you get a total of 1.) If you want to understand which features really drive a component, look for the ones with the largest absolute value. For component 1, the field goal percentage of the home team (FG_PCT_HOME) is most important, followed by the number of rebounds of the away team (REB_AWAY).

Recall from our discussion at the beginning of this chapter that each component attempts to capture the variance that remains after the previous component or components.

8. Let's show the five most important features for the first three components. The first component seems to be largely about the field goal percentage of the home team and the rebounding of each team. The second component does not seem very different, but the third one is driven by free throws made and attempted (FTM_HOME and FTA_HOME) and turnovers (TOV_HOME and TOV_AWAY):

```
components.pc1.abs().nlargest(5)
FG_PCT_HOME      0.38
REB_AWAY         0.37
DREB_AWAY        0.34
```

```
REB_HOME          0.33
FG_PCT_AWAY       0.30
Name: pc1, dtype: float64


components.pc2.abs().nlargest(5)
DREB_HOME         0.38
FG_PCT_AWAY       0.37
DREB_AWAY         0.32
REB_HOME          0.32
FG_PCT_HOME       0.29
Name: pc2, dtype: float64


components.pc3.abs().nlargest(5)
FTM_HOME          0.55
FTA_HOME          0.53
TOV_HOME          0.30
STL_AWAY          0.27
TOV_AWAY          0.26
Name: pc3, dtype: float64
```

9.  We can use the explained_variance_ratio_ attribute of the pca object
    to examine how much of the variance is captured by each component. The first
    component explains 14.5% of the variance of the features. The second component
    explains another 13.4%. If we use NumPy's cumsum method, we can see that the
    seven components explain about 65% of the variance altogether.

    So, there is still a fair bit of variance out there. We might want to use more
    components for any model we build:

```
np.set_printoptions(precision=3)


pipe1['pca'].explained_variance_ratio_
array([0.145, 0.134, 0.095, 0.086, 0.079, 0.059, 0.054])


np.cumsum(pipe1['pca'].explained_variance_ratio_)
array([0.145, 0.279, 0.374, 0.46 , 0.539, 0.598, 0.652])
```

10. We can plot the first two principal components to see how well they can separate home team wins and losses. We can use the `transform` method of our pipeline to create a DataFrame with the principal components and join that with the DataFrame for the target.

We use the handy `hue` attribute of Seaborn's `scatterplot` to display wins and losses. The first two principal components do an okay job of separating wins and losses, despite together only accounting for about 28% of the variance in our features:

```
X_train_pca = pd.DataFrame(pipe1.transform(X_train),
    columns=components.columns, index=X_train.index).
join(y_train)


sns.scatterplot(x=X_train_pca.pc1, y=X_train_pca.pc2,
hue=X_train_pca.WL_HOME)
plt.title("Scatterplot of First and Second Components")
plt.xlabel("Principal Component 1")
plt.ylabel("Principal Component 2")
plt.show()
```

This produces the following plot:

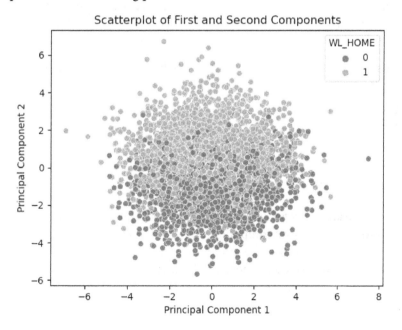

Figure 15.4 – Scatterplot of wins and losses by first and second principal components

11. Let's use principal components to predict whether the home team wins. We just add a logistic regression to our pipeline to do that. We also do a grid search to find the best hyperparameter values:

```
lr = LogisticRegression()

pipe2 = make_pipeline(OutlierTrans(2),
  SimpleImputer(strategy="median"), StandardScaler(),
  pca, lr)

lr_params = {
  "pca__n_components": randint(3, 20),
  "logisticregression__C": uniform(loc=0, scale=10)
}

rs = RandomizedSearchCV(pipe2, lr_params, cv=4,
  n_iter=40, scoring='accuracy', random_state=1)

rs.fit(X_train, y_train.values.ravel())
```

12. We can now look at the best parameters and score. As we suspected from an earlier step, the grid search suggests that our logistic regression model does better with more components. We get a very high score.

We discuss the hyperparameter *C* in detail in *Chapter 10, Logistic Regression*:

```
rs.best_params_
{'logisticregression__C': 6.865009276815837, 'pca__n_
components': 19}

rs.best_score_
0.9258345296842831
```

This section demonstrated how we can generate principal components from our dataset and how to interpret those components. We also looked at how to use principal components in a model, rather than the initial features. But we assumed that the principal components can be well described as linear combinations of features. This is often not the case. In the next section, we will use kernel PCA to handle nonlinear relationships.

# Using kernels with PCA

With some data, it is not possible to construct principal components that are linearly separable. This may not actually be easy to visualize in advance of our modeling. Fortunately, there are tools we can use to determine the kernel that will yield the best results, including a linear kernel. Kernel PCA with a linear kernel should perform similarly to standard PCA.

In this section, we will use kernel PCA for feature extraction with data on labor force participation rates, educational attainment, teenage birth frequency, and participation in politics by gender at the country level.

> **Note**
>
> This dataset on gender-based differences in educational and labor force outcomes is made available for public use by the United Nations Development Program at `https://www.kaggle.com/datasets/undp/human-development`. There is one record per country with aggregate employment, income, and education data by gender for 2015.

Let's start building the model:

1. We will import the same libraries we have been using plus scikit-learn's `KernelPCA` module. We will also import the `RandomForestRegressor` module:

```
import pandas as pd
import numpy as np
from sklearn.model_selection import train_test_split
from sklearn.preprocessing import MinMaxScaler
from sklearn.pipeline import make_pipeline
from sklearn.impute import SimpleImputer
from sklearn.decomposition import KernelPCA
from sklearn.ensemble import RandomForestRegressor
from sklearn.model_selection import RandomizedSearchCV
import seaborn as sns
import matplotlib.pyplot as plt
import os
import sys
sys.path.append(os.getcwd() + "/helperfunctions")
from preprocfunc import OutlierTrans
```

2. We load the data on educational and labor force outcomes by gender. We construct series for the ratio of female to male incomes, the years of education ratio, the labor force participation ratio, and the human development index ratio:

```
un_income_gap = pd.read_csv("data/un_income_gap.csv")
un_income_gap.set_index('country', inplace=True)
un_income_gap['incomeratio'] = \
  un_income_gap.femaleincomepercapita / \
    un_income_gap.maleincomepercapita
un_income_gap['educratio'] = \
  un_income_gap.femaleyearseducation / \
      un_income_gap.maleyearseducation
un_income_gap['laborforcepartratio'] = \
  un_income_gap.femalelaborforceparticipation / \
      un_income_gap.malelaborforceparticipation
un_income_gap['humandevratio'] = \
  un_income_gap.femalehumandevelopment / \
      un_income_gap.malehumandevelopment
un_income_gap.dropna(subset=['incomeratio'],
inplace=True)
```

3. Let's look at some descriptive statistics. There are a few missing values, particularly for genderinequality and humandevratio. Some features have much larger ranges than others:

```
num_cols = ['educratio','laborforcepartratio',
  'humandevratio','genderinequality',
  'maternalmortality','adolescentbirthrate',
  'femaleperparliament','incomepercapita']

gap_sub = un_income_gap[['incomeratio'] + num_cols]

gap_sub.\
  agg(['count','min','median','max']).T
```

|                     | count  | min  | median | max  |
|---------------------|--------|------|--------|------|
| incomeratio         | 177.00 | 0.16 | 0.60   | 0.93 |
| educratio           | 169.00 | 0.24 | 0.93   | 1.35 |
| laborforcepartratio | 177.00 | 0.19 | 0.75   | 1.04 |

| humandevratio | 161.00 | 0.60 | 0.95 | 1.03 |
| genderinequality | 155.00 | 0.02 | 0.39 | 0.74 |
| maternalmortality | 174.00 | 1.00 | 60.00 | 1,100.00 |
| adolescentbirthrate | 177.00 | 0.60 | 40.90 | 204.80 |
| femaleperparliament | 174.00 | 0.00 | 19.35 | 57.50 |
| incomepercapita | 177.00 | 581.00 | 10,512.00 | 123,124.00 |

4.  We should also look at some correlations:

```
corrmatrix = gap_sub.corr(method="pearson")

sns.heatmap(corrmatrix,
    xticklabels=corrmatrix.columns,
    yticklabels=corrmatrix.columns, cmap="coolwarm")
plt.title('Heat Map of Correlation Matrix')
plt.tight_layout()
plt.show()
```

This produces the following plot:

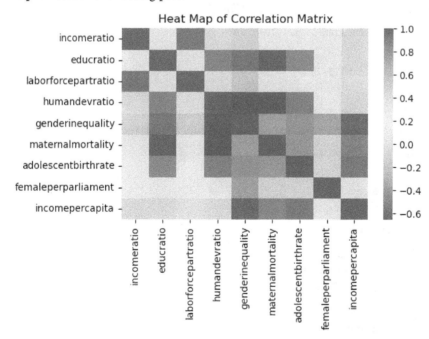

Figure 15.5 – Correlation matrix of NBA games data

`humandevratio` and `educratio` are highly correlated, as are `genderinequality` and `adolescentbirthrate`. We can see that `educratio` and `maternalmortality` are highly negatively correlated. It would be hard to build a well-performing model with all of these features given their high correlation. However, we might be able to reduce dimensions with kernel PCA.

5.  We create training and testing DataFrames:

```
X_train, X_test, y_train, y_test =  \
  train_test_split(gap_sub[num_cols],\
  gap_sub[['incomeratio']], test_size=0.2,
  random_state=0)
```

6.  We are now ready to instantiate the `KernelPCA` and `RandomForestRegressor` objects. We add both to a pipeline. We also create a dictionary with our hyperparameters for the kernel PCA and the random forest regressor.

    The dictionary has a range of hyperparameter values for the number of components, gamma, and the kernel to use with the kernel PCA. For those kernels that do not use gamma, those values are ignored. Notice that one option for kernel is the linear kernel.

    We discuss gamma in more detail in *Chapter 8, Support Vector Regression*, and *Chapter 13, Support Vector Machine Classification*:

```
rfreg = RandomForestRegressor()


kpca = KernelPCA()


pipe1 = make_pipeline(OutlierTrans(2),
  SimpleImputer(strategy="median"), MinMaxScaler(),
  kpca, rfreg)


rfreg_params = {
 'kernelpca__n_components':
    randint(2, 9),
 'kernelpca__gamma':
     np.linspace(0.03, 0.3, 10),
 'kernelpca__kernel':
     ['linear', 'poly', 'rbf',
      'sigmoid', 'cosine'],
 'randomforestregressor__max_depth':
```

```
        randint(2, 20),
    'randomforestregressor__min_samples_leaf':
        randint(5, 11)

}
```

7. Now let's do a randomized grid search with these hyperparameter values. The kernel for the PCA that gives us the best performance with the random forest regressor is polynomial. We get a good square for mean squared error, about 10% of the size of the mean:

```
rs = RandomizedSearchCV(pipe1, rfreg_params,
    cv=4, n_iter=40,
    scoring='neg_mean_absolute_error',
    random_state=1)


rs.fit(X_train, y_train.values.ravel())


rs.best_params_
{'kernelpca__gamma': 0.12000000000000001,
 'kernelpca__kernel': 'poly',
 'kernelpca__n_components': 4,
 'randomforestregressor__max_depth': 18,
 'randomforestregressor__min_samples_leaf': 5}


rs.best_score_
-0.06630618838886537
```

8. Let's look at other top-performing models. A model with an rbf kernel and one with a sigmoid kernel do nearly as well. The second and third best-performing models have more principal components than the best-performing model:

```
results = \
    pd.DataFrame(rs.cv_results_['mean_test_score'], \
        columns=['meanscore']).\
    join(pd.DataFrame(rs.cv_results_['params'])).\
    sort_values(['meanscore'], ascending=False)


results.iloc[1:3].T
```

| | | |
|---|---|---|
| meanscore | -0.067 | -0.070 |
| kernelpca__gamma | 0.240 | 0.180 |
| kernelpca__kernel | rbf | sigmoid |
| kernelpca__n_components | 6 | 6 |
| randomforestregressor__max_depth | 12 | 10 |
| randomforestregressor__min_samples_leaf 5 | | 6 |

Kernel PCA is a relatively easy-to-implement dimension reduction option. It is most useful when we have a number of highly correlated features that might not be linearly separable, and interpretation of predictions is not important.

## Summary

This chapter explored principal component analysis, including how it works and when we might want to use it. We learned how to examine the components created from PCA, including how each feature contributes to each component, and how much of the variance is explained. We went over how to visualize components and how to use components in subsequent analysis. We also examined how to use kernels for PCA and when that might give us better results.

We explore another unsupervised learning technique in the next chapter, k-means clustering.

# 16
# K-Means and DBSCAN Clustering

Data clustering allows us to organize unlabeled data into groups of observations with more in common with other members of the group than with observations outside of the group. There are a surprisingly large number of applications for clustering, either as the final model of a machine learning pipeline or as input for another model. This includes market research, image processing, and document classification. We sometimes also use clustering to improve exploratory data analysis or to create more meaningful visualizations.

K-means and **density-based spatial clustering of applications with noise (DBSCAN)** clustering, like **principal component analysis (PCA)**, are unsupervised learning algorithms. There are no labels to use as the basis for predictions. The purpose of the algorithm is to identify instances that hang together based on their features. Instances that are in close proximity to each other, and further away from other instances, can be considered to be in a cluster. There are a number of ways to gauge proximity. **Partition-based clustering**, such as k-means, and **density-based clustering**, such as DBSCAN, are two of the more popular approaches. We will explore those approaches in this chapter.

Specifically, we will go over the following topics:

- The key concepts of k-means and DBSCAN clustering
- Implementing k-means clustering
- Implementing DBSCAN clustering

# Technical requirements

We will mainly stick to the pandas, NumPy, and scikit-learn libraries in this chapter.

# The key concepts of k-means and DBSCAN clustering

With k-means clustering, we identify $k$ clusters, each with a center, or **centroid**. The centroid is the point that minimizes the total squared distance between it and the other data points in the cluster.

An example with made-up data should help here. The data points in *Figure 16.1* seem to be in three clusters. (It is not usually that easy to visualize the number of clusters, $k$.)

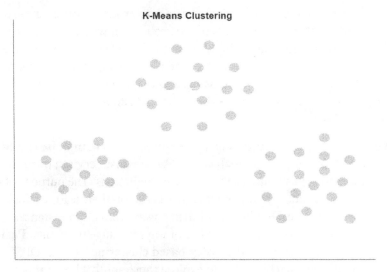

Figure 16.1 – Data points with three discernible clusters

We perform the following steps to construct the clusters:

1. Assign a random point as the center of each cluster.
2. Calculate the distance of each point from the center of each cluster.

3.  Assign data points to a cluster based on their proximity to the center point. These first three steps are illustrated in *Figure 16.2*. The points with an **X** are the randomly chosen cluster centers (with *k* set at 3). Data points that are closer to the cluster center point than to other cluster center points get assigned to that cluster.

Figure 16.2 – Random points assigned as the center of the cluster

4.  Calculate a new center point for the new cluster. This is illustrated in *Figure 16.3*.

Figure 16.3 – New cluster centers calculated

5.  Repeat steps 2 through 4 until there is not much change in the centers.

K-means clustering is a very popular algorithm for clustering for several reasons. It is quite intuitive and typically quite fast. It does have some disadvantages, however. It processes every data point as part of a cluster, so the clusters can be yanked around by extreme values. It also assumes clusters will have spherical shapes.

The evaluation of unsupervised models is less clear than with supervised models, as we do not have a target with which to compare our predictions. A fairly common metric for clustering models is the **silhouette score**. The silhouette score is the mean silhouette coefficient for all instances. The **silhouette coefficient** is as follows:

$$s_i = \frac{b_i - a_i}{max(a_i, b_i)}$$

Here, $b_i$ is the mean distance to all instances of the next closest cluster for the ith instance, and $a_i$ is the mean distance to the instances of the assigned cluster. This coefficient ranges from -1 to 1, with scores near 1 meaning that the instance is well within the assigned cluster.

Another metric to evaluate our clusters is the **inertia score**. This is the sum of squared distances between each instance and its centroid. This distance will decrease as we increase the number of clusters but there are eventually diminishing marginal returns from increasing the number of clusters. The change in inertia score with $k$ is often visualized using an **elbow plot**. This plot is called an elbow plot because the slope gets much closer to 0 as we increase $k$, so close that it resembles an elbow. This is shown in *Figure 16.4*. In this case, we would choose a value of $k$ near the elbow.

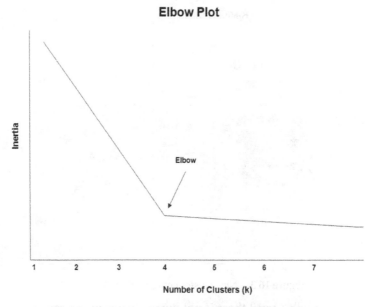

Figure 16.4 – An elbow plot with inertia and k

Another metric often used when evaluating a clustering model is the **Rand index**. The Rand index tells us how frequently two clusterings have assigned the same cluster to instances. Values for the Rand index will range between 0 and 1. We typically use an adjusted Rand index, which corrects for chance in the similarity calculation. Values for the adjusted Rand index can sometimes be negative.

**DBSCAN** takes a different approach to clustering. For each instance, it counts the number of instances within a specified distance of that instance. All instances within ε of an instance are said to be in that instance's ε-neighborhood. When the number of instances in an ε-neighborhood equals or exceeds the minimum samples value that we specify, that instance is considered a core instance and the ε-neighborhood is considered a cluster. Any instance that is more than ε from another instance is considered noise. This is illustrated in *Figure 16.5*.

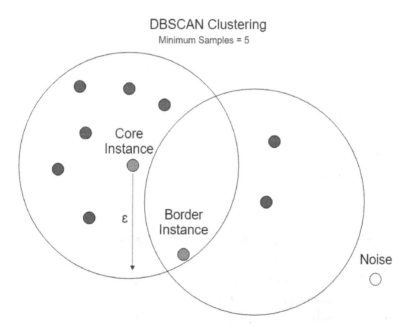

Figure 16.5 – DBSCAN clustering with minimum samples = five

There are several advantages of this density-based approach. The clusters do not need to be spherical. They can take any shape. We do not need to guess at the number of clusters, though we do need to provide a value for ε. Outliers are just interpreted as noise and so do not impact the clusters. (This last point hints at another useful application of DBSCAN: identifying anomalies.)

We will use DBSCAN for clustering later in this chapter. First, we will examine how to do clustering with k-means, including how to choose a good value for k.

# Implementing k-means clustering

We can use k-means with some of the same data that we used with the supervised learning models that we developed in earlier chapters. The difference is that there is no longer a target for us to predict. Rather, we are interested in how certain instances hang together. Think of how people arrange themselves in groups during a stereotypical high school lunch break and you kind of get a general idea.

We also need to do much of the same preprocessing work that we did with supervised learning models. We will start with that in this section. We will work with data on income gaps between women and men, labor force participation rates, educational attainment, teenage birth frequency, and female participation in politics at the highest level.

> **Note**
>
> The income gap dataset is made available for public use by the *United Nations Development Program* at https://www.kaggle.com/datasets/undp/human-development. There is one record per country with aggregate employment, income, and education data by gender for 2015.

Let's build a k-means clustering model:

1.  We load the familiar libraries. We also load the KMeans and silhouette_score modules. Recall that the silhouette score is often used to evaluate how good a job our model has done of clustering. We also load rand_score, which will allow us to compute the Rand index of similarity between different clusterings:

    ```python
    import pandas as pd
    from sklearn.preprocessing import MinMaxScaler
    from sklearn.pipeline import make_pipeline
    from sklearn.cluster import KMeans
    from sklearn.metrics import silhouette_score
    from sklearn.metrics.cluster import rand_score
    from sklearn.impute import KNNImputer
    import seaborn as sns
    import matplotlib.pyplot as plt
    ```

2.  Next, we load the income gap data:

    ```python
    un_income_gap = pd.read_csv("data/un_income_gap.csv")
    un_income_gap.set_index('country', inplace=True)
    un_income_gap['incomeratio'] = \
    ```

```
      un_income_gap.femaleincomepercapita / \
        un_income_gap.maleincomepercapita
un_income_gap['educratio'] = \
   un_income_gap.femaleyearseducation / \
       un_income_gap.maleyearseducation
un_income_gap['laborforcepartratio'] = \
   un_income_gap.femalelaborforceparticipation / \
       un_income_gap.malelaborforceparticipation
un_income_gap['humandevratio'] = \
   un_income_gap.femalehumandevelopment / \
       un_income_gap.malehumandevelopment
```

3.  Let's look at some descriptive statistics:

```
num_cols =
['educratio','laborforcepartratio','humandevratio',
   'genderinequality','maternalmortality','incomeratio',
   'adolescentbirthrate', 'femaleperparliament',
   'incomepercapita']

gap = un_income_gap[num_cols]

gap.agg(['count','min','median','max']).T
```

|                     | count  | min    | median    | max       |
|---------------------|--------|--------|-----------|-----------|
| educratio           | 170.00 | 0.24   | 0.93      | 1.35      |
| laborforcepartratio | 177.00 | 0.19   | 0.75      | 1.04      |
| humandevratio       | 161.00 | 0.60   | 0.95      | 1.03      |
| genderinequality    | 155.00 | 0.02   | 0.39      | 0.74      |
| maternalmortality   | 178.00 | 1.00   | 64.00     | 1,100.00  |
| incomeratio         | 177.00 | 0.16   | 0.60      | 0.93      |
| adolescentbirthrate | 183.00 | 0.60   | 40.90     | 204.80    |
| femaleperparliament | 185.00 | 0.00   | 19.60     | 57.50     |
| incomepercapita     | 188.00 | 581.00 | 10,667.00 | 23,124.00 |

4.  We should also look at some correlations. The education ratio (the ratio of female educational level to male educational level) and the human development ratio are highly correlated, as are gender inequality and the adolescent birth rate, as well as the income ratio and the labor force participation ratio:

```
corrmatrix = gap.corr(method="pearson")

sns.heatmap(corrmatrix,
  xticklabels=corrmatrix.columns,
  yticklabels=corrmatrix.columns, cmap="coolwarm")
plt.title('Heat Map of Correlation Matrix')
plt.tight_layout()
plt.show()
```

This produces the following plot:

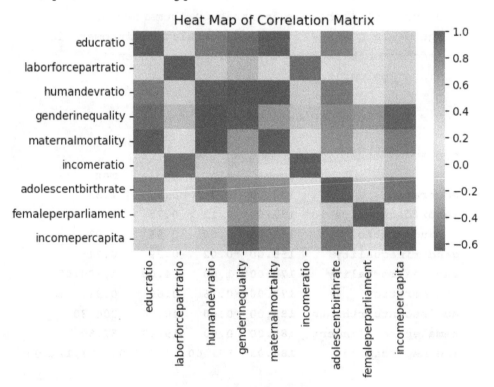

Figure 16.6 – A heat map of the correlation matrix

5.  We need to scale the data before running our model. We also use **KNN imputation** to handle the missing values:

```
pipe1 = make_pipeline(MinMaxScaler(), KNNImputer(n_
neighbors=5))


gap_enc = pd.DataFrame(pipe1.fit_transform(gap),
   columns=num_cols, index=gap.index)
```

6.  Now, we are ready to run the k-means clustering. We specify a value for the number of clusters.

    After fitting the model, we can generate a silhouette score. Our silhouette score is not great. This suggests that our clusters are not very far apart. Later, we will look to see whether we can get a better score with more or fewer clusters:

```
kmeans = KMeans(n_clusters=3, random_state=0)


kmeans.fit(gap_enc)
KMeans(n_clusters=3, random_state=0)


silhouette_score(gap_enc, kmeans.labels_)
0.3311928353317411
```

7.  Let's take a closer look at the clusters. We can use the `labels_` attribute to get the clusters:

```
gap_enc['cluster'] = kmeans.labels_


gap_enc.cluster.value_counts().sort_index()
0      40
1     100
2      48
Name: cluster, dtype: int64
```

8.  We could have used the `fit_predict` method instead to get the clusters, like so:

```
pred = pd.Series(kmeans.fit_predict(gap_enc))
pred.value_counts().sort_index()
0      40
1     100
```

```
2      48
dtype: int64
```

9. It is helpful to examine how the clusters differ in terms of the values of their features. Cluster 0 countries have much higher maternal mortality and adolescent birth rate values than countries in the other clusters. Cluster 1 countries have very low maternal mortality and high income per capita. Cluster 2 countries have very low labor force participation ratios (the ratio of female labor force participation to male labor force participation) and income ratios. Recall that we have scaled the data:

```
gap_cluster = gap_enc.join(cluster)

gap_cluster[['cluster'] + num_cols].groupby(['cluster']).
mean().T
```

| cluster | 0 | 1 | 2 |
|---|---|---|---|
| educratio | 0.36 | 0.66 | 0.54 |
| laborforcepartratio | 0.80 | 0.67 | 0.32 |
| humandevratio | 0.62 | 0.87 | 0.68 |
| genderinequality | 0.79 | 0.32 | 0.62 |
| maternalmortality | 0.44 | 0.04 | 0.11 |
| incomeratio | 0.71 | 0.60 | 0.29 |
| adolescentbirthrate | 0.51 | 0.15 | 0.20 |
| femaleperparliament | 0.33 | 0.43 | 0.24 |
| incomepercapita | 0.02 | 0.19 | 0.12 |

10. We can use the `cluster_centers_` attribute to get the center of each cluster. There are nine values representing the center for each of the three clusters, since we used nine features for the clustering:

```
centers = kmeans.cluster_centers_
centers.shape
(3, 9)
np.set_printoptions(precision=2)
centers
array([[0.36, 0.8 , 0.62, 0.79, 0.44, 0.71, 0.51, 0.33,
0.02],
       [0.66, 0.67, 0.87, 0.32, 0.04, 0.6 , 0.15, 0.43,
0.19],
       [0.54, 0.32, 0.68, 0.62, 0.11, 0.29, 0.2 , 0.24,
0.12]])
```

11. We plot the clusters by some of their features, as well as the center. We place the number for the cluster at the centroid for that cluster:

```
fig = plt.figure()
plt.suptitle("Cluster for each Country")
ax = plt.axes(projection='3d')
ax.set_xlabel("Maternal Mortality")
ax.set_ylabel("Adolescent Birth Rate")
ax.set_zlabel("Income Ratio")
ax.scatter3D(gap_cluster.maternalmortality,
   gap_cluster.adolescentbirthrate,
   gap_cluster.incomeratio, c=gap_cluster.cluster,
cmap="brg")
for j in range(3):
   ax.text(centers2[j, num_cols.
index('maternalmortality')],
   centers2[j, num_cols.index('adolescentbirthrate')],
   centers2[j, num_cols.index('incomeratio')],
   c='black', s=j, fontsize=20, fontweight=800)
plt.tight_layout()
plt.show()
```

This produces the following plot:

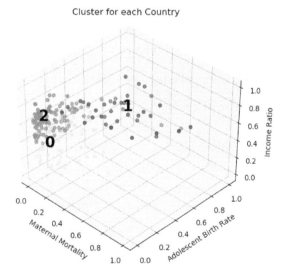

Figure 16.7 – A 3D scatter plot of three clusters

We can see here that the cluster 0 countries have higher maternal mortality and higher adolescent birth rates. Cluster 0 countries have lower income ratios.

12. So far, we have assumed that the best number of clusters to use for our model is three. Let's build a five-cluster model and see how those results look.

The silhouette score has declined from the three-cluster model. That could be an indicator that at least some of the clusters are very close together:

```
gap_enc = gap_enc[num_cols]


kmeans2 = KMeans(n_clusters=5, random_state=0)


kmeans2.fit(gap_enc)


silhouette_score(gap_enc, kmeans2.labels_)
0.2871811434351394


gap_enc['cluster2'] = kmeans2.labels_


gap_enc.cluster2.value_counts().sort_index()
0    21
1    40
2    48
3    16
4    63
Name: cluster2, dtype: int64
```

13. Let's plot the new clusters to get a better sense of where they are:

```
fig = plt.figure()
plt.suptitle("Cluster for each Country")
ax = plt.axes(projection='3d')
ax.set_xlabel("Maternal Mortality")
ax.set_ylabel("Adolescent Birth Rate")
ax.set_zlabel("Income Ratio")
ax.scatter3D(gap_cluster.maternalmortality,
    gap_cluster.adolescentbirthrate,
    gap_cluster.incomeratio, c=gap_cluster.cluster2,
    cmap="brg")
```

```
for j in range(5):
  ax.text(centers2[j, num_cols.
index('maternalmortality')],
  centers2[j, num_cols.index('adolescentbirthrate')],
  centers2[j, num_cols.index('incomeratio')],
  c='black', s=j, fontsize=20, fontweight=800)


plt.tight_layout()
plt.show()
```

This produces the following plot:

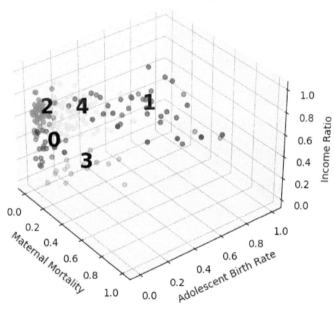

Figure 16.8 – A 3D scatter plot of five clusters

14. We can use a statistic called the Rand index to measure the similarity between the clusters:

```
rand_score(kmeans.labels_, kmeans2.labels_)
0.7439412902491751
```

15. We tried three-cluster and five-cluster models, but were either of those a good choice? Let's look at scores for a range of *k* values:

```
gap_enc = gap_enc[num_cols]

iner_scores = []
sil_scores = []
for j in range(2,20):
  kmeans=KMeans(n_clusters=j, random_state=0)
  kmeans.fit(gap_enc)
  iner_scores.append(kmeans.inertia_)
  sil_scores.append(silhouette_score(gap_enc,
    kmeans.labels_))
```

16. Let's plot the inertia scores with an elbow plot:

```
plt.title('Elbow Plot')
plt.xlabel('k')
plt.ylabel('Inertia')
plt.plot(range(2,20),iner_scores)
```

This produces the following plot:

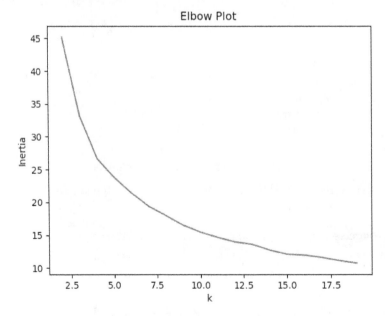

Figure 16.9 – An elbow plot of inertia scores

17. We also create a plot of the silhouette scores:

```
plt.title('Silhouette Score')
plt.xlabel('k')
plt.ylabel('Silhouette Score')
plt.plot(range(2,20),sil_scores)
```

This produces the following plot:

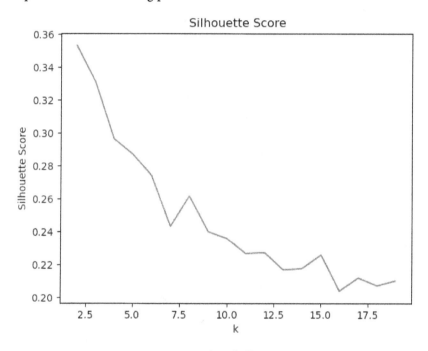

Figure 16.10 – An plot of silhouette scores

The elbow plot suggests that a value of $k$ around 6 or 7 would be best. We start to get diminishing returns in inertia at $k$ values above that. The silhouette score plot suggests a smaller $k$, as there is a sharp decline in silhouette scores after that.

K-means clustering helped us make sense of our data on the gap between women and men in terms of income, education, and employment by country. We can now see how certain features hang together, in a way that the simple correlations we did earlier did not reveal. This largely assumed, however, that our clusters have a spherical shape, and we had to do some work to confirm that our value of $k$ was the best. We will not have any of the same issues with DBSCAN clustering, so we will try that in the next section.

# Implementing DBSCAN clustering

DBSCAN is a very flexible approach to clustering. We just need to specify a value for ε, also referred to as **eps**. As we have discussed, the ε value determines the size of the ε-neighborhood around an instance. The minimum samples hyperparameter indicates how many instances around an instance are needed for it to be considered a core instance.

> **Note**
>
> We use DBSCAN to cluster the same income gap data that we worked with in the previous section.

Let's build a DBSCAN clustering model:

1.  We start by loading familiar libraries, plus the DBSCAN module:

    ```
    import pandas as pd
    from sklearn.preprocessing import MinMaxScaler
    from sklearn.pipeline import make_pipeline
    from sklearn.cluster import DBSCAN
    from sklearn.impute import KNNImputer
    from sklearn.metrics import silhouette_score
    import matplotlib.pyplot as plt
    import os
    import sys
    sys.path.append(os.getcwd() + "/helperfunctions")
    ```

2.  We import the code to load and preprocess the wage income data that we worked with in the previous section. Since that code is unchanged, there is no need to repeat it here:

    ```
    import incomegap as ig
    gap = ig.gap
    num_cols = ig.num_cols
    ```

3.  We are now ready to preprocess the data and fit a DBSCAN model . We have chosen an eps value of 0.35 here largely through trial and error. We could have also looped over a range of eps values and compared silhouette score:

    ```
    pipe1 = make_pipeline(MinMaxScaler(),
        KNNImputer(n_neighbors=5))
    ```

```
gap_enc = pd.DataFrame(pipe1.fit_transform(gap),
  columns=num_cols, index=gap.index)


dbscan = DBSCAN(eps=0.35, min_samples=5)


dbscan.fit(gap_enc)


silhouette_score(gap_enc, dbscan.labels_)
0.31106297603736455
```

4.  We can use the labels_ attribute to see the clusters. We have 17 noise instances, those with a cluster of -1. The remaining observations are in one of two clusters:

```
gap_enc['cluster'] = dbscan.labels_


gap_enc.cluster.value_counts().sort_index()
-1      17
 0     139
 1      32
Name: cluster, dtype: int64


gap_enc = \
  gap_enc.loc[gap_enc.cluster!=-1]
```

5.  Let's take a closer look at which features are associated with each cluster. Cluster 1 countries are very different from cluster 0 countries in maternalmortality, adolescentbirthrate, and genderinequality. These were important features with the k-means clustering as well, but there is one fewer cluster with DBSCAN and the overwhelming majority of instances fall into one cluster:

```
gap_enc[['cluster'] + num_cols].\
  groupby(['cluster']).mean().T
```

| cluster | 0 | 1 |
| --- | --- | --- |
| educratio | 0.63 | 0.35 |
| laborforcepartratio | 0.57 | 0.82 |
| humandevratio | 0.82 | 0.62 |
| genderinequality | 0.40 | 0.79 |
| maternalmortality | 0.05 | 0.45 |
| incomeratio | 0.51 | 0.71 |

| | | |
|---|---|---|
| adolescentbirthrate | 0.16 | 0.50 |
| femaleperparliament | 0.36 | 0.30 |
| incomepercapita | 0.16 | 0.02 |

6. Let's visualize the clusters:

```
fig = plt.figure()
plt.suptitle("Cluster for each Country")
ax = plt.axes(projection='3d')
ax.set_xlabel("Maternal Mortality")
ax.set_ylabel("Adolescent Birth Rate")
ax.set_zlabel("Gender Inequality")
ax.scatter3D(gap_cluster.maternalmortality,
    gap_cluster.adolescentbirthrate,
    gap_cluster.genderinequality, c=gap_cluster.cluster,
    cmap="brg")
plt.tight_layout()
plt.show()
```

This produces the following plot:

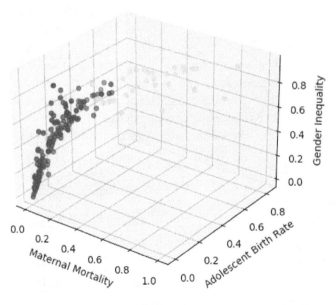

Figure 16.11 – A 3D scatter plot of the cluster for each country

DBSCAN is an excellent tool for clustering, particularly when the characteristics of our data mean that k-means clustering is not a good option; for example, when the clusters are not spherical. It also has the advantage of not being influenced by outliers.

# Summary

We sometimes need to organize our instances into groups with similar characteristics. This can be useful even when there is no target to predict. We can use the clusters created for visualizations, as we did in this chapter. Since the clusters are easy to interpret, we can use them to hypothesize why some features move together. We can also use the clustering results in subsequent analysis.

This chapter explored two popular clustering techniques, k-means and DBSCAN. Both techniques are intuitive, efficient, and handle clustering reliably.

# Index

Packt.com

Subscribe to our online digital library for full access to over 7,000 books and videos, as well as industry leading tools to help you plan your personal development and advance your career. For more information, please visit our website.

## Why subscribe?

- Spend less time learning and more time coding with practical eBooks and Videos from over 4,000 industry professionals

- Improve your learning with Skill Plans built especially for you

- Get a free eBook or video every month

- Fully searchable for easy access to vital information

- Copy and paste, print, and bookmark content

Did you know that Packt offers eBook versions of every book published, with PDF and ePub files available? You can upgrade to the eBook version at packt.com and as a print book customer, you are entitled to a discount on the eBook copy. Get in touch with us at customercare@packtpub.com for more details.

At www.packt.com, you can also read a collection of free technical articles, sign up for a range of free newsletters, and receive exclusive discounts and offers on Packt books and eBooks.

# Other Books You May Enjoy

If you enjoyed this book, you may be interested in these other books by Packt:

**Applied Machine Learning Explainability Techniques**

Aditya Bhattacharya

ISBN: 9781803246154

- Explore various explanation methods and their evaluation criteria
- Learn model explanation methods for structured and unstructured data
- Apply data-centric XAI for practical problem-solving
- Hands-on exposure to LIME, SHAP, TCAV, DALEX, ALIBI, DiCE, and others
- Discover industrial best practices for explainable ML systems
- Use user-centric XAI to bring AI closer to non-technical end users
- Address open challenges in XAI using the recommended guidelines

**Hyperparameter Tuning with Python**

Louis Owen

ISBN: 9781803235875

- Explore manual, grid, and random search, and the pros and cons of each
- Understand powerful underdog methods along with best practices
- Explore the hyperparameters of popular algorithms
- Discover how to tune hyperparameters in different frameworks and libraries
- Deep dive into top frameworks such as Scikit, Hyperopt, Optuna, NNI, and DEAP
- Get to grips with best practices that you can apply to your machine learning models right away

# Packt is searching for authors like you

If you're interested in becoming an author for Packt, please visit authors. packtpub.com and apply today. We have worked with thousands of developers and tech professionals, just like you, to help them share their insight with the global tech community. You can make a general application, apply for a specific hot topic that we are recruiting an author for, or submit your own idea.

# Share Your Thoughts

Now you've finished *Data Cleaning and Exploration with Machine Learning*, we'd love to hear your thoughts! Scan the QR code below to go straight to the Amazon review page for this book and share your feedback or leave a review on the site that you purchased it from.

https://packt.link/r/1-803-24167-5

Your review is important to us and the tech community and will help us make sure we're delivering excellent quality content.